Engine Service
AUTOMOTIVE MACHINING
AND
ENGINE REPAIR

by
GARY LEWIS
DE ANZA COLLEGE
Cupertino, California
2007

ENGINE SERVICE
Automotive Machining and Engine Repair

Copyright
1980, 1986, 1994 and 2001
ISBN 0-13-277849-1

Revised Edition
2007
ISBN 0-9787415-0-1

Update 29
January 2017

Website
www.enginebooks.net

Purchase
at
www.AERA.org

On the Author

Gary Lewis began developing his interest in all things mechanical at an early age repairing his old cars and fabricating farm equipment. He gained formal training in an apprenticeship machining aircraft engine parts. After accumulating a few more years experience, he returned to college and earned Baccalaureate and Master's degrees from California State University San Jose.

While in his senior year at CSU San Jose, Mr. Lewis accepted a part time teaching assignment in machine tools at De Anza College in Cupertino, California. In 1969, he was hired full time and began building nationally recognized programs in Machine Tool and Automotive Technologies at De Anza College. He has also consulted in developing programs at other colleges and for private industry.

In 1972, following his love of engines, Mr. Lewis began developing an Automotive Machining program at De Anza. With this assignment, he learned that he had to develop his own teaching materials with an appropriate mix of machining and engine technologies. Other materials at the time were either entirely academic or limited to parts replacement and assembly. Slowly, the materials developed for instruction in this program accumulated to become the current textbook.

In 1985, he gained recognition for presenting to the Automotive Engine Rebuilders Association process technology that he developed for stress relieving and straightening aluminum cylinder heads. This process had been developed over time beginning in the mid-seventies coinciding with huge increases in import overhead cam engines. Since the presentation in 1985, the process has become a standard for the industry.

PREFACE

The information in this text is presented from the view of a machine shop serving automotive technicians. Basic procedures are selected from those with wide industry acceptance and represent well balanced and competent "state of the art" practice. While skills of technicians and machinists sometimes overlap in basic engine repair areas such as valve grinding and engine overhaul, the complexity of today's engine systems clearly calls for specialization in each of the two areas. However, it has been my experience in the teaching of apprentice technicians and machinists, that a well-rounded knowledge of engine principles and related services directly meets the needs of both.

There is also a significant population of automotive restorers and street rodders that will find the information in this text useful. While some do their own assembly and installations, their interest is far more than casual and nearly all sublet machining. Just like those in the field, they too can benefit from the content of this text.

The content of this text addressing performance engines is essential to custom engine builders and the principles introduced are not widely shared in the field. Considering the size of the custom and performance engine market, it is certainly advisable that technicians and machinists be prepared to properly advise customers.

While emphasis is upon "how to" information and procedures, appropriate attention is given to theory, especially in difficult to understand areas. The importance of job sequences, routines, and planning are also emphasized, as these are the key to productivity. Considering the cost of shop time, the need for productivity cannot be ignored.

The engine rebuilding and machining content of this text bridges the gap that typically exists between textbook coverage and service manuals used by technicians. After mastering the content of this text, the student apprentice is better attuned to all-important details and can make on-the-job judgments that eliminate the oversights associated with inexperience.

Each chapter has review test questions at the end. Studying the content of this text and completion of the review quizzes prepare students and apprentices for Automotive Engine Rebuilders Association machinist certification examinations and for Automotive Service Excellence engine repair certification.

ACKNOWLEDGMENTS

The author expresses sincere gratitude to the following companies and corporations: Chrysler Corporation, General Motors Corporation, Kwik-Way Manufacturing Company, Petersen Machine Tools, Sunnen Products Company, Fel-Pro Incorporated, Lock-N-Stitch, Maintenance Welding Alloys, and Extrude Hone Corporation.

Sincere gratitude is also given to the following shop owners and individuals: Ron Rosa of Clarke's Auto Parts and Machine, Jan Huff of Penniman and Richards Machine Shop, Ted Yamashiro of Techcraft Machine Shop, Jim Andersen of Maintenance Welding Alloys, Bud Riebhoff, automotive machinist, Marc Vertin of Chevron-Texaco, and Guymond Louie of Copyrite, San Ramon, CA.

Particular gratitude is expressed for the consultation of two faculty members at De Anza College. One is Dema Elgin of Elgin's Cams in Santa Rosa, California. His knowledge and experience in performance engine preparation and related teaching at De Anza College in Cupertino, California, were of immeasurable value in reviewing the content and in writing the Performance Engines chapter of this text. The same appreciation is extended to Dave Capitolo at De Anza College, formerly of Snap On tools. Dave's experience with computer diagnostics was invaluable in updating the engine diagnosis content.

Special thanks are also due my wife, Edjie, for her support during this long project.

CONTENTS

Chapter-Page

1   SHOP SAFETY AND HAZARDOUS WASTE MANAGEMENT

Lifting Hazards   1-1
Accidents During Installation   1-2
Respiratory Hazards in the Shop   1-3
Substance Abuse in the Workplace   1-4
General Safety Guidelines   1-4
Handling Hazardous Materials   1-8
Best Management Practices   1-10
Summary   1-12

Review Questions   1-14

2   FUNDAMENTALS OF MACHINING

Machining Processes   2-1
Tool Materials   2-4
Cutting Tool Glossary   2-6
Single Pointed Tools   2-6
Milling Cutters   2-9
Drills, Reamers and Other Drilling Tools   2-11
Grinding and Honing   2-15
Speeds and Feeds   2-16
Machine Installation and Set-Up   2-18
Correcting Alignments Between Centers   2-20
Tramming Spindles   2-23
Summary   2-24

Review Questions   2-25

3   MEASURING TOOLS

Understanding Specifications and Tolerances   3-1
Calculating Thermal Expansion   3-3
Comparing Units of Measurement   3-4
Using Micrometers   3-5
Making Transfer Measurements   3-8
Using Dial Indicators   3-10
Using Dial Bore Gauges   3-10
Using Calipers   3-12
Checking Alignments   3-12
Measuring Surface Finishes   3-13

Measuring Thicknesses of Castings   3-14
Summary   3-15

Review Questions   3-16

4   FASTENERS

Determining the Strength of Fasteners   4-1
Comparing Clamping Force and Torque   4-3
Identifying Threads   4-3
Using Pipe Threads and Fittings   4-4
Removing Broken Fasteners   4-5
Installing Helicoils   4-6
Removing Broken Tools   4-8
Summary   4-8

Review Questions   4-9

5   ENGINE THEORY

The Four-Stroke Cycle   5-1
Compression Ignition Engines   5-3
Variable Valve Timing and Valve Action   5-6
Valve Train Configurations   5-7
Valve Lifters and Lash Compensators   5-9
Engine Oiling   5-12
Engine Oils   5-15
Engine Measurements   5-21
Fits and Clearances   5-23
Cooling System Operation   5-23
Combustion Efficiency   5-25
Summary   5-27

Review Questions   5-28

6   ENGINE DIAGNOSIS

Looking for Signs of Engine Wear   6-2
Checking the Block Assembly   6-2
Testing Power Balance   6-3
Testing Compression   6-4
Testing Cylinder Leakage   6-5
Checking Valve Timing   6-5
Testing Manifold Vacuum   6-7
Testing Exhaust Back Pressure   6-9
Testing With a Scan Tool; An Introduction   6-10

Diagnosing Engine Noises 6-12
Testing Engine Oil Pressure 6-14
Testing Cooling systems 6-15
Summary 6-16

Review Questions 6-17

7   ENGINE DISASSEMBLY

Hints for Disassembly in the Chassis 7-1
Disassembling Cylinder Heads 7-2
Numbering Connecting Rods 7-4
Ridge Reaming 7-5
Removing Piston and Rod Assemblies 7-6
Removing the Timing Chain and Sprockets 7-7
Removing the Crankshaft 7-8
Removing Cams and Lifters from Pushrod Engines 7-8
Removing Camshaft Bearings 7-9
Removing Oil Plugs and Core Plugs 7-9
Summary 7-11

Review Questions 7-12

8   CLEANING ENGINE PARTS

Using Solvent and Cold Solutions 8-2
Cleaning in Hot Tanks 8-2
Degreasing in Ovens 8-3
Using Airless Shot Blasters 8-4
Bead Blasting 8-4
Blasting With Baking Soda 8-5
Blasting With High Pressure Water 8-6
Tumbling Small Parts 8-6
Using Hand and Power Tools 8-7
Removing Rust and Scale 8-8
Working Under Regulations 8-8
Summary 8-10

Review Questions 8-11

9   INSPECTING VALVE TRAIN COMPONENTS

Determining Valve Guide Wear 9-1
Checking Valves 9-2
Checking Natural Gas Valve Trains 9-4
Testing Valve Springs 9-5

Inspecting Camshafts, Lifters, and Followers 9-6
Inspecting Rocker Arms and Pushrods 9-10
Checking Timing Chains and Gears 9-12
Checking Cylinder Head Castings 9-13
Summary 9-15

Review Questions 9-16

10 INSPECTING ENGINE BLOCK COMPONENTS

Measuring Cylinder Wear 10-1
Measuring Piston Clearance 10-3
Checking Pistons 10-3
Checking Piston Pin Clearances 10-5
Checking Cylinder Block Flatness 10-6
Measuring Main Bearing Bores 10-6
Checking the Crankshaft 10-7
Measuring Connecting Rod Bores 10-9
Summary 10-10

Review Questions 10-11

11 CRACK DETECTION AND REPAIR

Using Dry Magnetic Particle Inspection 11-1
Using Wet Magnetic Particle Inspection 11-2
Using Dye Penetrants 11-3
Pressure Testing Castings 11-3
Using Crack Repair Pins 11-4
Stop Drilling 11-11
Welding Head and Block Castings 11-12
Sealing Castings 11-14
Summary 11-15
Review Questions 11-17

12 RECONDITIONING VALVE TRAIN COMPONENTS

Removing and Replacing Valve Guides 12-1
Knurling Valve Guides 12-3
Fitting Oversized Valve Stems 12-6
Replacing Integral Valve Guides 12-6
Refacing Valves and Valve Stems 12-12
Grinding Valve Seats 12-14
Cutting Valve Seats 12-19
Installing Valve Seats 12-21

Fitting Valve Seals   12-24

Replacing Rocker Arm Studs   12-27

Correcting Installed Spring Height   12-28

Correcting Installed Stem Height   12-29

Refacing Rocker Arms   12-31

Straightening Aluminum Heads   12-31

Correcting Overhead Camshaft Centerlines   12-35

Regrinding Camshafts, Lifters, and Followers   12-37

Summary   12-41

Review Questions   12-42

13   RECONDITIONING ENGINE BLOCK COMPONENTS

Honing Cylinders for Overhaul   13-1

Piston Inspection and Knurling   13-2

Reboring and Honing Cylinders   13-4

Sleeving Cylinders   13-10

Line Boring and Honing   13-11

Fitting Piston Pins   13-14

Resizing Connecting Rod Housing Bores   13-17

Assembling and Aligning Pistons and Connecting Rods   13-21

Regrinding and Polishing Crankshafts   13-26

Resurfacing Flywheels and Replacing Ring Gear   13-35

Overhauling Oil Pumps   13-38

Summary   13-41

Review Questions   13-43

14   RESURFACING CYLINDER HEADS AND BLOCKS

Comparing Resurfacing Machines   14-1

General Precautions   14-3

Correcting V-Block Intake Manifold Alignment   14-6

Determining V-Block Ratios   14-9

Resurfacing Overhead Cam Cylinder Heads   14-11

Resurfacing Diesel Cylinder Heads   14-12

Resurfacing Air Cooled Cylinder Heads   14-13

Summary   14-14

Review Questions   14-15

15 ENGINE BALANCING

Weighing Pistons and Connecting Rods   15-3
Balancing Connecting Rods   15-4
Balancing Pistons and pins   15-4
Balancing Crankshafts   15-5
Balancing Flywheels and Clutches   15-10
Balancing Torque Converters   15-10
Balancing with Heavy Metal   15-11
Suggestions for Minimum Balancing   15-12
Summary   15-13

Review Questions   15-14

16 ENGINE ASSEMBLY

Cleaning and Deburring for Assembly   16-1
Assembling Cylinder Heads   16-2
Installing Core Plugs   16-3
Installing Camshaft Bearings and Camshaft   16-4
Installing Oil Plugs   16-7
Sealing Rotating Shafts; the Basics   16-8
Fitting the Rear Main Seal   16-10
Installing the Main Bearings and Crankshaft   16-12
Setting Valve Timing   16-14
Installing Piston Rings   16-18
Installing Piston and Connecting Rod Assemblies   16-19
Assembling Cylinder Heads to Engine Blocks   16-21
Installing Rocker Arms   16-22
Adjusting Valves   16-23
Installing the Oil Pump   16-24
Pre-Oiling the Engine   16-25
Installing Timing Covers   16-26
Hints on Gaskets, Seals, and Sealants   16-27
Using an Assembly Checklist   16-28
Testing in a Run-In Stand   16-31
Assembling Flywheels and Flexplates   16-32
Attaching Bellhousings   16-33
Summary   16-34

Review Questions   16-35

17 PREPARING PERFORMANCE ENGINES

Improving Efficiency   17-2
Improving Flow Through Ports   17-3
Reducing Restriction at the Valves   17-7

Flow Testing   17-11
Improving Flow through Manifolds   17-14
Extrude Honing   17-16
Dealing with Tumble and Swirl   17-16
Synchronizing Valve Opening and Piston Travel   17-17
Maximizing Cylinder Pressure   17-18
Selecting Camshafts   17-24
Selecting Camshafts for Forced Induction   17-27
Matching Intake Airflow to the Engine   17-29
Matching Exhaust Systems to the Engine   17-31
Running Computer Simulations   17-34
Tuning Performance Engines   17-35
Project; Prepare a Performance Engine   17-36
Studying Sport Compact Engines   17-47
Summary   17-51

Review Questions   17-52

18   ENGINE INSTALLATION AND BREAK-IN

Removing and Installing the Engine   18-1
Inspecting and Servicing the Cooling System   18-3
Preparing for Emissions Testing   18-5
Making a Final Inspection   18-8
Starting and Breaking-In the Engine   18-8
Following Up on the Installation   18-9
Summary   18-9

Review Questions   18-10

KEY TO REVIEW QUESTIONS K-1

APPENDIX REFERENCE TABLES

Torque Recommendations, English   A-1
Torque Recommendations, Metric   A-1
Pipe Plug Torque Recommendatons   A-2
Torque Conversions   A-3
Decimal Equivalents for Drills   A-4
Tap Drill Sizes, English   A-5
Tap Drill Sizes, Metric   A-6
Helicoil Tap Drill Sizes, English and Metric   A-7
English-Metric Conversions   A-8
Metric-English Conversions   A-9
Conversion Factors   A-10

Chapter 1

SHOP SAFETY AND HAZARDOUS WASTE MANAGEMENT

Upon completion of this chapter, you will be able to:
* Identify lifting hazards and know how to avoid them.
* Recognize and avoid common accidents that occur in automotive repair and machine shops.
* Minimize exposure to respiratory hazards.
* Explain the scope of substance abuse problems for employees and the employer.
* Observe safe practices and follow correct procedures for shop environments.
* Describe the intent of the Resource Recovery and Conservation Act.
* Describe the type of information to be found in Material Safety Data Sheets.
* Define best management practices for the handling of hazardous wastes.

INTRODUCTION

Discussions of safety are often anticipated with a degree of boredom. Everyone understands the necessity, but unless an accident temporarily raises consciousness, the information is only of secondary interest. It does however help to focus on the specific hazards encountered by heavy-duty mechanics and automotive machinists.

A major new area of concern is the handling of hazardous materials. Besides regulatory compliance, nearly everyone in the industry must now consider issues of health and safety. Many materials used in shops pose both immediate and long-term risks that machinists and technicians are only now learning about. Because all employees have a role in regulatory compliance, and since their health and safety are at stake, learning about hazardous materials and wastes is a necessity.

LIFTING HAZARDS

Too many technicians and machinists experience back pain and related complications. While associated with lifting heavy weights, this is not the cause of back pain for many individuals. Strains frequently result from lifting lighter weights or pulling wrenches from positions that make the back vulnerable to injury.

Avoid working while bent over or reaching overhead (see Fig.1-1 and 2). For engine installation, work as much as possible at workbench height. With some thought, one can assemble engines complete with accessories before installation in the chassis. When working over the fenders, consider raising the car to a comfortable work height.

Fig.1-1 Reduce work done bent over

Fig.1-2 Reduce work done overhead

Most obviously, get help from your coworkers when handling anything heavy. What is considered "heavy" varies not only with individual size and strength, but also with the position the technician is working in. For example, lifting a 75-pound crankshaft from a workbench into a machine is less strain than lifting a 40-pound cylinder head across an engine compartment during installation. It is easy to say "lift with your legs" but how is this done when it is necessary to reach across or around an engine in the chassis? As mentioned, many technicians and machinists have already experienced back pain and will gladly help with the expectation that such help will be returned in kind.

If back pain is already a reality, stretching and strengthening exercises both relieve pain and decrease vulnerability to injury. Stretching hamstrings and strengthening abdominal muscles are commonly recommended. Check with your doctor and get an opinion on exercises for your particular case.

ACCIDENTS DURING INSTALLATION

Accidents frequently result from failure to observe basic procedures. This is especially true for under hood fires. When removing an engine, begin by disconnecting the battery cable to ground and plugging any disconnected fuel lines. These steps prevent fuel leakage and limit sources of ignition. All shop personnel need to know the location and use of fire extinguishers and keep one on hand when starting engines for the first time. Even without fuel leaks, something as simple as crossed ignition wires can start fires in the manifold that readily engulf the engine compartment.

Prepare for an unusual phenomenon. When a fire does start, the technician on the job often goes into a temporary state of shock. While I have no explanation, I know from experience that it is often someone else in the area that first takes action.

Remote starter switches also can lead to accidents because they bypass transmission neutral safety switches. Unless the transmission is in neutral or park, engines can start in gear and drive the vehicle forward. Injury accidents are not uncommon and technicians will tell stories of crushed toolboxes, workbenches, and other close calls that they have personally witnessed. Place the transmission in neutral or park and set the parking brake before connecting a remote starter switch.

It is common to install engines after storing the vehicle for two weeks or more. During this time, the battery often discharges and is not recharged until installation of the rebuilt engine. The technician is sometimes in a hurry at this point and then recharges the battery at the maximum charging rate. The problem with this practice is that batteries produce flammable hydrogen gas during heavy charging. Given any source of ignition, such as arcing at a loose battery cable or other electrical connection, the hydrogen gas ignites and blows the battery apart. Avoid this hazard by trickle charging the battery two or three days ahead of time. Of course, check for loose battery cables and other electrical connections before attempting to crank the engine.

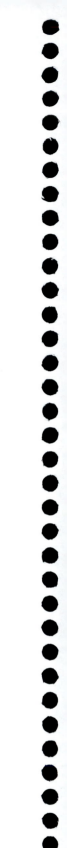

RESPIRATORY HAZARDS IN THE SHOP

In the author's opinion, too many technicians take hazards in their work environment for granted. They routinely work around gasoline and use cleaning solvent or other special purpose chemicals, and even when such products are known to affect health, they do not think of them as really "dangerous." It is in our own best interest to learn which materials are dangerous even when they do not cause immediate injury.

Avoid inhaling any of these products. While this seems obvious, how often do technicians and machinists use compressed air to dry solvent or chemicals from parts? How often do technicians and machinists handle worn flywheels and clutches or brake parts contaminated with asbestos friction material? We must consider the long range, and sometimes unknown, consequences of working around chemicals, petroleum products, and friction materials and exercise greater caution.

Begin by checking the "MSDS," or Material Safety Data Sheet, for each product used in the shop. Second, find alternative products to replace chemicals that are particularly hazardous. For example, avoid cleaning products containing chlorinated hydrocarbons under any circumstances. Chlorinated hydrocarbons are carcinogens and accumulate in your system over time. Many technicians and machinists are unaware of potential hazards and accept unknown risks to their health by continuing to use them.

Other steps required to reduce exposure are especially difficult because they involve changing work habits. For example, when using compressed air to dry parts, chemicals become airborne and subject everyone in the immediate area to respiratory hazards. Why not let the parts drain and blow passages clear later? Why not use hot water as a rinse? Habits are most difficult to change when there is no immediate hazard, but we must consider the cumulative effects of exposure over time.

A few basic practices reduce respiratory risks. First, always work in well-ventilated areas or outside if the quantity of vapor is too great for interior ventilation systems to handle. Second, wear facemasks appropriate for either vapor or particulate hazards (see Figs. 1-3 and 4). Particulate hazards include grinding dust and asbestos from clutch or brake parts brought into the shop for machining. While inexpensive disposable paper masks are adequate for some dust, be sure to wash asbestos from clutches or brake parts before handling or machining them.

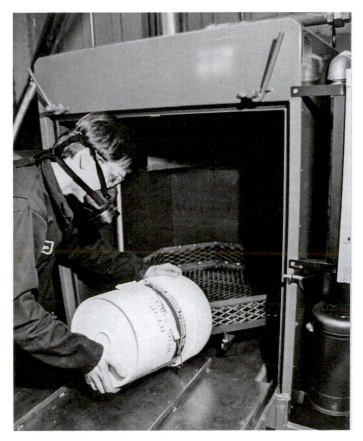

Fig.1-3 Respirator mask with filter

Fig.1-4 A disposable face mask for dust protection

SUBSTANCE ABUSE IN THE WORK-PLACE

This is one unpopular subject to deal with both as a safety concern and as something that adds to overhead cost. According to some sources, two thirds of those in the 18 to 25 year old age group use "recreational" drugs and over 10 percent use them on the job. Some drugs, marijuana for example, have a half-life of 72 hours and can affect weekend users through the following Monday or Tuesday.

We cannot ignore the potential hazard to workers and their co-workers that these numbers suggest. The problem being sufficiently widespread, many employers have substance abuse programs for employees. In industries where the public safety is the issue, such as in transportation industries, random drug testing is possible and sometimes mandatory.

Of particular concern to employers is statistical data suggesting that "abusers" are four times more likely to have accidents and five times more likely to file disability claims. Because of these claims, these individuals increase the employer's cost of insuring for liability and disability. Consider that these overhead expenses come from the owner's profits and there is every motive not to employ these individuals.

Co-workers must consider the possibility that someone under the influence endangers everyone. Such behavior cannot be ignored and individuals should be encouraged, perhaps as a condition of employment, to seek counseling.

GENERAL SAFETY GUIDELINES

Each piece of equipment has safety guidelines prepared for operators. Become familiar with these guidelines. Listed below are general safety rules that apply in any shop environment:

1. Keep floors clean to prevent slipping on fluids or tripping over tools, parts, or air hoses.

2. Wear safety glasses continuously while on the job. It is not always possible to anticipate when hazards will arise.

3. Learn the location of fire extinguishers in the shop and how to use them. To be sure they are ready for use; check their state of charge every few months.

4. Connect exhaust pipes from running engines to outside ventilation when working inside (see Fig.1-5). Avoid breathing exhaust when working outside.

5. 5.Hold a shop towel over work or holes to capture material blown loose with compressed air (see Fig.16). Keep in mind that chips and dirt are directed across the shop unless precautions are taken.

Fig.1-5   Connecting vehicle exhaust to ventilation system

Fig.1-7   Pumping fuel from one tank to another using a transfer pump

Fig.1-6   Using a shop towel to trap chips when using an air hose

6. Do not siphon fuel. Drain tanks by transferring (pumping) fuel into another tank or approved gas can (see Fig.1-7). Add fuel only from approved gas cans.

7. Check all fuel lines and electrical connections before cranking engines.

8. Do not run on shop floors. The possibility of slipping and injury is very high.

9. Leave keys in the ignition of cars in the shop. This enables moving cars quickly in an emergency.

10. Rinse batteries with clear water before handling. Wear gloves or wash your hands after handling to limit exposure to acids. Wash yourself and your clothing immediately in clear water should you spill or come in contact with battery acid.

11. Switch battery chargers off and wear safety glasses before connecting or disconnecting charger cables. Because battery charging produces explosive hydrogen gas, trickle charging is preferred to fast charging.

12. Keep clear of exhaust manifolds, fans, and belts or pulleys when working in the engine compartment. Watch especially for electric fans because they can start without warning.

13. Wear goggles and gloves for protection when welding or using an oxy-acetylene torch (see Fig.1-8). Remember that electric welding exposes others in the area to arc welding flash and that they may not be aware that work pieces are still hot. Set up safety screens (see

Fig.1-9) and warn co-workers that the flash can burn eyes. Write "hot" on welded parts with soapstone or chalk.

Fig.1-8   Welding gloves left, gas welding goggles center and arc welding helmet right

Fig.1-9   A screen to protect others from welding flash

14. Roll up long sleeves or wear short sleeves when operating machinery to reduce chances of clothing getting caught in rotating machinery. Wristwatches and rings are unacceptable for the same reasons. Short sleeve coveralls do not get caught in machinery and still protect street clothing when working on other jobs.

15. Flush skin and clothing with water if exposed to chemicals or fluids (see Fig.1-10). If eyes are exposed, flush with water for 15 minutes and have them checked by a doctor or nurse (see Fig.1-11). While any sensation of pain may be low level, if you wait to check for damage, the damage may be permanent.

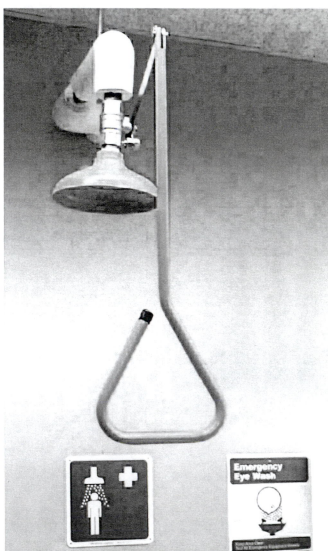

Fig.1-10   An emergency safety shower

Fig.1-11   An emergency eyewash station

16. Wear only work shoes with oil-resistant soles and safety toes. Tennis type shoes do not provide any protection from hazards and are unacceptable.

17. Only one operator switches machinery on or off. When working with others, remember that there must be only one operator and that all others must stand clear on start up.

18. Check for proper lift points before jacking up a car or raising it on a hoist (see Fig.1-12). When using a floor jack, place safety stands under the car before continuing work (see Fig.1-13).

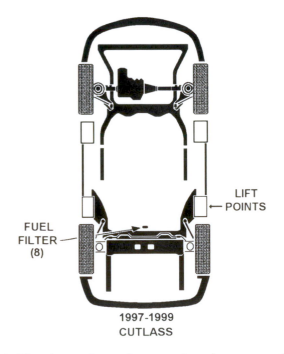

Fig.1-12   A service reference showing proper lift points

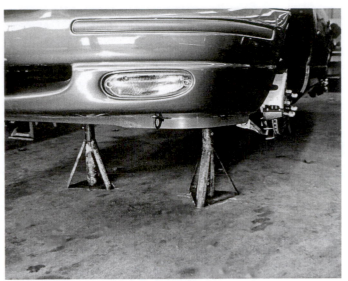

Fig.1-13 Safety stands under a vehicle

19. Lift heavy objects from the floor using your legs. Do this by bending low at the knees with the back kept straight (see Fig.1-14). If you think that the object exceeds 75 pounds, get help from a co-worker.

Fig.1-14 Lifting with legs bent and back straight

20. Check that hood supports are secure before going to work in the engine compartment. If in doubt, prop open hood with a rod or, on major jobs, remove the hood entirely.

21. Check that transmissions are in neutral or park and that the parking brake is set before cranking any engine. Remember remote starter switches bypass neutral safety switches.

22. Immediately report even minor injuries so they may be checked and treated if necessary.

23. Close drawers after removing tools from tool chests so that others do not catch themselves on open drawers as they pass by. To prevent sudden weight shifts and tipped over tool chests, close and lock drawers before rolling them around the shop.

HANDLING HAZARDOUS MATERIALS

Over the years, technicians have generally taken the variety of chemicals and chemical byproducts in their workplace for granted. While it is easy to view efforts at controlling these materials simply as regulatory meddling, our personal health and our environment are at stake.

Hazardous materials are defined in regulations as ignitable, reactive, corrosive or toxic. While specific parameters for each of these characteristics are detailed in the Resource Conservation and Recovery Act (RCRA), defining characteristics include the following:

1. Ignitable materials have a flash point below 140 degrees F or ignite spontaneously or by friction.

2. Corrosive materials have a pH below 2.5 or above 12.5 and corrode steel at a rate of .250in or more per year at a temperature of 128 degrees F.

3. Reactive materials are unstable, change violently, form explosive mixtures or generate toxic fumes.

4. Toxic materials are poison if ingested or inhaled or are carcinogenic.

Information describing the contents, hazards, and first aid is included in Material Safety Data Sheets or MSDSs. All vendors are required to make available MSDSs for the products they sell. Under "Right to Know" requirements included in regulations, shops must maintain a file of MSDSs for employee reference. If uncertain of product contents, or if considering new products, check the information contained in the MSDSs. While form varies, the primary sections of an MSDS contain the following:

Section 1
Who makes the product and their address, emergency phone number, trade name and the product formula (unless proprietary).

Section 2 Hazardous Ingredients
Chemical identification of components and exposure limit guidelines.

Section 3 Physical Data
Appearance and odor under normal conditions, specific gravity, boiling point, vapor pressure, vapor density, and evaporation rate.

Section 4 Fire and Explosion Data
Indicates what kind of fire extinguishers to use, the flash point, special fire Fighting procedures, and any special dangers.

Section 5 Reactivity Data
What the chemicals react with, and if they do react, what might happen. It also tells what situations to avoid so there are no unexpected chemical reactions.

Section 6 Health Hazard Data
How the chemical can enter the body through inhalation, eyes, or skin contact and what symptoms might exist. This section also gives emergency first aid procedures.

Section 7 Spill or leak Procedure

Includes what to do if there is a spill or leak, equipment and procedure for cleaning up the spill, method of disposal/special precautions.

Section 8 Special Protection

Safety measures for adequate protection such as a respirator, gloves, ventilation, eye protection, or protective clothing for safe handling of different chemicals.

Section 9 Special Precautions - Transport and Handling

Special handling requirements in relation to temperature, open spark humidity, and special containers. To forewarn users of hazards, chemical products have color bars (blue, red, yellow, and white) or National Fire Protection Agency labels (see Fig.1-15). Although not as detailed as an MSDS, these labels warn of potential hazards. Note the scale numbered from 1 to 4 for hazard levels under health, fire, and reactivity, 4 being most dangerous and 1 least dangerous. Should there be any number greater than 2, check the MSDS prior to using the product.

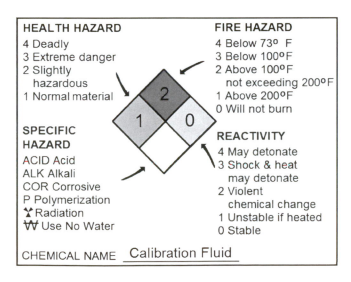

Fig.1-15 Hazard levels on NFPA label

There are also specialized storage requirements that are a further incentive to eliminate or reduce inventories on hand. For example, all hazardous materials must be stored where there is no possibility of leaking chemicals into storm drains or ground water. Corrosive materials must also be stored separately from flammables. These requirements essentially mean that shops must provide separate "double containment" storage units to protect against leakage, explosion or fire.

In addition, many materials become hazardous waste once removed from use. Examples include hot tank chemicals, drain oil and coolant from engines. Also included are materials such as blast media from dust collectors that are not hazardous when new but become contaminated in use. These wastes must also be properly contained, stored, labeled, dated, and hauled within specified time limits. When hauled, an Environmental Protection Agency (EPA) identification number and manifest are required which assigns liability to the shop indefinitely.

In simplified terms, these regulations encourage business and industry to reduce worker and environmental exposure to hazardous materials. Regulatory compliance requirements are met by following one or more of the following four general alternatives:

1. **Eliminate** Replace products, and sometimes processes, with alternatives that produce less waste. An example in engine building would be the replacement of hot tanks with baking ovens for degreasing engine parts.

2. **Reduce** Switch to less hazardous products or reduce wastes by taking measures to extend the useful life of products in use. For example, replace the caustics used for cleaning iron and steel only with detergents. While efficiency in some cases may be reduced slightly, one tank and product will clean aluminum as well as iron and steel. Oil skimmer and sludge separation systems can also be added to hot tanks to extend the useful life of products.

3. **Recycle** Find a recycling stream for used products that otherwise become hazardous waste. Examples of these include cleaning solvents, oils, and coolants, all of which can be recycled and exchanged for products equivalent to new.

4. **Dispose of Properly** When the process generates waste that cannot be recycled, the waste must then be hauled and disposed of according to regulations. Examples of this include hot tank chemicals and sludge that can only be removed by licensed haulers, using an EPA manifest, and delivered to a licensed disposal facility.

To be safe and comply with regulations, much more than management effort is required. Note the role of employees in following these guidelines and for the handling of hazardous materials. Without employee cooperation, little chance of remaining safe or maintaining compliance exists.

BEST MANAGEMENT PRACTICES

The measures required to manage hazardous wastes are referred to as Best Management Practices. Regulatory compliance requires the efforts of both management and employees. Aside from regulations, worker health and safety and the health of the environment require that these efforts be conscientious. While state and local requirements vary, following the guidelines below assures safe handling and greatly diminishes the likelihood of being found in noncompliance with regulations.

Work Stations
1. Change fluids or perform cooling systems services and repairs only in covered workstations. Working outdoors increases the possibility of fluids getting into storm drains or ground water.
2. Take all necessary precautions to capture fluids and wipe up spills prior to mopping floors. Since mop water is disposed of in sewers, using mops to clean up excessive quantities of oil adds to wastewater treatment problems. Use absorbents and then wipe with shop towels prior to mopping. Of course, the absorbents require handling as a hazardous waste and we depend upon commercial laundries to properly

handle the oily towels.
3. Drain and recycle all fluids from engine and transmission cores before shipping or storage. This reduces the potential for spilled oil getting into storm drains or ground water.
4. Dump mop water only into drains directed to sewers.

Shop Towels
1. Use cotton shop towels. Once used, disposable towels become hazardous waste.
2. Store cotton shop towels in fireproof metal cans after use (see Fig.1-16). While working, technicians can collect used towels in a plastic bag in their toolboxes.

Fig.1-16 Fire proof storage for shop towels

3. Launder cotton shop towels through a commercial laundry service. They are equipped to properly handle water discharges after washing.

Oils, Fluids, and Coolants

1. Drain engine oils and transmission fluids into double containment units (see Fig.1-17). Cover the primary container when not in use and keep the secondary side clean.

Fig.1-17   Double containment for drain oil

2. Capture coolant and store in a separate double containment unit (see Fig.1-18).

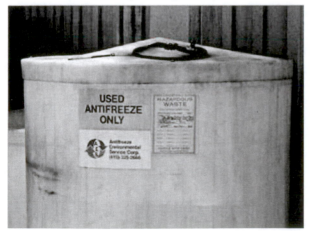

Fig.1-18   Double containment unit to capture and recycle coolant with labels

3. Recycle oils and coolants to conserve resources and to avoid hazardous waste disposal handling charges.

4. In case of a spill, use absorbents and/or wipe with shop towels before mopping. For small spills, wiping with shop towels is preferred since used absorbent materials require handling as hazardous waste.

5. Report spills of more than one pound (estimated) to management. Regulatory agencies generally require that management keep a log of such spills.

Oil Filters

1. Drain used filters thoroughly by puncturing the case and placing them upright in a bucket or barrel. Crushing them is even better since more can be collected per barrel thereby reducing the cost of hauling (see Fig.1-19).

Fig.1-19   An oil filter crusher

2. Collect drained or crushed filters in a closed top barrel and store in a dry place. These materials are recyclable.

Solvent

1. Work only in approved solvent cleaning stations that meet requirements for vapor control and fire safety. This typically means a limited sized opening over the reservoir and a lid equipped with a fusible link that drops over the tank in case of fire (see Fig.1-20).

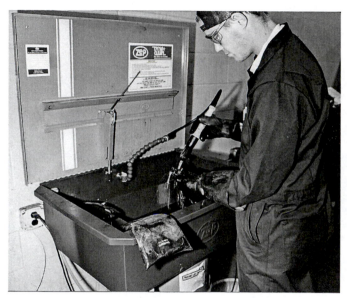

Fig.1-20   A solvent cleaning station with vapor control and fire safety lid

2. Add no hazardous chemicals such as gasoline or cleaners, especially those containing chlorinated hydrocarbons, to the solvent. Residual oil and brake fluid are not considered contaminants in this case.

3. Report spills of more than one pound (estimated) to management, as a log of such spills is generally required.

Friction Materials

1. Assume that all friction materials including clutches and brakes contain asbestos.

2. Wash all flywheels, clutches and brake drums or rotors before machining them and wash hands after handling. This reduces the possibility of inhaling or ingesting asbestos.

3. Place all used clutches and other friction materials in a closed barrel and store in a dry place. These materials are recyclable or can often be returned to rebuilders as cores. Other means of disposal depend upon local regulators.

Batteries

1. To reduce handling, charge batteries while still in the vehicle if possible.

2. If removed from the vehicle, place them in a plastic secondary container (non-metallic) to protect against spills containing acid and lead (see Fig.1-21).

Fig.1-21 Battery temporarily placed in a plastic bin

3. Store battery cores separate from other materials in a secondary containment unit (see Fig.1-22). Battery cores are recyclable.

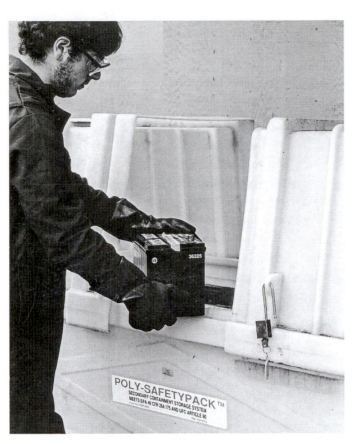

Fig.1-22 Double containment for battery cores

SUMMARY

Workplace safety is both a personal and a business priority. Employees want to avoid injury and employers want an accident-free workplace out of interest for their employees and because accidents directly affect operating cost. Practices associated with hazardous materials and the management of wastes are integral parts of both shop safety and shop management. All employees need to know at least the following:

1. Basic safety guidelines and the particular hazards present in their shop environment.

2. Potential hazards associated with hazardous materials and best management practices for handling, containing, and storing hazardous materials and hazardous wastes.

3. The information available on Material Safety Data Sheets.

4. How to read National Fire Protection Agency labels and assess potential hazards.

5. The location and proper use of safety equipment and fire extinguishers.

Because workplace activities and equipment vary, the content of this chapter may not be comprehensive in terms of covering any particular shop situation. For example, in regard to engine building alone, the workplace could be machine shop, an automotive repair facility, or an engine rebuilder also doing engine installations. Ultimately, workplace safety requires conscientious attitudes toward safety and constant vigilance for workplace hazards.

Chapter 1

SHOP SAFETY AND HAZARDOUS WASTE MANAGEMENT

Review Questions

1. Technician A says that when lifting heavy objects, keep your back straight. Technician B says lift with your legs. Who is right?

   a. A only   c. Both A and B
   b. B only   d. Neither A or B

2. Technician A says that to avoid back injuries, lift with your back straight and legs bent. Technician B says avoid working in awkward positions that strain your back. Who is right?

   a. A only   c.   Both A and B
   b. B only   d.   Neither A or B

3. Technician A says that hazards to the back include working while bent over fenders or overhead. Technician B says do not lift engine or transmission parts. Who is right?

   a. A only   c.   Both A and B
   b. B only   d.   Neither A or B

4. Technician A says that when removing engines, disconnect or remove the battery. Technician B says plug disconnected fuel lines. Who is right?

   a. A only   c. Both A and B
   b. B only   d. Neither A or B

5. In case of an engine fire,
   a. check with supervisor
   b. call for help
   c. use the nearest extinguisher
   d. leave the building

6. Technician A says that before using a remote starter switch, place manual transmissions in neutral. Technician B says first place automatic transmissions in park. Who is right?

   a. A only   c. Both A and B
   b. B only   d. Neither A or B

7. Technician A says that heavy battery charging produces explosive hydrogen gas. Technician B says that there is no hazard of explosion when charging sealed batteries. Who is right?

   a. A only   c. Both A and B
   b. B only   d. Neither A or B

8. Machinist A says that respiratory hazards in machine shops include airborne solvent or vapor. Machinist B says that friction material from clutches and flywheels are hazards. Who is right?

   a. A only   c. Both A and B
   b. B only   d. Neither A or B

9. Clean up spilled oil
   a. at the next break
   b. when finished with the present job
   c. at the end of the day
   d. immediately

10. For safety in a machine shop environment, wear safety glasses when
    a. the supervisor is present
    b. doing hazardous work
    c. instructed to do so
    d. on the job

11. Machinist A says that all shop personnel need to know the location of eye wash stations. Machinist B says that everyone needs to know the location of fire extinguishers. Who is right?

    a. A only
    b. B only
    c. Both A and B
    d. Neither A or B

12. If you get solvent or other liquids in your eyes,
    a. flush eyes with water for 15 minutes
    b. wait for professional care
    c. check first aid kit for recommendations
    d. check with supervisor

13. Technician A says that when installing engines; check all fuel and electrical connections before cranking. Technician B says put transmissions in neutral or park before cranking. Who is right?

    a. A only
    b. B only
    c. Both A and B
    d. Neither A or B

14. Technician A says that when working on running engines, hazards include hot cooling and exhaust system components. Technician B says that fans and belts are hazards. Who is right?

    a. A only
    b. B only
    c. Both A and B
    d. Neither A or B

15. When using welding equipment, wear
    a. goggles and gloves
    b. safety glasses
    c. sleeves rolled up
    d. coveralls

16. Technician A says that if materials are ignitable or corrosive, they are classified as hazardous. Technician B says that they are hazardous if reactive or toxic. Who is right?

    a. A only
    b. B only
    c. Both A and B
    d. Neither A or B

17. Technician A says that all required information pertaining to hazards is on product labels. Technician B says that complete information is on the MSDS for the product. Who is right?

    a. A only
    b. B only
    c. Both A and B
    d. Neither A or B

18. Technician A says that under RCRA guidelines, one should eliminate or reduce the use of hazardous materials where possible. Technician B says that one can also recycle wastes or properly dispose of them. Who is right?

    a. A only
    b. B only
    c. Both A and B
    d. Neither A or B

19. Technician A says that secondary containment units must be leak tight and hold more than the primary containers. Technician B says that old asbestos friction materials must be stored in double contained systems. Who is right?

    a. A only
    b. B only
    c. Both A and B
    d. Neither A or B

20. Technician A says that used oil filters may be crushed, contained, and hauled away for scrap metal. Technician B says that thoroughly draining them is acceptable for recycling. Who is right?

    a. A only
    b. B only
    c. Both A and B
    d. Neither A or B

FOR ADDITIONAL STUDY

1. Draw a map of your shop space and locate all fire extinguishers, eye wash stations, and first aid kit.

2. Locate the Material Safety Data Sheets for materials and products in your shop. Select any one of these sheets and list the health hazard data and any required special protection.

3. Find the National Fire Protection Agency label for cleaning solvent, or any other cleaning product, and list the hazards and the required protective measures.

4. Make a list of the respiratory hazards in your shop.

5. List the characteristics that make a material hazardous.

Chapter 2
FUNDAMENTALS OF MACHINING

Upon completion of this chapter, you will be able to:

- Identify the different machining processes used in engine repair and rebuilding.
- List the advantages and disadvantages of different cutting materials.
- Identify the terms used to describe the geometry of cutting tools.
- Explain the effects of each single pointed tool angle on cutting action and surface finish.
- Describe the differences between, and uses of, drills, reamers, countersinks, and spot-facing.
- Describe the abrasives and explain the cutting action of grinding wheels and hones.
- Determine correct cutting speeds for tools and materials and calculate machining RPM.
- Check machinery installations for level and twist.
- Explain the required alignment for machining between centers and list examples in automotive machining.
- Check alignment of vertical spindle machines such as used in milling, drilling, and guide and seat machining.

INTRODUCTION

Machining, in the simplest possible terms, involves precise metal removal to bring parts to dimensional specifications. For automotive machinists, this is usually from castings or forgings coming from engine assemblies. Because automotive machinists rework parts as opposed to manufacturing them, they typically learn about engines and engine assembly and how to operate machine tools that perform very specific operations. Common operations among these are:

1. Boring cylinders
2. Honing cylinders
3. Grinding crankshafts
4. Grinding camshafts
5. Grinding or milling head and block surfaces
6. Grinding valves and valve seats
7. Grinding flywheel surfaces
8. Drilling and reaming for valve guides.

Training in this field generally lacks answers to "why" tools work the way they do. Eventually, most automotive machinists find themselves in front of a lathe or milling machine and wishing they knew a little more about the fine points of machining. This is because lathes and milling machines are general machine tools not designed for specific automotive operations. Machinists in the manufacturing field, while learning their trade, first learn how these machines setup, operate, and how cutting tools perform but often nothing at all about the function of the machined parts. This situation is the exact reverse of automotive machinist training.

MACHINING PROCESSES

All machine tool processes must meet certain basic requirements. There must be relative motion between the work piece and the cutting tool and the tool material must be harder than the work piece. Described below are the basic machining processes that automotive machinists must be familiar with.

1. **Drilling and Related Operations** - Drills, reamers, and counterbores are all end cutting tools used for a variety of miscellaneous operations in drilling or milling machines and with portable drill motors (see Fig.2-1). Drills are used to rough holes to size such as when removing broken fasteners or preparing to tap threads. Reamers follow drilling to improve size control and surface finishes such as when reaming valve guides. Counterbores, for example, are used to "spot-face" castings under the heads of capscrews or to cut off rocker stud bosses when converting to screw-in

studs. One of the more routine operations for automotive machinists is probably drilling and reaming for "false" valve guide bushings.

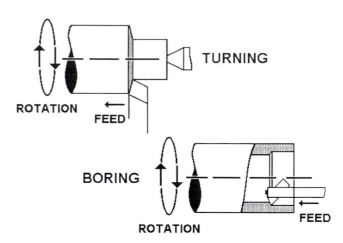

Fig.2-2   Relative motion when turning or boring

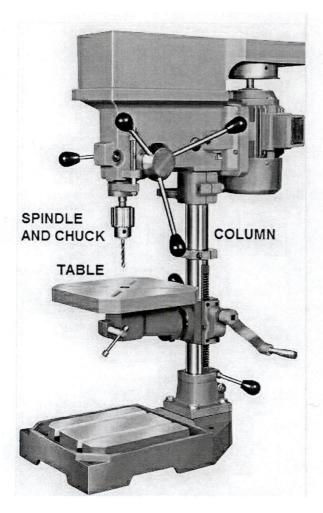

Fig.2-1   Features of a basic drilling machine

2. **Turning and Boring** - Turning is done in engine lathes with single pointed tools. Boring refers only to turning operations done on inside diameters (see Fig.2-2). Automotive machinists routinely bore cylinders using a boring bar but will use an engine lathe only for special operations or tool making (see Fig.2-3).

Fig.2-3   Features of an engine lathe

3. **Milling** - Milling is done in machine tools with vertical or horizontal spindles. For automotive machinists, experience is likely limited to vertical spindle machines (see Fig.2-4). Chip removal is done using end mills for miscellaneous odd jobs and face mills for the routine surfacing of cylinder heads and blocks (see Figs. 2-5 and 6).

Fig.2-4   Features of a vertical milling machine

Fig.2-5   Milling cutters used in vertical mills

Fig.2-6   A face milling cutter assembly

4. **Grinding** - Grinding is a form of abrasive machining where chip removal is done by millions of abrasive grains bonded into grinding wheels. Stock removal is generally small and the required precision often high. The machine tools include vertical spindle surfacing machines and horizontal spindle cylindrical grinders. Common vertical spindle operations include surfacing cylinder heads, blocks, and flywheels (see Fig.2-7). Horizontal spindle grinders are largely limited to crankshaft grinding and small machine tools such as valve grinders (see Fig.2-8). These machine tools in automotive machine shops have special features to make them more suitable to the particular work done.

Fig.2-7   Vertical spindle flywheel grinder

Fig.2-8   Horizontal spindle crankshaft grinder

5. **Broaching** - Although the name is borrowed for some automotive equipment, there is no true broaching done in small automotive machine shops. Broaches remove chips as progressively deeper cutting teeth are forced across or through the work (see Fig.2-9). Many cylinder head and block castings are surfaced by broaching in manufacturing. Key slots through gears and sprockets are examples of internal broaching.

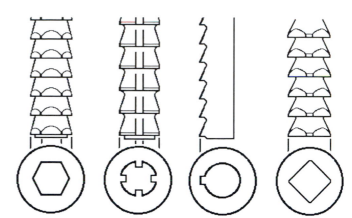

Fig. 2-9   Broaching operations

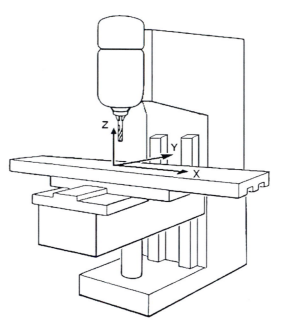

Fig.2-10   X, Y, and Z-axis

TOOL MATERIALS

Drills, reamers, and milling cutters are commonly made of HSS or high-speed steel. This refers to the ability of the tool material to withstand the heat generated at higher cutting speeds. Alloying elements such as tungsten, vanadium, and cobalt enhance the hardness of carbon steels not only at room temperature but also at the higher temperatures encountered in machining. While plain high carbon tools are also very hard, they soften and fail at higher cutting speeds.

Tungsten carbide tools are also commonly used in automotive machining for boring tools and the individual cutting tools in face mills used for surfacing. The key ingredients are the carbides of metallic tungsten produced in electric furnaces. The carbides are mixed in varying percentages with cobalt powder, compressed into pellets of the appropriate shapes, and "sintered," or bonded together in furnaces. The finished material is brazed or clamped to steel tool holders. Tungsten carbides are very heat resistant and operate at cutting speeds at least triple those of HSS tools.

There is a trade-off in that carbides are not particularly resistant to shock. Heat resistance and shock resistance can be balanced somewhat by varying the percentage of cobalt binder relative to carbides; the higher the percentage of cobalt binder, the greater the resistance to shock but the lower the resistance to heat. This is especially important in automotive machining because many operations, such as surfacing cylinder heads and blocks, involve "interrupted" cuts. As cutting tools pass across cylinders, combustion chambers, and bolt holes, the cut is interrupted and the tools must withstand the shock.

Advances in this tool technology exist in the form of "CBN" or cubic boron nitride and "PCD" or polycrystalline diamond tool materials. These cutting tools operate at approximately 3,000 surface feet per minute or ten times the cutting speed of carbides. In automotive surfacing operations, CBN tools are used for iron castings and PCD tools for aluminum castings.

Grinding wheels for automotive machining applications generally use aluminum oxide or silicon carbide abrasives and have "vitrified" glass bonds. Aluminum oxide wheels are used for grinding steel and nodular iron such as found in crankshafts. Silicon carbide wheels are used for grinding iron such as cylinder head or block surfaces and integral iron valve seats.

These wheels are manufactured by carefully sorting the abrasives by grit size, mixing them with silica sand in varying percentages, compressing the mixture into wheel shaped molds and curing the wheels at temperatures sufficient to melt the sand.

The result is an even distribution of abrasives bonded in place with posts of glass or vitrified silica sand.

A third composition of grinding wheel is the silicon carbide "green grit" wheel used to grind carbide tools. These wheels are mounted in tool grinders and are available in different grit sizes for the rough or finish grinding of carbide tools but are not suitable for grinding steel or iron (see Fig.2-11). A standard wheel marking code describes some possible variations in wheel composition (see Fig.2-12). Although most machine shops buy wheels for specific purposes, crankshaft grinding for example, the machinists can study the code and order wheels with a different composition if performance problems are encountered.

Fig.2-11 A tool grinder for single pointed tools

ABRASIVE	GRAIN SIZE	GRADE	STRUCTURE	BOND
A	36	L	S	V
Aluminum Oxide (A)	1 Coarse ↓ 70 Fine	A Soft ↓ Z Hard	1 Dense ↓ 15 Open	V-Vitrified S-Silicate R-Rubber B-Resinoid
Silicon Carbide (C)				

Fig. 2-12 The industry standard wheel marking code

CUTTING TOOL GLOSSARY

Below is a short glossary of special terms necessary to follow any discussion of cutting tools.

1. "Depth of cut" refers to how much of the work the tool engages. For example, when surfacing a cylinder head, if the depth of cut is .005in (.15mm), then .005in will be removed from the deck. When turning or boring, we normally set the tool according to the change in diameter and the depth is half that change (see Fig.2-13).

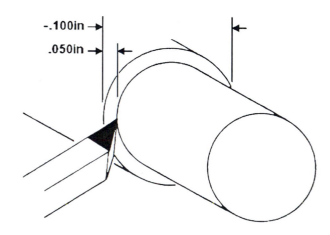

Fig. 2-13 Depth of cut versus change in diameter

2. "Feed rate" refers to the rate that the tool advances during the cut. When boring, the feed rate is in terms of inches per revolution

(see Fig.2-14). In milling machines, the feed rate depends upon the RPM and number of cutting teeth and is in inches per minute.

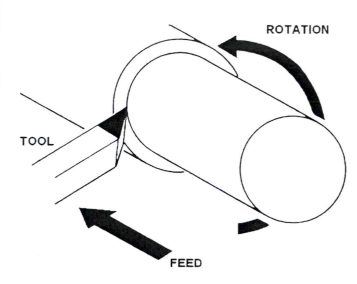

Fig. 2-14 Tool feed per revolution

3. "Cutting speed" refers to the number of feet per minute that the tool sees relative to the work. Cutting speeds vary according to the material and with the cutting tool's tolerance for heat.

4. "RPM" is not a new term but it is important to understand that for machining, it is calculated according the material cutting speed and tool material. If the tool is rotating such as in a boring bar, RPM is in reference to the tool. If the work rotates such as in a lathe, RPM is in reference to the work. The larger the diameter is at a given RPM, the greater the number of surface feet per minute.

SINGLE POINTED TOOLS

Single pointed tools have cutting edges and other angles that influence machining efficiency. There are standardized terms that define the angles and features of single pointed tools. The example below is for a general purpose HSS lathe tool (Fig. 2-15). It is important to understand the purpose of each angle and how it affects cutting efficiency.

Side rake...12 Degrees
Back rake 12
End cutting edge angle 30
Side cutting edge angle 15
End relief.......................................10
Side relief......................................10
Nose radius ....1/32in.

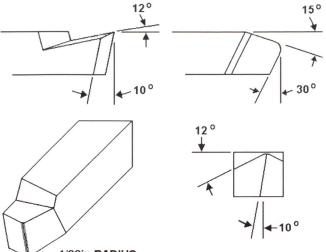

Fig. 2-15 Sample tool angles for a general purpose HSS lathe tool

To begin, the back and side rake angles together form the surface that the chips pass over as they come off the work (see Fig.2-16). The greater these angles, the more free the cutting action. Reducing rake angles somewhat forces the chip to turn and break off and reduces the tendency for the tool to "dig" in.

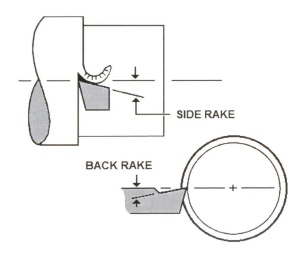

Fig.2-16 Back and side rake angles

The relief angles expose the nose and cutting edge of the tool to the work (see Fig.2-17). Without sufficient side relief, the tool cannot feed into the cut as the work rotates because the side of the tool bumps into the work. With too much relief, there is too little support and mass under the cutting edge and the resultant tool wear and overheating reduces longevity.

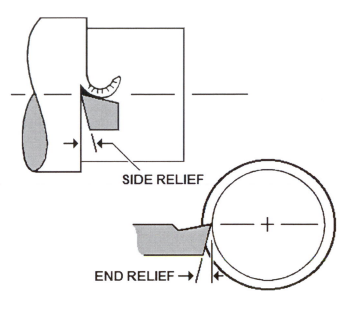

Fig.2-17 Side and end relief angles

The nose radius is the single greatest factor affecting surface finish. Unless other factors interfere such as a lack stiffness or rigidity, a larger radius produces a better finish. However, a larger radius increases the power required and the tendency for the work and the tool to vibrate or "chatter." While good surface finishes are not always required, using a larger radius enables improved surface finishes at higher feed rates.

Also important is the difference between the angle ground on the tool and the angle the chip sees at the work. For example, a 10 degree back rake angle when turning an outside diameter is one thing, but quite different when boring on the inside diameter (see Fig.2-18). Basically, back rake angles are greater for boring than for turning.

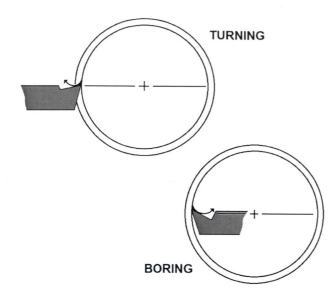

TURNING

BORING

Fig.2-18   Differences in apparent rake angle

Because the work and cutting tool deflect under the force of the cut, setting the cutting tool to the correct height relative to the center of rotation is important. The tool is set on center for turning but above center for boring (see Fig.2-19). With the tool set at the proper height, deflection causes the tool to slightly disengage from the cut. If set incorrectly, the depth of cut increases with deflection and the very force causing the deflection increases even further.

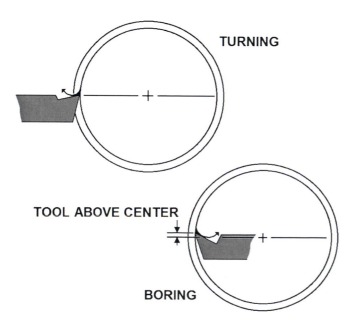

TURNING

TOOL ABOVE CENTER

BORING

Fig.2-19   Tool setting relative to the centerline

The angle that the side cutting edge is presented to the work is important for two reasons. First, the greater the angle, the greater is the tendency for the work to deflect away from the tool. Second, the smaller the angle, the thicker the chip and the more concentrated the tool wear (see Fig.2-20). Presenting the side cutting edge to the work at an angle of 15 to 25 degrees allows the chip to thin somewhat and reduces the forces of deflection against the work.

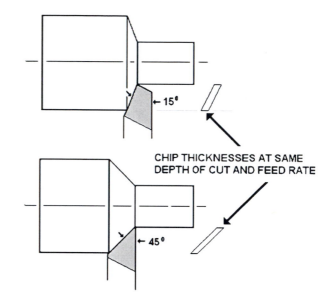

15°

CHIP THICKNESSES AT SAME
DEPTH OF CUT AND FEED RATE

45°

Fig.2-20   Lead angle and its effect on chip thickness

The recommendations in the following table eliminate experimentation in the search for angles that give the best performance. Any variations from these will be minor and made only to adjust for particular variables on the job.

Recommended Angles for HSS Single Point Tools

Material	Side Relief	End Relief	Back Rake	Side Rake
Alloy, stainless, High-carbon steel	7-9	6-8	5-7	8-10
SAE steels:				
1020-1040	8-10	8-10	10-12	10-12
1045-1090	7-9	8-10	10-12	10-12
1300s	7-9	7-9	12-14	14-16
2315-2340	7-9	7-9	8-10	10-12
2345-2350	7-9	7-9	6-8	8-10
3115-3130	7-9	7-9	8-10	10-12
3135-3140	7-9	7-9	8-10	8-10
3250-6145	7-9	7-9	6-8	8-10
Aluminum	12-14	8-10	30-35	14-16
Brass	10-12	8-10	0	0
Bronze	8-10	8-10	0	2-4
Cast iron, gray	8-10	6-8	3-5	10-12
Copper, soft	12-14	12-14	14-16	18-20

For SAE steels in the codes above, the first digit is the alloying element, the second is the percent of alloy, and the last digits are "points" carbon. To determine points, 30 points are equal to .30 percent. The codes 1020 through 1090 refer to plain carbon steel with no alloying elements and 20 to 90 points carbon. Complete breakdowns are found in "Machinery's Handbook" and other technical references.

Cutting oil extends tool longevity. The oil cools the tool and prevents chips from bonding to the cutting edges and effectively dulling them. Cast iron machines freely without cutting oil but aluminum requires it. All threading operations require cutting oil.

MILLING CUTTERS

End mills are cutting tools that remove chips from the end or sides of the tool. "Two flute" end mills have two cutting edges that cross over the centerline of the tool and penetrate a work piece directly from the end similar to a drill. End mills with more cutting edges will not cut from the end because the cutting edges do not go to the center of the cutter (see Fig.2-21). End mills are used by automotive machinists in vertical milling machines for a variety of odd jobs and tool making projects. Typically, these jobs are not in the planned workload, but if the shop has the machine, a million and one uses will be found.

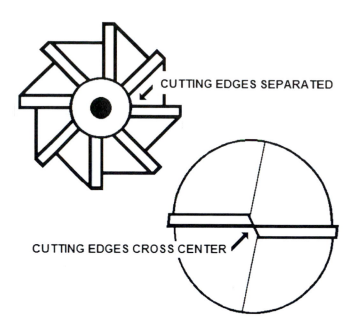

CUTTING EDGES SEPARATED

CUTTING EDGES CROSS CENTER

Fig.2-21   Two-lip and multi-flute end mills

As mentioned earlier, automotive machinists use face mills for surfacing operations in automotive machine tools. A face-milling cutter is really a large hub carrying 10 or more single pointed tools (see Fig.2-22). Each individual tool is set an equal distance from the face of the hub. When dull, brazed carbide tools must be removed, resharpened, and reset to equal heights. Inserted carbides are indexed to expose fresh cutting edges but may still require checking for equal heights.

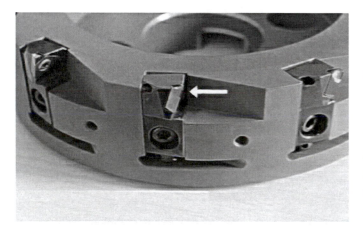

Fig.2-22   Individual single point tools in face mill assembly

A notable difference in automotive surfacing machines is that the axis of the vertical spindle of the machine is tilted slightly (see Fig.2-23). The amount of tilt is approximately .004in (.10mm) across a 13in (33cm) diameter face mill or grinding wheel. This means that there is a slight arc to surfaces or a "hollow cut" as viewed from the end. The arc however is less than .0005in (.013mm). The tilt in the adjustment reduces the total force downward on the work when cutting. Evidence of this adjustment is that the tool marks made by the cutter are from the leading edge of the cutter only. If the pattern suggests that both leading and trailing edges of the cutter are contacting the surface, the cutter is probably dull and deflecting under the force of the cut or the work piece is moving in the set-up.

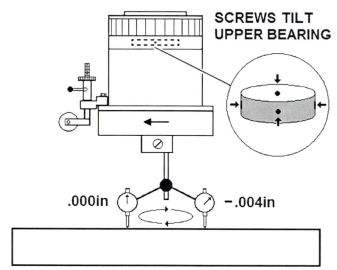

SCREWS TILT UPPER BEARING

.000in   −.004in

Fig.2-23   The tilted axis in surfacing machines

One caution with milling cutters is to avoid "climb milling." If cutting from the side of an end mill, and the cutter rotation is in the same direction as the feed, the cutter pulls the work deeper into the cut (see Fig.2-24). The danger is that the cutter may break or pull the work loose from the setup.

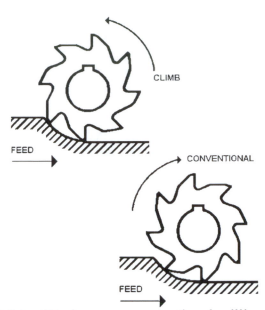

Fig.2-24   Climb versus conventional milling

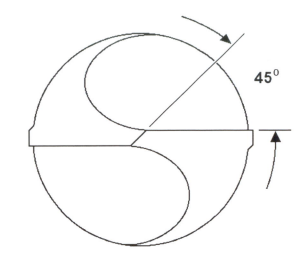

Fig.2-25   The chisel edge at 45 degrees between cutting edges

As mentioned with single pointed tools, cutting oil extends tool longevity. This is especially important in regard to milling cutters because they cannot be sharpened off-hand in a pedestal grinder as are HSS lathe tools or drills. Resharpening milling cutters requires a tool and cutter grinder and most shops sublet the work. It pays to operate these cutters at proper speeds, feeds, and with cutting oil to extend their use to maximum limits.

DRILLS, REAMERS, AND OTHER DRILLING TOOLS

Drills cut from the end only. "Flutes" spiral around the outside of the drill and lift chips out of the drilled hole. The edges of the flutes are ground cylindrically and therefore will not cut. To cut efficiently, the length of the two cutting edges, and their angles, must be equal. On a properly ground drill, the cutting edges form a 45-degree angle between them when viewed from the end (see Fig.2-25). This angle between the cutting edges is called a "chisel edge".

The "web" runs the length of the drill and gradually thickens, as the drill becomes shorter (see Fig.2-26). At the point of the drill, the web forms the chisel edge where cutting is least efficient. For this reason, small diameter pilot holes are drilled before drilling with large diameter drills. Also, as the drill becomes shorter after repeated sharpening, the web thickens and the need for a pilot hole is even greater.

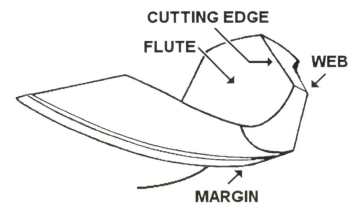

Fig.2-26 The web of a drill forming the chisel edge

Both drills and reamers are available in fractional, letter, and number or wire sizes. Of course, metric drills are also available although most shops find that the selection of sizes in inches is adequate. Drills have straight shanks through 1/2in diameters and Morse taper shanks for larger diameters (see Fig.2-27). This taper is 5/8in per foot but the drive tang is what allows these drills to transmit greater torque than possible with straight shanks. Without the drive tang on taper shank drills they would slip in the machine spindle. Removing taper shank drills from a machine spindle requires driving a tapered "drift" through a slot in the spindle (see Fig.2-28).

Fig.2-28   A drift for taper shank tools

Center drills are short and rigid and therefore do not drift off location when used to create pilot holes (see Fig.2-29). The 60-degree angle produced by the center drill also prepares the work to run on centers in a lathe or grinder (see Fig.2-30).

Fig.2-27   The Morse taper on a drill shank

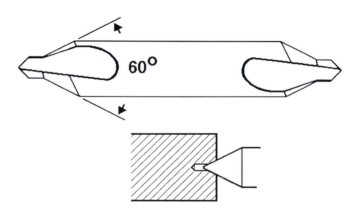

Fig.2-29   A 60-degree center drill and center

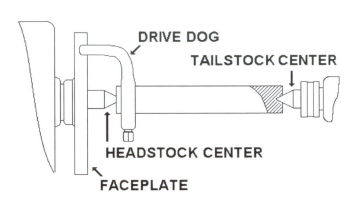

Fig.2-30 A set-up for turning or grinding between centers

"Core" drills are common in automotive shops. They do not cut at the center, but they will enlarge existing holes (see Fig.2-31). The drills for false valve guide bushings are a good example of core drills. The term "core" refers to the use of sand cores to form holes during the casting process.

Fig.2-31 Core drill (L) and core reamer (R)

There are both hand and machine reamers. It is important to recognize the difference because the hand reamer will break and rip the work out of its set-up if run in a machine. All reamers are end

cutting tools and, like drills, the margins on the sides are ground cylindrically. Hand reamers cut on the end for a distance approximately equal to the diameter (see Fig.2-32). The shank is square on the end to provide for a tap handle. Machine reamers cut on a short 45-degree bevel at the end of the flutes (see Fig.2-33).

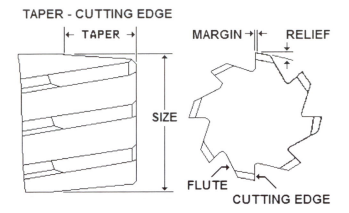

Fig.2-32 The cutting edges of a hand reamer

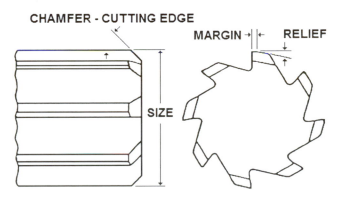

Fig.2-33 The cutting edges of a machine reamer

It is important to drill approximately 5 percent undersize for machine reaming. This amounts to drilling .025in (.64mm) undersize for a 1/2in (13mm) diameter reamer. Considering that reaming is done to improve finishes and for better size control, the chip load should be kept small. Be aware that because drills and reamers are end cutting, they follow the alignment of existing holes. If the hole- location is wrong, it can be moved with an end mill but not with another drill.

Hand reamers should remove only .003 to .005in (.10mm), certainly not more than .010in (25mm). Frequently the amount allowed for hand reaming depends upon the finish size desired and the drill selection on hand. Countersinks are used in automotive machining to bevel the edges or "chamfer" bolt holes. The included angle of countersinks is typically 82 degrees (see Fig.2-34). This angle fits rivet heads, and unless specified otherwise, countersinks will likely come with this angle.

Fig.2-35 Spotfacing head bolt holes with a counterbore

Fig.2-34 A countersink

Although intended for a variety of other jobs, machinists frequently use counterbores to remove rocker stud bosses and spotface head bolt holes (see Fig.2-35). Counterbores come with straight or tapered shanks and, for larger sizes, have removable pilots (see Fig.2-36). To spotface a head bolt hole, for example, select a counterbore diameter matching the original spotfaced area and a pilot that fits the head bolt hole. It is best to modify the counterbores used for spotfacing by grinding the corners at a 45 degree angle to prevent forming sharp corners under head bolts and causing stress concentration.

Fig.2-36 A counterbore and removable pilot

Except for cast iron, use cutting oil where possible to extend tool life. While drills can be sharpened off-hand in a pedestal grinder, reamers and counterbores require sharpening in a tool and cutter grinder.

2-14

GRINDING AND HONING

Chip removal in abrasive machining is somewhat different than with milling cutters or single pointed tools. Exposed abrasive grains on the wheel face remove chips but without the precise geometry of other cutting tools. Instead, each grain removes microscopic chips until it dulls and it then breaks out of the wheel face under the pressure of the cutting action. If these grains do not break away, it becomes necessary to "dress" the wheel with a diamond or star wheel dresser. Although these dressers work differently, both break-up the grinding wheel faces and expose fresh abrasive grains.

A diamond dresser is much harder than aluminum oxide or silicon carbide and removes dull material from the wheel face with minimal wear. The diamond should remove only .001 to .002in (.04mm) from the wheel face per pass and, if available, use coolant when dressing (see Fig.2-37). The diamond is fed across the wheel face quickly when a coarse dress is desired for roughing. The depth is reduced to .0005in and fed slowly across the wheel for finish grinding. Coolant flushes loose grit from the wheel and cools the diamond in its setting. Also, the diamond should be rotated occasionally in its mount to distribute wear around the point. Dressing with a diamond both sharpens the wheel and trues the face.

A star wheel dresser works on a different principle. The star wheels are steel and rotate on an axle. By pressing the dresser into the face of the wheel, the star wheels rotate and hammer on the wheel face thereby loosening dull material (see Fig.2-38). This action sharpens the wheel but does not true it. When using a star wheel dresser, there must be sufficient force against the wheel face to force rotation. With too little force, the star wheels grind away.

Fig.2-38   A star wheel dresser

To determine if the grinding wheel is cutting properly, watch for "hard" or "soft" grinding action. "Hard" action is caused by failure of the abrasive grains to wear away and expose sharper abrasives. Stock removal rates are low and the work tends to burn because of the friction of dull abrasives. The grinding action can be corrected by dressing the wheel, but if dressing is required too frequently, the wheel selection is probably incorrect for the job. "Soft" action is caused by too rapid wheel wear. Excessive wheel wear releases grit that jams between the work and the wheel and produces scored surface finishes. Causes include too coarse a wheel dress and wheel diameters worn too far undersize. In either of these cases, each abrasive overloads at normal stock removal rates and breaks away under the pressure. Unless spindle speeds can be increased to correct surface speed, under-

Fig.2-37   Positioning of diamond for wheel
dressing and truing

sized grinding wheels must be replaced. This is a common occurrence in valve grinders when wheels wear 20 percent or more undersize.

While surface speeds are much lower, honing abrasives work on the same principles. Some wear is necessary to keep the abrasives sharp and RPM must be adjusted for diameter to keep surface speeds in the correct range. Honing machines have pressure adjustments that can reduce or increase the wear rate of the abrasives and thereby affect cutting action.

Coolants prevent burning the work and flush loose abrasives out of the operation.

Grinding or honing oils flush loose material but also lubricate the cutting action. Coolants and grinding oils should not be allowed to carry excessive grit or surface finishes will be affected. At the same time, hazardous waste disposal requirements encourage filtering and reusing coolants and oils as long as possible.

Water based coolants require rust inhibiting and anti-bacterial additives to extend usable life. It also helps to separate solids and remove oils from water-based coolant. Solids are sometimes filtered but usually allowed to settle out in coolant tanks (seeFig.2-39). Skimmers are used to remove oil from both coolants and, when cold, cleaning solutions (Fig.2-40).

Fig.2-40 "Skimming" oil from a cold jet spray machine

SPEEDS AND FEEDS

Having made many references to cutting speeds and surface speeds, we can now discuss them in precise terms. Any cutting action generates heat, and the harder the material, the greater the heat. Using more heat resistant tool materials such as carbide increase the allowable surface speeds. Materials with hardness levels above Rockwell 40C will probably require machining by grinding.

The following table lays out the required information and formulas in an organized manner. Study the table and work through the sample problems to develop a full understanding. As much as possible, the calculations should be performed in your head and the machinery adjusted to the nearest value.

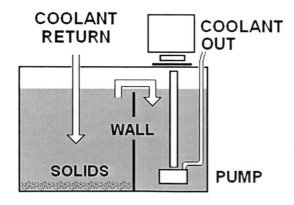

Fig.2-39 Separating solids in the coolant tank

Speeds and Feeds

Material Cutting Speeds in Surface Feet per Minute

Material	Turning/Drilling	Milling
Tool steels	50	40
Cast iron	60	50
Mild steels 1020-1040	100	80
Brass and soft bronze	200	160
Aluminum and magnesium	300	200

Note: These recommendations are for HSS tools. Triple these speeds for carbide tools. Run CBN at 3,000 feet per minute.

Formulas for Machining Speeds

Operation	Formula	Notes
Drilling, milling	RPM = 4 x CS ÷ D	Cutter diameter
Turning, boring	RPM = 4 x CS ÷ D	Work diameter
Grinding	RPM = 4 x CS ÷ D	Use 5,500 SFPM
Grinding	SFPM = RPM x D ÷ 4	

Note (1) Operate reamers, counterbores, and countersinks at no more than half of drilling RPM.

Note (2) The formula above is accurate within 5% (it is simplified from RPM = 12 x CS ÷ 3.1416 X D).

Feed Rates

Operation	Formula or Rate
Turning, boring	.002 to .006 per revolution
Drilling	.001 per 1/16in. of diameter
Reaming	.002 per cutting edge per revolution
Milling	IPM = RPM x No. Teeth x Chip Load

Note: IPM equals inches per minute and chip loads are in the table below.

Chip Loads for Milling Cutters

Cutter	Chip Load Per Tooth
End mills	.001 to .005in.
Face mills	.010 to .012in.

Note: Chip load is equivalent to the thickness of each chip. Add 50% for iron and aluminum.

As mentioned, some practice is required to apply these formulas correctly. The following sample problems are typical of those encountered in an automotive machine shop.

Sample Problems

1. To make a driver, it is necessary to turn down a piece of 1in diameter mild steel stock in a lathe. What RPM and feed rate should be used?

$$RPM = 4 \times CS \div D$$
$$= 4 \times 100 \div 1$$
$$= 400 \div 1$$
$$= 400$$

Feed Rate: Use .002in per revolution for finishing and up to .006in per revolution for roughing.

2. A 3/4in diameter 4-lip end mill is being used to cut down iron rocker stud bosses on an iron head. What RPM should be used?

$$RPM = 4 \times CS \div D$$
$$= 4 \times 50 \div .750$$
$$= 200 \div .750$$
$$= 267$$

3. 3A 10in diameter face mill with 10 carbide cutters is used to face iron engine blocks. What RPM and feed rate should be used?

$$RPM = 4 \times CS \div D$$
$$= 4 \times 50 \div 10$$
$$= 200 \div 10$$
$$= 20 \times 3 \text{ (for carbide)}$$
$$= 60$$

Feed Rate = RPM x No. Teeth x Chip Load
$$= 60 \times 10 \times .018in$$
$$= 10.8 \text{ IPM}$$

4. A 7in valve grinder wheel is worn down to a 5in diameter. The drive motor runs at 3150 RPM and is not adjustable. Does this account for the scored finish?

$$SFPM = RPM \times D \div 4$$
$$= 3150 \times 5 \div 4$$
$$= 3,938$$

Surface speed is 72 percent of normal (3,938/5,500) and causes "soft" grinding action. Replacing the wheel will restore the surface speed and the grinding action.

5. A 1/2in hole is being drilled in a 3in diameter piece of mild steel in a lathe. What RPM should the work run at?

$$RPM = 4 \times CS \div D$$
$$= 4 \times 100 \div .5$$
$$= 400 \div .5$$
$$= 800$$

Machining at optimum speeds and feeds improves productivity and extends tool life. However, this is sometimes not possible because of the lack of rigidity in machine spindles and in some work pieces or work holding systems. This lack of rigidity causes vibration or "chatter" in machining that can be seen in poor surface finishes and in reduced tool life. Eliminating chatter requires that speeds be reduced and/or that feed rates be increased.

MACHINE INSTALLATION AND SET-UP

Without proper installation, the best machine tools are not likely to produce the precision that they are capable of. This is especially true for large machinery such as surfacing machines and crankshaft grinders.

Moving this machinery is the first problem. Machinery jacks with a very low profile are available for jacking the machine up off the floor (see

Fig.2-41). From the start, keep in mind that hands should never be placed under machinery. Place jacks at every corner and raise the machine evenly to prevent tilting off-balance. If the machine is too low for the jacks, raise the machine with long machinery pinch bars (see Fig.2-42). The machine will lift easily but this is obviously not a one-man job.

With the machine elevated approximately 6in (15cm) off of the floor, place machinery dollies under each corner, lower the machine onto the dollies, and remove the jacks (see Fig.2-43). These dollies steer at every corner and the machine will maneuver into very tight spaces. Push the machine into position steering the dollies as necessary.

Fig.2-41   Low profile machinery jacks

Fig.2-43   Dollies roll under the machine after raising it with machinery jacks

Once in place, raise the machine again with the machinery jacks and remove the dollies. Place pads or machinery mounts under the corners of the machine and lower the machine off the jacks. Check the Machine Instruction Manual for installation instructions and recommended mounting points. Next, using a precision level on the machinery ways or table, level the machine (see Fig.2-44). This may require shims or the machine or machine mounts might be equipped with adjusting screws. These levels are precise within .0005in (.013mm) per foot.

Fig.2-42   Pinch bars lifting a machine to clear machinery jacks

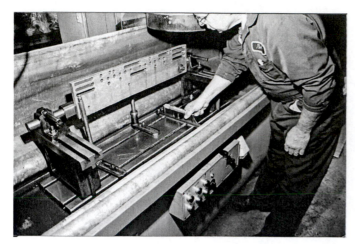

Fig.2-44   Leveling the base or table

Leveling is especially important for machine tools such as guide and seat machines that use a level to set-up work pieces for machining. After leveling the table, use the level that comes with the machine to check the alignment of the spindle with the table (see Fig.2-45).

Fig.2-45   Checking spindle alignment with the machine base using a level in the spindle

Twist also must be removed from machine tools. Place the precision level on the machine assembly with the longest axis of travel (see Fig.2-46). Note the level reading and move the assembly the length of its travel. If the machine is without twist, the reading of the level will not change. If the reading does change, locate the lowest corner of the machine and raise it to eliminate the twist.

Fig.2-46   Checking for twist in a machine base

Remember that we are advising customers to resurface heads with as little as .002in (.05mm) warpage. Unless surfacing machines are properly installed, these heads will not be flat within .002in after resurfacing. The equivalent is also true for other machining operations.

CORRECTING ALIGNMENTS BETWEEN CENTERS

Machine tools such as cylindrical grinders and lathes are capable of machining shafts between centers on the X-axis. The center at the left end is in the "workhead" and the center on the right end is in the "tailstock" (see Fig.2-47). A variation on this in crankshaft grinders is the clamping of shafts in chucks at both headstock and tailstock ends (see Fig.2-48). Should the headstock and tailstock be misaligned in these machines, machining without taper or an out-of-round condition is difficult.

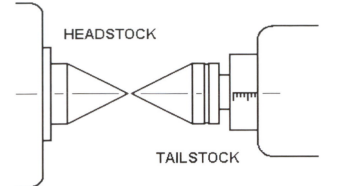

Fig.2-47   Alignment of centers between the work head and the tailstock

Fig.2-48   A crankshaft grinder with chucks at both headstock and tailstock

In an engine lathe, where the shaft is driven by the headstock and runs on a center in the tailstock, the shaft will machine with a taper unless the tailstock is brought back into alignment with the headstock. One way of making this correction is to first place a shaft equal in diameter along its length between centers. Second, take a dial indicator reading against the side of the shaft at the left, or headstock end (see Fig.2-49). Third, take another indicator reading at the right, or tailstock end. The two indicator readings should be the same. If unequal, adjust the tailstock position in or out along the Y-axis as required (see Fig.2-50).

Keep in mind that a .001in (.025mm) offset of the tailstock causes .002in (.05mm) taper.

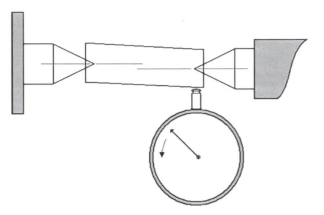

Fig.2-49   Comparing lathe headstock and tailstock alignment with a dial indicator

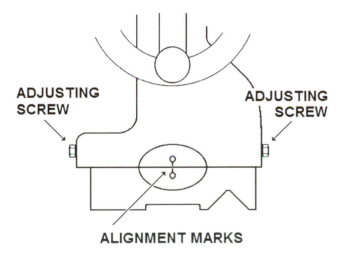

Fig.2-50   Adjusting the tailstock alignment in an engine lathe

In a crankshaft grinder, the shaft is mounted between centers or in chucks at both ends. As in the engine lathe, misalignment of the headstock and tailstock causes taper but because the shaft is clamped rigidly at both headstock and tailstock ends, the shaft will "buck" as it rotates causing out-of-round journals. To check alignment, indicate in two centers of the same diameter at both the head stock and tailstock ends (see Fig.2-51). Next, check against the sides of the centers both ends with a dial indicator (see Fig.2-52). Because there is wear along the underside of the tailstock, indicator readings should also be taken along the topsides of the centers at each end (see Fig.2-53). The same checks can be made against a true shaft mounted between the two centers. The tailstock may be adjusted side-to-side as in the engine lathe or, depending upon the construction of the machinery, it may be necessary to shim under the tailstock to tilt it into position.

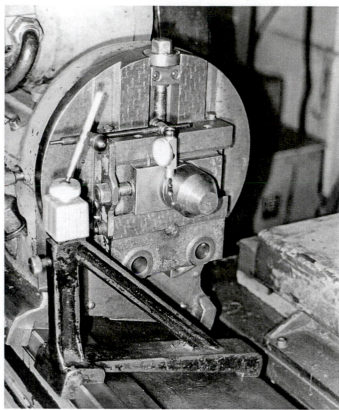

Fig.2-52   Comparing headstock and tailstock centers from the side

Fig.2-51   Indicating-in headstock and tailstock centers in a crankshaft grinder

Fig.2-53   Comparing headstock and tailstock centers from the top

With extreme wear, it will be found that tailstock alignment is lost each time it is moved along the ways. Correcting this condition may require truing the ways or bed of the machine.

TRAMMING SPINDLES

"Tramming" refers to setting the vertical spindles, or Z-axis, of milling and drilling machines square to the worktables. Unless square, milling cutters will tip and not machine surfaces flat or machine vertical and horizontal surfaces square to each other. In both milling and drilling machines, holes drilled or bored into the work will not be square to surfaces.

Detect errors by placing an indicator in the spindle and rotating the spindle by hand around the machine table (see Fig.2-54). If square, there will be no variation in indicator readings. The spindles in vertical milling machines are adjustable and can be tilted front to rear or side-to-side to correct errors. This is a common adjustment in a milling machine following a set up in which the spindle was tilted to machine at angles. These adjustments are also graduated with witness marks for each degree of rotation but understand that these graduations are not sufficiently accurate for many jobs (see Fig.2-55). Errors in drilling spindles typically require adjusting or shimming the base of the vertical column supporting the spindle (see Fig.2-56).

Fig.2-55 Witness marks for position of the alignment in a milling machine

Fig.2-56 Correcting spindle alignment in this machine required shimming under the column

Fig.2-54 Checking for out-of-square spindle vertical spindle of a milling machine

SUMMARY

While many automotive machinists frequently begin their apprenticeship with some experience in automotive repair, many possess little or no knowledge of machining. Knowledge of machining fundamentals removes much of the mystery surrounding why and how castings and forgings are machined or reconditioned. In addition, this assists machinists in the selection and use of tools and in troubleshooting machining problems.

Operations such as milling, grinding, honing, drilling and reaming, or machining with single pointed tools are basic to the engine building process. Machinists must know what RPM or speed and rates of feed are correct for these processes. They must machine flat or round surfaces as required and hold alignment with other related engine components. They must also produce the required surface finishes to assure engine sealing or to minimize wear.

Many problems in meeting flatness requirements, or alignment, can be traced to poor set-up of work pieces in machinery or even to improper installation of the machinery on the shop floor. Other problems with surface finishes can be traced to poor cutting tool performance. To identify and remedy these problems requires knowledge of the fundamentals presented in this chapter.

Chapter 2
MACHINING PRINCIPLES
Review Questions

1. Machinist A says that cutting tools cannot be harder than the material machined. Machinist B says that there must be relative motion between the work and cutting tools for chips to form. Who is right?

    a. A only   c. Both A and B
    b. B only   d. Neither A or B

2. The cutting action of grinding wheels is provided by
    a. vitrified glass or sand
    b. aluminum oxide
    c. silicon carbide
    d. aluminum oxide or silicon carbide

3. Machinist A says that high carbon steel has sufficient hardness for cutting tools but cutting speeds are limited. Machinist B says that HSS tools retain hardness better at high temperatures. Who is right?

    a. A only   c. Both A and B
    b. B only   d. Neither A or B

4. Machinist A says that as the percentage of cobalt increases in carbide tools so does resistance to heat. Machinist B says that as the percentage of carbide increases in carbide tools so does resistance to shock. Who is right?

    a. A only   c. Both A and B
    b. B only   d. Neither A or B

5. Machinist A says that relative to HSS, carbide tools run one-third as fast. Machinist B says that HSS better resists shock. Who is right?

    a. A only   c. Both A and B
    b. B only   d. Neither A or B

6. When turning or boring, a .025in depth of cut changes the diameter _____ in
    a. .0125
    b. .025
    c. .0375
    d. .050

7. Machinist A says that feed rates for milling cutters are in inches per revolution. Machinist B says that feed rates for boring or turning are in inches per minute. Who is right?

    a. A only   c. Both A and B
    b. B only   d. Neither A or B

8. As diameter increases, _____ RPM to maintain correct cutting speeds.
    a. maintain
    b. reduce
    c. increase
    d. adjust according to materials

9. Machinist A says that without sufficient clearance on single point tools, tools overheat. Machinist B says that this limits feed rates. Who is right?

    a. A only   c. Both A and B
    b. B only   d. Neither A or B

10. Machinist A says that rake angles direct chips off of the work. Machinist B says that rake angles are smallest for boring tools. Who is right?

    a. A only   c. Both A and B
    b. B only   d. Neither A or B

11. Machinist A says that for turning, set tools below center. Machinist B says that for boring, set tools on center. Who is right?

    a. A only
    b. B only
    c. Both A and B
    d. Neither A or B

12. Machinist A says that increasing the lead angle on a single pointed tool increases chip thickness and reduces tool life. Machinist B says that increasing the lead angle reduces deflection between the tool and the work. Who is right?

    a. A only
    b. B only
    c. Both A and B
    d. Neither A or B

13. Machinist A says that a smaller nose radius on a single pointed tool improves surface finishes. Machinist B says that reduced feed rates improves surface finishes. Who is right?

    a. A only
    b. B only
    c. Both A and B
    d. Neither A or B

14. Machinist A says that all end-milling cutters cut from the sides. Machinist B says that only two-lip end mills bore from the end. Who is right?

    a. A only
    b. B only
    c. Both A and B
    d. Neither A or B

15. Machinist A says that in climb milling, the cutter rotates in the same direction as the work feeds. Machinist B says that in conventional milling, the cutter rotates opposite the direction of the work feed. Who is right?

    a. A only
    b. B only
    c. Both A and B
    d. Neither A or B

16. Machinist A says that in climb milling, the cutter tends to pull the work into the cut. Machinist B says that in conventional milling, cutting from the sides of spiral fluted cutters can unscrew them from collets. Who is right?

    a. A only
    b. B only
    c. Both A and B
    d. Neither A or B

17. The spindle axis of a surfacing machine is
    a. square with the machine
    b. square with the work
    c. square with the work travel
    d. tilted in the direction of travel

18. Machinist A says that drills and reamers are ground to cut from the ends and not the sides. Machinist B says that these tools follow the alignment and positioning of existing or predrilled holes. Who is right?

    a. A only
    b. B only
    c. Both A and B
    d. Neither A or B

19. Machinist A says that the cutting action of drills is most efficient at the chisel edge. Machinist B says that the chisel edge widens as the drill becomes shorter. Who is right?

    a. A only
    b. B only
    c. Both A and B
    d. Neither A or B

20. Machinist A says that center drills hold location better because of their short length. Machinist B says that center drills form a 60-degree included angle for setting up shafts between centers. Who is right?

    a. A only
    b. B only
    c. Both A and B
    d. Neither A or B

21. Machinist A says that a machine reamer forms chips along cutting edges tapered 1/16in per foot. Machinist B says that hand reamers form chips along cutting edges on the 45-degree angle on the end. Who is right?

 a. A only
 b. B only
 c. Both A and B
 d. Neither A or B

22. Prior to reaming, holes are bored or drilled _____ undersize.
 a. .012in
 b. .02in
 c. 1/64 - 1/32in
 d. 5%

23. Drilling operations such as reaming, counterboring, and counter-sinking are done at _____ drilling speed.
 a. half
 b. full
 c. double
 d. triple

24. Hole alignments or positions are changed using
 a. drills
 b. reamers
 c. hones
 d. boring tools or end mills

25. Machinist A says that grinding coolants flush away grit and cool both the wheel and the work. Machinist B says that cutting oils lubricate the cutting action and prevent chip welding on cutting edges. Who is right?

 a. A only
 b. B only
 c. Both A and B
 d. Neither A or B

26. Machinist A says that evidence of soft grinding action is a burned surface on the work. Machinist B says that evidence of hard grinding action is a scored finish. Who is right?

 a. A only
 b. B only
 c. Both A and B
 d. Neither A or B

27. Machinist A says that to reduce tool chatter, decrease speeds. Machinist B says that to reduce chatter, increase feeds. Who is right?

 a. A only
 b. B only
 c. Both A and B
 d. Neither A or B

28. For drilling a 3/8 inch hole in steel, what RPM is correct?
 a. 1060
 b. 530
 c. 150
 d. 100

29. Machinist A says that if the headstock and tailstock are misaligned in a lathe, it cuts on a taper. Machinist B says that because the crank is clamped at both ends in a crank grinder, it grinds out-of-round if centers are misaligned. Who is right?

 a. A only
 b. B only
 c. Both A and B
 d. Neither A or B

30. Machinist A says square milling and drilling spindles with tables or they bore at angles. Machinist B says square milling spindles or they mill surfaces out of square. Who is right?

 a. A only
 b. B only
 c. Both A and B
 d. Neither A or B

FOR ADDITIONAL STUDY

1. Compare the advantages of tungsten carbide tools to machining with high-speed steel tools. What are the advantages to tungsten carbide?

2. You are milling the deck surface of and engine block. What can you do to improve the surface finish?

3. You have just resurfaced a cylinder head and the surface is twisted instead of flat. What could cause this?

4. The surface of the cylinder head casting under the head bolts is crushed. What process is used to correct this condition?

5. A valve is being refaced in a grinder and the surface is scored. What causes this?

6. A cylinder head must be drilled and then reamed to .500in for a valve guide bushing. After drilling, how much can be removed by reaming?

7. A hole drilled through a valve guide, in a guide and seat machine, is out of alignment with the original. What causes this?

8. The centers in a crankshaft grinder are running true but when a crankshaft is placed between these centers, the main journals run-out. What causes this?

9. What is the difference between a drill and a core drill?

10. What is the effect of increasing the radius at the point of single pointed tools?

Chapter 3

MEASURING TOOLS

Upon completion of this chapter, you will be able to:

- Explain the logic behind the writing of specifications and tolerances for engine parts.
- Calculate thermal expansion and contraction of iron and aluminum engine parts.
- Read micrometer scales in metric and inch units.
- List the differences between direct, transfer, and comparative measuring tools and measurements.
- Explain how correct alignments are maintained in the set-up of engine castings for machining.
- Describe surface scales and methods of measurement.
- Explain how casting thickness is measured and the needs for such measurements.

INTRODUCTION

Current trends in the development of automotive engines require that technicians and machinists measure more thoroughly and more closely. First, more engines use precision, thin wall castings to reduce engine weight. This means that cylinder oversizes are generally smaller and machining operations on castings, such as for false valve guide bushings or valve seat inserts, require more care because of the thin walls. Second, more engines use overhead camshaft and multi-valve cylinder heads that require inspection techniques unfamiliar to many technicians. Third, there are a number of light duty diesel engines that require potentially new and very precise inspection techniques.

Most important, manufacturers have improved "in-process" gauging systems in engine plants that make it possible to measure all major engine parts, mark them according to size, and selectively assemble them to optimum clearances. If we are to rebuild to comparable standards, very thorough and precise measurements and assembly are necessary.

UNDERSTANDING SPECIFICATIONS AND TOLERANCES

Most technicians and machinists understand the need to follow specifications. What is perhaps not so well understood is that the OEM (original equipment manufacturer) specifications may apply only to OEM parts. For example, when replacing an OEM cast aluminum piston with a forged aluminum replacement, the specified clearance for the forged piston must be used (see Fig.3-1). There is little doubt about the quality of the forged replacement, but specified clearance likely differs from the factory manual. Note the enlarged piston pin bosses in the forged piston. These pin bosses lower the temperature at the head of the piston by drawing heat into the lower part of the piston; therefore, the piston shape and clearance must be changed. Most importantly, be aware that specifications for all replacement parts and OEM manuals may be required.

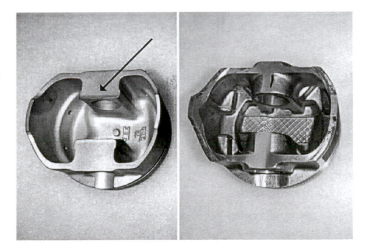

Fig.3-1   The forged piston (L) has larger pin bosses than the cast piston (R)

Diameters of crankshafts and housing bores for engine bearings are specified in the form of tolerance limits. The limits are the minimum and maximum allowable diameters. For example:

Crankshaft diameter	1.9995/2.0005in
Housing bore diameter	2.1245/2.1250in

Clearances also are stated as minimums and maximums but clearance is the difference in diameters between the shaft diameters and bearing inside diameters. To calculate the bearing inside diameter, subtract the thickness for each bearing insert from the housing bore. Because the housing bore is in terms of tolerance limits, this must be done twice. To calculate maximum bearing inside diameter, subtract the bearing thicknesses from the maximum housing bore as follows:

Maximum housing bore	+ 2.1250in
Bearing thickness	- .0615
Bearing thickness	- .0615
Maximum bearing inside diameter	2.0020

To calculate the minimum bearing inside diameter, subtract the bearing thicknesses from the minimum housing bore as follows:

Minimum housing bore	+ 2.1245in
Bearing thickness	- .0615
Bearing thickness	- .0615
Minimum bearing inside diameter	2.0015

To find minimum clearance, subtract the maximum crankshaft diameter from the minimum bearing inside diameter as follows:

Minimum bearing inside diameter	+ 2.0015in
Maximum crankshaft diameter	- 2.0005
Minimum clearance	.0010

To find maximum clearance, subtract the minimum crankshaft diameter from the maximum bearing inside diameter as follows:

Maximum bearing inside diameter	+ 2.0020in
Minimum crankshaft diameter	- 1.9995
Maximum clearance	.0025

These calculations show that tolerance limits and clearance combine in a very logical way. While manuals do not explain the logic behind specifications, we should understand them and be prepared to apply them in different ways. If we look in a bearing shop manual from a bearing manufacturer, the specifications (in inches) appear in a form similar to the following:

Part Number	Standard Shaft Dia.	Bearing OD or; Housing Bore	Oil Clearance	Bearing Thickness	Overall Length
1234AB	1.9995/ 2.0005	2.1245/ 2.1250	.0010- .0025	.0615	.842

Valve and valve guide specifications also are stated in terms of tolerance limits and clearance with one difference; the valve guide inside diameter is frequently not given. With experience, we learn what valve guide diameters are common but, if we understand the logic behind the specifications for valve stem diameter and clearance, we can quickly figure out the appropriate guide diameter. Do this by adding the minimum specified clearance to the maximum valve stem diameter, or the reverse, by adding the maximum specified clearance to the minimum valve stem diameter. For example:

Valve stem diameter	.3405 - .3415in
Stem-to-guide clearance	.0015 - .0025

Add the minimum clearance to the maximum valve stem diameter:

Minimum clearance	.0015in
Maximum stem diameter	+.3415
Valve guide diameter	.3430

Or add the maximum clearance to the minimum valve stem diameter:

Maximum clearance	.0025in
Minimum stem diameter	+.3405
Valve guide diameter	3430

Occasionally, these calculations do not match the specifications because tolerance limits for diameters are too great and original equipment assembly was "selective", that is, individual parts were sorted and matched by size to achieve optimum fit. In such cases, it is wise to check diameters and clearances by calculation before machining. While selective fitting is an economical means of improving quality in large volume production, when building one engine at a time, we must study the tolerances and machine engines and parts on hand to fit.

In most cases, following a prescribed procedure yields the desired results and by studying the procedure, we can usually judge what the tolerances really are. For example, a piston manufacturer may specify boring and honing a cylinder to the nominal standard bore diameter plus an oversize such as plus .020 or .030in (.50 or .75mm). The expectation is that pistons will fit in the oversize cylinders with the minimum specified clearance or, in this example, up to .0005in (.013mm) more. To see how this works, begin by finding the oversized cylinder bore:

Standard cylinder bore	4.000in
Oversize	+.030
Oversized cylinder bore	4.030

Next, measure the pistons and record the largest and smallest diameters. The difference between piston diameters, if within specifications, is no more than .0005in. Remember also that following recommended procedure should give minimum clearance up to plus .0005in. For example:

Largest piston	4.0290in
Smallest piston	- 4.0285
Difference in diameters	.0005

From these calculations, we find two things pertaining to tolerances. One, the piston diameters must be within 4.0285 to 4.0290in and two, clearance must be within .0010 to .0015in. Again, the optimum fit for pistons in production is typically minimum clearance. In this case, this would mean the optimum piston diameter would be 4.0290in with .001in clearance. The 4.0285in diameter piston and .0015in clearance would be acceptable. This is an example of "unilateral" tolerance, that is, the pistons can be smaller but not larger or clearance could be less than allowed with the potential for damage.

"Bilateral" tolerance means that tolerances permit a variation over and under a specified size. Such specifications typically apply to individual parts with no reference to how they might fit in assembly.

CALCULATING THERMAL EXPANSION

In machining or assembling engines, keep in mind that we are building the engine at room temperature, but it operates at widely varying temperatures. Exhaust valves operate at 1,400 degrees F plus. Pistons run up to 600 degrees F at the head or 300 degrees F at the pin bores. Cylinder head temperatures run a cycle from approximately 180 to 250 degrees F depending on engine load.

Because of these temperature variations, it is extremely important to stay within specifications. A seemingly insignificant error at room temperature may cause noise or seizure at operating temperature. Keep in mind that because of customer complaints, even minor noises require repair under warranty and therefore minor variations from specifications are potentially as expensive as major engine failures.

Metals expand and contract as temperatures cycle upward and downward. More importantly, different metals have different expansion rates, called coefficients of expansion. Engines use parts made of iron, steel, and aluminum. Iron and steel, because they have the same expansion rate, are not a problem. Aluminum though has double the expansion rate of iron and steel. Precise coefficients of expansion are as follows:

Iron and steel   .000006 per in per degree F
Aluminum   .000012

Calculate thermal expansion by multiplying the coefficient times the temperature change times the diameter or thickness. For example, a steel piston pin would grow with heat as follows:

.000006in x 300 degrees F increase x .900 diameter = .0016in growth

Consider what happens to different metals assembled together when the temperature rises. For example, consider a steel piston pin fitted in an aluminum piston. Unless restrained by steel reinforcements, the piston would expand at twice the rate as the piston pin. A clearance of .0005in (.015mm) would change to at least .0019in (.035mm) when the operating temperature rises 250 degrees F. The calculations are as follows:

Piston pin clearance at room temperature
=.9005in Bore - .9000in Pin = .0005in

Piston pin expansion
= .000006in x 250 degrees F x .9000in
= .0013in

Piston pin diameter at 250 degrees F
= .9000in + .0013in = .9013in
Piston pin bore expansion

= .000012in x 250 degrees F x .9005in
= .0027in

Piston pin bore diameter at 250 degrees F
= .9005in + .0027in = .9032in

Piston pin clearance at 250 degrees F
= .9032in Bore - .9013in Pin
= .0019in

These calculations show that piston pin clearances could triple in warm engines. An error in machining, although insignificant at room temperature, also would triple when the engine is hot. Because metal parts also contract on cooling, too little clearance at room temperature could cause a piston to bind on a piston pin at extremely low temperatures.

Consider also that engine parts heat up and expand during machining. Because parts must be measured at room temperature to be sure that sizes are correct, it is necessary to allow for contraction on cooling. With experience, this can be allowed for in machining. For example, cylinders honed .0005in oversize will be on size 20 minutes later. Machinists also allow for cooling by "rough" machining parts slightly undersize, letting them cool, and then "finish" machining them to exact size. Handle measuring tools with the same consideration for heat as engine parts. Tools should be kept at room temperature and not held in hands for long periods or even carried in pockets. Also be sure to check or calibrate tools at room temperature only.

COMPARING UNITS OF MEASUREMENT

One problem that technicians and machinists must deal with is the use of both English and metric measuring systems. Domestic manufacturers are changing over to metric units but the aftermarket will continue dealing with English units for years to come. While many manuals include "dual notation" of specifications in both systems, when one set of specifications is converted to the other system it invariably will contain errors. It is best to use metric specifications and tools when original specifications are in metric units and the reverse when original specifications are in English units.

Measuring in inches is confusing to a technician or machinist who first learns the metric system. This is because specifications are frequently in fractions of inches such as 1/8, 1/16, 1/32, or 1/64in. Furthermore, these specifications are nearly always decimal equivalents of the fractions. For example, a specified cylinder bore diameter of 3 7/8in will be measured as 3.875in. Another source of confusion is that units of length are not directly related to units of volume or weight.

The metric system on the other hand, is simple. The basic unit of length is the meter (M) and sub-units are 1/10th (dm), 1/100th (cm), and 1/1000th (mm) parts of a meter. For engine specifications, the millimeter (mm) or 1/1000th a meter, is the basic unit. It in turn breaks down into sub-units of 1/10, 1/100, or possibly 1/1000th of a millimeter. The names, relationships, and equivalents in inches are as follows:

Meter	M	1	39.37in
Decimeter	dm	1/10	3.937
Centimeter	cm	1/100	.3937
Millimeter	mm	1/1000	.03937

As mentioned, for engine specifications, the basic unit is the millimeter. A millimeter breaks down into sub-units as follows:

Millimeter	mm	1	.03937in
.1mm	1/10		.003937
.01mm	1/100		.0003937
.001mm	1/1000		.00003937

Unlike measuring in inches, metric units of length, volume, and weight are related. Using one centimeter (cm) as a basic unit of length, a unit of volume would be one cubic centimeter (cc or cc), and a unit of weight would be one gram (gm). Water is the base for the system of weights; one cc of water weighs one gram. For larger units of volume, use liters. One liter is equal to 1,000cc and one kilogram (Kg) of water. Conversely, one cubic centimeter also may be called a milliliter (ml). The units of volume and weight are as follows:

Cubic centimeter	1cc (or cc)	gram
Milliliter	1cc (or cc)	gram
Liter	1,000cc	kilogram

Occasionally, it is necessary to convert measuring units to the other system. An easy number to recall for this purpose is 25.4, the number of millimeters per inch. To convert inches to millimeters, multiply inches by 25.4. To convert millimeters to inches, divide millimeters by 25.4. Tables also may be used to make conversions (see Appendix A). Some examples of conversions are:

3.875in x 25.4 = 98.43mm
84mm ÷ 25.4 = 3.307in

Units of volume, usually engine displacement, may be converted by recalling the number of cubic centimeters per cubic inch, 16.387. Convert cubic inches to cubic centimeters by multiplying by 16.387. Convert cubic centimeters to cubic inches by dividing by 16.387. Some examples of conversions are:

110in3 x 16.387 = 1,803cc
1,600cc ÷ 16.387 = 98in3

Many people in technical fields wonder why we are not fully metric by now. In the author's opinion, the only answer is that measuring units are not all there is to manufacturing systems. Also involved are specifications for fits, clearances, threads, and other standards for manufacturing. While time is necessary for domestic and international manufacturers to develop mutually acceptable manufacturing standards, competitive international business no doubt creates pressure to accelerate the process.

USING MICROMETERS

Outside micrometers are the most frequently used measuring tools. They make direct measurements of diameter, length, and thickness (see Fig.3-2). Such measurements are "linear" measurements because they are along a straight line. Micrometer scales are dependent upon the pitch of the micrometer screw. An inch micrometer screw has a pitch of 1/40in or .025in, and a metric micrometer screw has a pitch of .50mm. One full revolution of the thimble is equal to the pitch of the micrometer screw (see Fig.3-3).

Fig.3-2   Measurements of diameter, as shown, are along a straight line

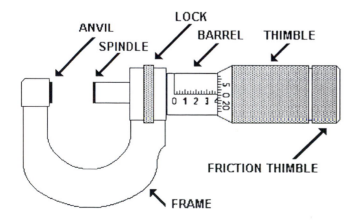

Fig.3-3   The major parts of a micrometer

An inch micrometer has a "resolution limit" of .0001in. This means that the smallest unit on its scale is .0001in and that the micrometer cannot be read closer. Read the micrometer by adding the .100in and .025in units on the barrel, the last unit of .001in on the thimble, and the fourth decimal place units of .0001in from the vernier scale (see Fig.3-4).

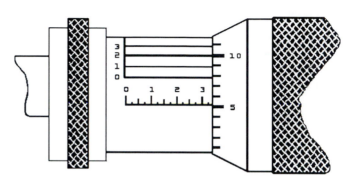

Fig.3-4   Micrometer reading in inches .3302in

A metric micrometer has a resolution limit of .01mm without the vernier scale and a resolution limit as small as .001mm, depending on the units of the vernier scale. Read the micrometer by adding the millimeters and half-millimeters on the barrel, the last unit of .01mm on the thimble, and the third decimal place units of .001mm from the vernier scale (see Fig.3-5).

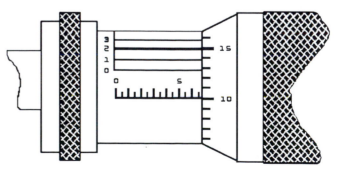

Fig.3-5 A micrometer reading in millimeters 6.602

Depth micrometers are ideal for such measurements as deck clearance (see Fig.3-6). Note that depth micrometer scales are numbered in the reverse direction from an outside micrometer. Also read depth micrometers by adding barrel and thimble units, but remember to note the reverse directions of the scales (see Fig.3-7).

Outside micrometers should be regularly calibrated with gauge blocks (see Fig.3-8). Read the micrometer over the gauge block and, if the reading is incorrect, calibrate by turning the barrel in the frame of the micrometer until the reading agrees with the size of the gauge block (see Fig.3-9). The gauge blocks are accurate within .000004in (.0001mm).

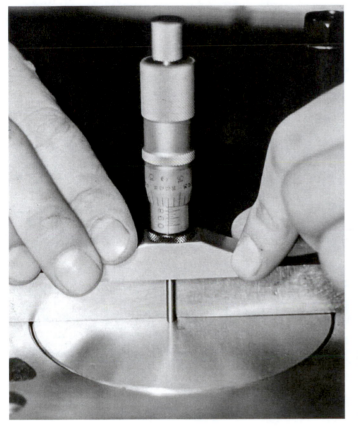

Fig.3-6   Measuring deck clearance with a depth mike

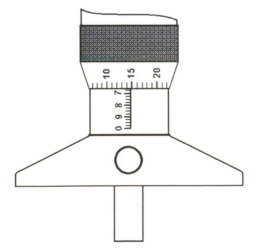

Fig.3-7   Depth micrometer reading .665in

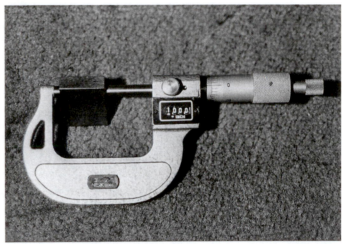

Fig.3-8   Checking micrometer calibration with a gauge block

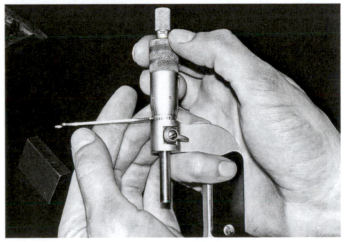

Fig.3-9   Calibrating a micrometer by turning the barrel with a hooked spanner

Take three precautions when calibrating micrometers. First, be sure that both the micrometer and gauge blocks are at room temperature. Second, clean the faces of the micrometer spindle and anvil. Third, use the friction thimble or ratchet stop when calibrating so that readings will not vary with different users providing they use the friction thimble or ratchet.

Check depth micrometer calibration in one of two ways. For the 0 to 1in measuring rod, wipe the depth mike clean and read it over a surface plate (see Fig.3-10). The depth mike should read zero. With a 1 to 2in measuring rod, wipe the depth mike clean and read it over a 1in gauge block on a surface plate and the depth mike should then read zero. Repeat these steps for each additional measuring rod. A worn micrometer reads incorrectly even though calibrated. To find wear, check the calibration at one end of the micrometer scale and then take readings of different gauge blocks along the length of the scale. For example, a worn 0 to 1in micrometer, if calibrated, will read "0" at the lower end of the scale but might read .5005in when measuring a .5000in gauge block.

MAKING TRANSFER MEASUREMENTS

A group of measurements called transfer measurements is commonly used in servicing engines. A common measurement involves using an inside micrometer and an outside micrometer to measure cylinder wear (see Fig.3-11). First, adjust the inside micrometer to the cylinder size by feel. Then find the diameter of the cylinder by measuring over the inside micrometer with an outside micrometer (see Fig.3-12). This is a transfer measurement because the size of the cylinder transfers to the outside micrometer. The accuracy of these measurements is dependent on the skill of the technician or machinist.

Fig.3-11 "Miking" cylinder diameter with an inside micrometer

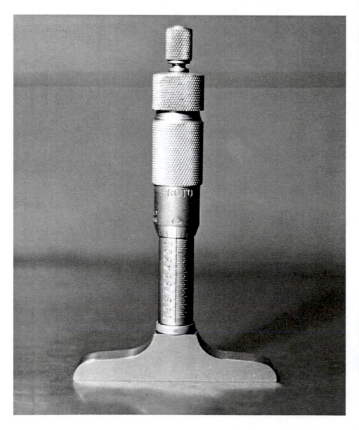

Fig.3-10 Checking calibration of a depth mike on a surface plate

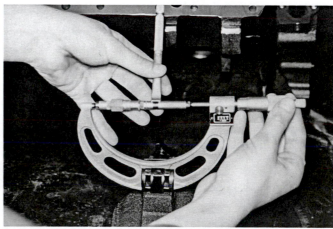

Fig.3-12 Measuring over an inside micrometer with an outside mike

Using the inside micrometer this way is convenient because it would otherwise require calibration with each change of micrometer extensions (see Fig.3-13). Without calibration, an inside micrometer can measure cylinder roundness, taper, or size differences, but it cannot accurately measure diameters unless calibrated. To calibrate an inside micrometer, attach the extensions required and compare the reading of the inside micrometer to the reading of an outside micrometer. Then adjust the inside micrometer to agree with the outside micrometer.

Fig.3-14 Measuring a valve guide with a telescoping gauge

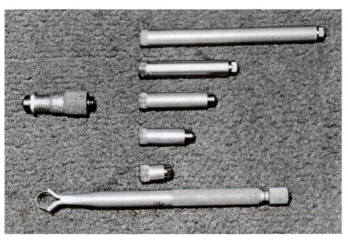

Fig.3-13 An inside micrometer with extensions

Other tools used with outside micrometers to make transfer measurements include telescoping gauges and split-ball or small-hole gauges (see Figs. 3-14 and 15). Transfer measurements with these tools are as accurate as those made with inside and outside micrometers.

Fig.3-15 Measuring a valve guide with a small hole gauge or "split-ball" gauge

USING DIAL INDICATORS

Dial indicators are used for many inspections of engine parts. Specifications for such inspections are generally in terms of total indicator reading or "TIR." For example, checking straightness of a crankshaft may be done in V-blocks with a dial indicator (see Fig.3-16). In making the inspection, be aware that conditions other than straightness add to indicator readings. If there is an out-of-round condition on a main journal or if journals are not concentric with each other, the TIR will be high.

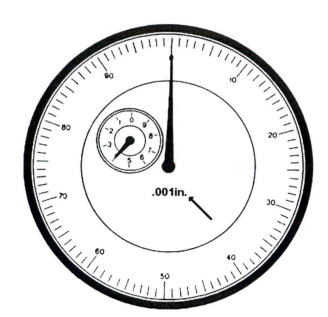

Fig.3-17   The resolution limit of an indicator printed on the face

Fig.3-16   Checking a crankshaft in V-blocks with a dial indicator

Most dial indicators have the resolution limit printed on the face (see Fig.3-17). This is the smallest unit that can be read with the particular indicator. Many indicators also have revolution counters for making readings requiring a longer range of indicator travel (see Fig.3-18). These count each full revolution of the indicator needle around the indicator face.

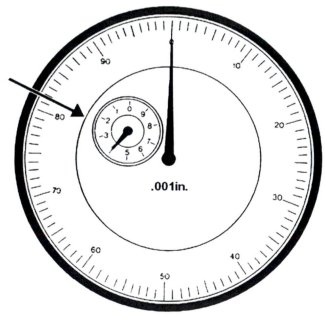

Fig.3-18   A revolution counter on a long stroke dial indicator

USING DIAL BORE GAUGES

Dial bore gauges are really adaptations of dial indicators. A dial bore gauge for measuring cylinders requires a mechanism to allow the dial indicator to read changes in cylinder diameter (see Fig.3-19). The cylinder diameter is not read on the dial indicator. Instead, the difference between the setting of the dial bore gauge and the cylinder diameter is read on a dial indicator. For example, for a cylinder rebored to 3.430in, a dial bore gauge set to this size should read zero. If the cylinder is .001in oversize, the dial indicator will read plus .001in (see Fig.3-20).

The key to accuracy is the tool used to set the dial bore gauge to size. Tool setting is done with an outside micrometer or with a setting fixture designed for this purpose (see Figs. 3-21 and 22). While the setting fixture shown works on a micrometer principle, it is more massive, remains in calibration, and the dial bore gauge fits perfectly into a "nest" when it is set to size.

Fig.3-19 Measuring cylinder diameter with a dial bore gauge

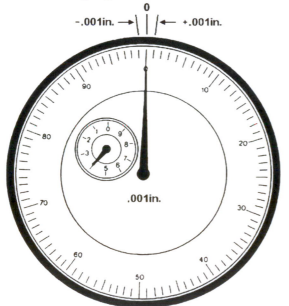

Fig.3-20 Zero means the cylinder is on size; plus or minus means over or under

Fig.3-21 Setting a dial bore gauge to size with a micrometer

Fig.3-22 A Sunnen dial bore gauge setting fixture

This method of gauging is "comparative" because it requires comparing size to a standard, the standard being a setting fixture. This method is most accurate because of the accuracy of standards and because measurements do not depend as heavily on feel or the skills of the technician or machinist.

USING CALIPERS

Dial or digital calipers are used for measurements of outside dimensions, inside dimensions, and depth (see Fig.3-23). The lack of rigidity however limits accuracy to approximately .002in (.05mm) even though the limit of resolution is much smaller.

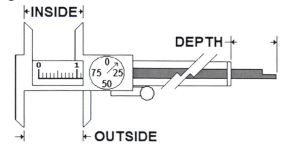

Fig.3-23 A dial caliper measuring outside, inside, and depth

Most machinists use dial or digital electronic calipers because they are quicker and easier to read. With dial calipers, read inches and hundreds of thousandths from the main scale and thousandths the dial (see Fig.3-24). Besides reading directly in decimal units, digital calipers switch to metric units with the push of a button (see Fig.3-25). Keep in mind that while these tools are easier to read, they still lack basic rigidity and are not necessarily accurate to their limit of resolution.

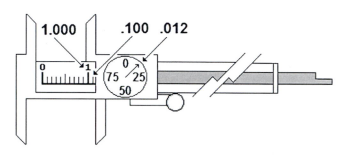

Fig.3-24 A dial caliper reading in inches

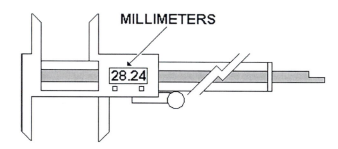

Fig.3-25 A digital vernier caliper reading in millimeters

CHECKING ALIGNMENTS

Many important engine alignments usually go unmeasured. This does not mean that the work is sloppy or that there is no concern over alignments. Instead, correcting alignments are part of routine machining and process planning.

For example, it is important to keep block decks parallel to the crankshaft centerline or pan rails. This alignment could be measured, but it is convenient to position blocks for resurfacing using the crankshaft centerline or pan rails as a reference. In this way, parallelism will be corrected automatically during resurfacing. It is also important to keep cylinders perpendicular (90 degrees) to the crankshaft centerline. Positioning cylinder boring bars using the crankshaft centerline and pan rails as a reference does this. It also can be done using the deck as a reference providing deck parallelism checks out or is resurfaced parallel to the crankshaft centerline.

MEASURING SURFACE FINISHES

While automotive machine shops are not typically equipped to measure surface finishes, all must follow set procedures that yield the desired results. Consider the following routine operations and their importance to engine longevity:

1. Crankshaft journals are ground and polished to produce surface finishes of 10 RA micro inches. Failure to obtain proper finishes will lead to abrasion, wear, and premature bearing failures.
2. Head and block deck surfaces are resurfaced in grinding and milling machines to obtain flatness and surface finishes between 20 and 60 RA micro inches. Depending on the gasket, too coarse a finish will not seal and too fine a finish will lack the "tooth" necessary to grip gaskets.
3. Cylinder honing produces surface finishes matching piston ring requirements. Typically, moly rings require 10-15 RA and iron or chrome rings require a 25-30 RA microinch finish. Too coarse or too fine a finish causes wear or failure to seat rings.

Machinists must consider what is necessary to obtain these finishes and the consequences of running engines with incorrect surface finishes. Essentially, their experience on the job teaches them the necessity of following procedures and obtaining predictable results. While not a substitute for actually measuring cylinder wall finishes, they know that honing equipment manufacturers often run tests and recommend abrasives and procedures that reproduce the specified surface finishes. This gives them a place to start. Following are key

elements of the procedure for finish honing one particular engine:

Engine Specifications:

Bore diameter	3.875in
Cylinder length	6in

Basic Machine Set-Up:

Stone overstroke at each end	7/16in
Rotational speed (RPM)	155
Strokes per minute (SPM)	61

Finishing for Iron/Chrome Rings:

Minimum stock removal	.003in
Grit size (Sunnen EHU 525)	220
Feed rate per load meter reading	45%
Approximate surface finish	34RA

Finishing for Moly Rings:

Stock removal	.0005in
Grit size (Sunnen JHU 820)	400
Feed rate per load meter reading	25%
Approximate surface finish	15RA

Currently, engine manufacturers increasingly use low-tension or reduced friction piston rings. Typically this means thinner rings and increased use of plasma moly ring faces. Shops now find that generally finer finishes and improved cylinder roundness are required. First, moly coated rings cannot withstand the abrasion of rough finishes, and second, thin rings do not apply sufficient force against rough or out-of-round walls to seat properly. Although extra work is required, these rings, especially the thin rings, significantly reduce frictional losses and thereby increase output.

Measuring surface finishes requires an instrument called a "Profilometer." A stylus similar to a phonograph needle moves across the surface at a rate of 1/8in per second and the surface finish reads on an analog or digital meter (see Fig.3-26). On surfaces such as honed cylinder walls, note that finish readings will vary upward and downward as the stylus travels over the crosshatch pattern produced by honing. Expect finishes to vary within a range of readings, and not read a constant number.

Fig.3-26 Using a profilometer to measure surface finishes

Arithmetic Average (AA) and Roughness Average (Ra) are the two most common surface finish scales. The Root Mean Square (RMS) scale is similar but is now considered obsolete. The differences are in the methods of calculating the average deviations between high and low points along the surface profile. All units are in microinches (millionths) and are similar within an approximate 11 percent range.

A third "Rz" scale discloses extremes in surface imperfections. It is a calculation of the mean distance between the five highest peaks and the five lowest valleys within the test range. These extremes are lost in averages within other measuring scales. Because of the extremes, specified numerical values are likely to be 5 to 10 times greater than Ra values. Using cylinder head and block surfaces as an example, a 30 Ra specification also allows 150 Rz.

A fourth "W" scale is used to detect valley to peak "wave" height. In milling for example, a typical limit would be 800 microinches with peak spacing not less than .100in apart and 500 microinches with peak spacing between .030 and .100in apart. Special instrumentation is required for these measurements.

MEASURING THICKNESSES OF CASTINGS

Wall thicknesses in newer engines are thinner than in the past and checking castings prior to boring oversize or removing major amounts of metal is sometimes advised. Besides basic thickness, castings are subject to variations in manufacturing called "core shifts." That is, sand cores forming the interior of hollow castings shift in position during the pouring of the casting resulting in walls that are thick in some spots but thin in others (see Fig.3-27).

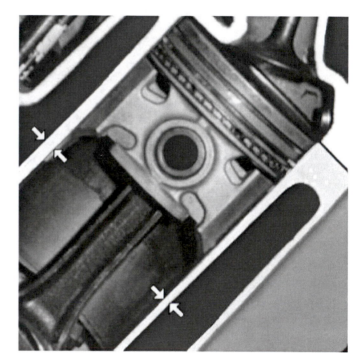

Fig.3-27 Varying cylinder thickness due to a "core shift"

Core shifts make themselves known in various ways. First, engine failures result from cracks in thin cylinders. Machinists may break into water jackets while boring for valve seat inserts. Most common is the difficulty encountered in obtaining roundness during cylinder honing. In this last instance, cylinders flex and allow the cylinder hone to pass by without removing the normal amount of material. While not all engines have these problems, they have become sufficiently common to cause machinists to find ways of inspecting casting thickness.

Ultrasonic testing is the accepted way of precisely measuring wall thicknesses. The tester transmits a sound wave through the casting wall and the time required to pass through the wall translates into a measurement of thickness (see Figs. 3-28 and 29). This is possible because the sound waves do not travel through the air on the other side of the wall. Where core shifts are involved, instead of a consistent .160in wall thickness, thicknesses ranging from .060 to .260in (1.5 to 6.6mm) are found.

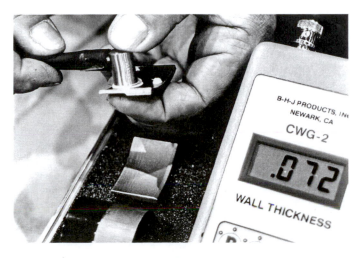

Fig.3-28 Checking calibration of the ultrasonic tester

Fig.3-29 Measuring casting thickness ultrasonically

SUMMARY

Engine machining and assembly requires close adherence to specifications and tolerances to obtain proper performance and longevity. Assumptions that parts are within specifications are unsafe. Technicians and machinists must have skills in reading and using micrometers, dial indicators, and vernier calipers. It is also necessary that technicians and machinists understand how systems of tolerances work and how errors stack-up and lead to out-of-tolerance assemblies.

Although it is confusing, there is also a continuing need to work with both English and metric systems of measurement. As part of this need, it is necessary to read measuring tools in both systems and to convert measurements. It is also helpful to understand the structure and standards of both systems.

Since engines are assembled at room temperature but run at temperatures three or more times higher, an understanding of thermal expansion and contraction is necessary. Such understanding helps in more precise machining, more precise assembly, and in making judgments pertaining to the accuracy of measurements.

Chapter 3

MEASURING

Review Questions

1. Tolerance limits are used in the automotive industry for
   a. piston pin clearance
   b. valve stem to guide clearance
   c. bearing bore and crank journal diameters
   d. piston clearance

2. Machinist A says that tolerance limits specify high and low limits of clearance. Machinist B says that they specify high and low limits of size. Who is right?

   a. A only
   b. B only
   c. Both A and B
   d. Neither A or B

3. Machinist A says that precision assemblies are most economically manufactured by reducing tolerances on individual parts. Machinist B says that such assemblies are more economical if selectively fitted. Who is right?

   a. A only
   b. B only
   c. Both A and B
   d. Neither A or B

4. An example of unilateral tolerance is
   a. plus .001in, minus .000in
   b. plus or minus .001in
   c. plus .001in
   d. minus .001in

5. An example of bilateral tolerance is
   a. plus .001in, minus .000in
   b. plus or minus .001in
   c. plus .001in
   d. minus .001in

6. A piston is specified to fit a 4.030 inch bore with minimum clearance or as much as .0005in more. The tolerance on piston diameter is
   a. plus .000, minus .0005in
   b. plus .0005, minus .0000in
   c. plus or minus .0005in
   d. .0005in

7. Machinist A says that aluminum expands .000006in per inch per degree F. Machinist B says that iron or steel expands .000012in per inch per degree F. Who is right?

   a. A only
   b. B only
   c. Both A and B
   d. Neither A or B

8. Machinist A says that if the temperature of a one-inch diameter piston pin increases 100 degrees F, the pin grows .0006in. Machinist B says that the pin bore grows as much as .0012in. Who is right?

   a. A only
   b. B only
   c. Both A and B
   d. Neither A or

9. To convert specifications given in millimeters to inches, the specification must be
   a. divided by 25.4
   b. multiplied by 25.4
   c. multiplied by 2.54
   d. divided by 2.54

10. Machinist A says that one cc of water weighs one gram. Machinist B says that one cc equals one milliliter. Who is right?

   a. A only
   b. B only
   c. Both A and B
   d. Neither A or B

11. One millimeter equals
    a. 1/100 of a meter
    b. 1/10 of a meter
    c. .003937in
    d. .03937in

12. One inch equals _____ cm.
    a. 2.54
    b. 2.45
    c. 25.4
    d. 24.5

13. Machinist A says that if a one-inch mike reads zero when the spindle and anvil are in contact, the mike is unworn. Machinist B says that the mike is in adjustment. Who is right?

    a. A only   c. Both A and B
    b. B only   d. Neither A or B

14. A ratchet stop or friction thimble on a micrometer is used to
    a. calibrate without regard to individual feel
    b. calibrate to individual feel
    c. prevent damage to the screw thread
    d. prevent damage to anvil or spindle

15. Machinist A says that transfer measurements are made with micrometers and telescoping gauges. Machinist B says that transfer measurements are made with inside and outside micrometers. Who is right?

    a. A only   c. Both A and B
    b. B only   d. Neither A or B

16.   Select the nearest 1/1000in feeler gauge for a .30mm valve clearance adjustment.
    a.   .002in
    b.   .004in
    c.   .008in
    d.   .012in

17. Machinist A says that transfer measuring is most dependent on skill. Machinist B says that comparative methods are more reliable. Who is right?

    a. A only   c. Both A and B
    b. B only   d. Neither A or B

18. Machinist A says that the resolution limit of a dial indicator is printed on the face of the indicator. Machinist B says that the resolution limit is .001in. Who is right?

    a. A only   c. Both A and B
    b. B only   d. Neither A or B

19. The reading of a dial indicator from high to low is called the
    a. travel
    b. reading
    c. total reading
    d. TIR

20. Machinist A says that using a dial bore gauge with a setting fixture to check cylinders is "comparative" measuring. Machinist B says that using a dial bore gauge in this way is "transfer" measuring. Who is right?

    a. A only   c. Both A and B
    b. B only   d. Neither A or B

21. The most reliable method of measuring clearances is
    a. direct
    b. transfer
    c. interpolative
    d. comparative

22. A dial bore gauge reading of cylinder diameter, if not zero, is
    a. over or under specified diameter
    b. over or under the setting of the dial bore gauge
    c. over or under the piston diameter
    d. the actual cylinder diameter as measured

23. The lack of rigidity in vernier calipers limits accuracy to approximately _____ in.
    a. .0001
    b. .0005
    c. .001
    d. .002

24. Machinist A says that vernier calipers measure both inside and outside diameters. Machinist B says that measure depth. Who is right?

    a. A only   c. Both A and B
    b. B only   d. Neither A or B

25. Machinist A says that ultrasonic testers measure surface finishes. Machinist B says that profilometers measure casting thickness. Who is right?

    a. A only   c. Both A and B
    b. B only   d. Neither A or B

FOR ADDITIONAL STUDY

1. A piston is 2.9985in in diameter and requires .0015in clearance. To what diameter should the cylinder be honed?

2. A 20 inch long aluminum cylinder head increases in temperature 200 degrees F. How much longer is it?

3. A 20 inch long iron engine block increases in temperature 200 degrees F. How much longer is it?

4. An 85mm piston requires .03mm clearance but your dial bore gauge reads in inches. What diameter should the cylinder be in inches?

5. An inch micrometer with a vernier scale has a resolution limit of _____ in.

6. A metric micrometer with no vernier scale has a resolution limit of _____ mm.

7. How are block decks kept parallel to crankshaft centerlines when resurfacing?

8. How are cylinders kept perpendicular to crankshaft centerlines when boring?

9. Core shifts in castings are found by measuring wall thicknesses with a(n) _____ .

10. Surface finishes are measured with a _____ .

Chapter 4

FASTENERS

Upon completion of this chapter, you will be able to:
- List the factors that affect fastener strength.
- Identify fastener grades.
- Explain the relationship between clamping force and torque.
- Explain the different methods for obtaining the required clamping force in assembly.
- Identify threads based upon diameter, pitch, and grade.
- Explain the purpose of pipe threads and requirements for correct assembly.
- List the steps for removing broken fasteners.
- Compare options for the repair of damaged threads.
- List the steps in installing Helicoils.
- Describe methods used to remove broken tools from castings.

INTRODUCTION

No single area of knowledge is taken for granted as much as that of threaded fasteners. Fasteners clamp, adjust, and plug all parts of engine assemblies and yet technicians regularly encounter problems with these devices. A thorough understanding of fastener fundamentals will decrease the frequency of such difficulties as stripped threads and inaccurate clamping force in assembly.

DETERMINING THE STRENGTH OF FASTENERS

Threaded fasteners, specifically called machine screws, capscrews, or bolts, stretch when tightened. This stretching helps keep fasteners tight even during heating and cooling cycles of cylinder heads and blocks. If over-tightened, the fasteners stretch beyond the point where they remain elastic; that is, they become permanently stretched.
Fasteners stretch as turning torque increases to the elastic limit, beyond which the torque required to

stretch the fasteners decreases until failure occurs (see Fig.4-1). Below elastic limits, fasteners can be loosened and tightened repeatedly and still return to their original lengths.

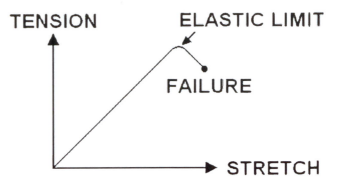

Fig.4-1 Fastener stretch, elastic limit, and failure

All have experienced the sensation of tightening a bolt and suddenly sensing that it had become easier to turn. This occurs when fasteners exceed elastic limits and permanently stretch. Remove a fastener at this point and inspection generally shows a "necking down" of the diameter (see Fig.4-2). Continue tightening and the fastener breaks.

Fig.4-2   A fastener "necked" down after exceeding the elastic limit

The cross-sectional area of a fastener is the largest single factor in determining strength. The diameter of the cross-section of an externally threaded fastener is equal to the minor diameter of the thread (see Fig.4-3). The cross-sectional area of an externally threaded fastener decreases as the distance between threads, or the pitch, increases (see Fig.4-4). This is because thread depth is calculated by multiplying the pitch by a thread system constant. Because cross-sectional area decreases as pitch increases, coarse-threaded external fasteners, or those having larger pitches, are weaker than fine-threaded external fasteners, or those having smaller pitches.

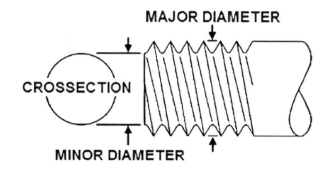

MAJOR DIAMETER

CROSSECTION

MINOR DIAMETER

Fig.4-3   The cross-sectional area of an external thread

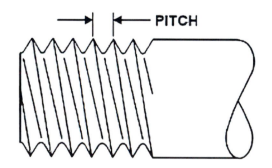

PITCH

Fig.4-4   Thread pitch

The cross-sectional area of the thread determines the strength of internal threads (see Fig.4-5). This is the reverse of the external thread. As thread pitch increases, the cross-section of internal threads increase and the cross-section of externally threaded fasteners decrease. Therefore, as pitch increases, the internal thread in a nut or in a casting becomes stronger, and the externally threaded

fastener used with it becomes weaker. Threaded holes in low strength materials, such as cast iron or aluminum engine castings, are nearly always coarse for increased strength. To increase strength for a coarse externally threaded fastener requires using the next larger diameter.

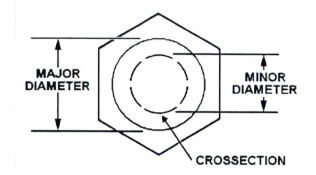

MAJOR DIAMETER   **MINOR DIAMETER**

CROSSECTION

Fig.4-5   The cross-sectional area of an internal thread

Besides cross-sectional areas, fastener strength varies with materials and heat treatment processes. Lower strength fasteners are made of low carbon mild steel. Higher strength fasteners have increased carbon content, making heat treating processes possible, and possibly alloying metals in varying percentages. The Society of Automotive Engineers (SAE) has developed a system of grading fastener strength. The head of the fastener is marked with a code to identify the grade. The SAE grade is equal to the number of marks on the head of the fastener plus two. Metric grades are numbered and are also on the head of the fastener (see Fig.4-6).

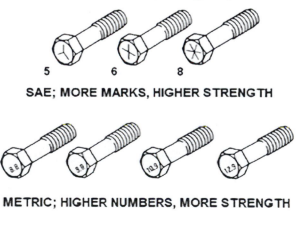

5   6   8

SAE; MORE MARKS, HIGHER STRENGTH

METRIC; HIGHER NUMBERS, MORE STRENGTH

Fig.4-6   Fastener grade markings on capscrews

COMPARING CLAMPING FORCE AND TORQUE

The purpose in tightening fasteners is to obtain clamping force on cylinder head gaskets, bearings in their housings, and other critical engine parts. It is the clamping force that is important, and tightening fasteners is only the means of obtaining it.

Common practice has been to tighten critical fasteners with a torque wrench. To obtain the engineered clamping force, threads must be cleaned, lubricated with engine oil, and tightened with a torque wrench to specifications. Just be sure to pull the torque wrench handle slowly and smoothly as jerking motions cause false indications of tightening torque

Fasteners are also tightened by counting the degrees of rotation. This method is more precise because variables such as cleanliness or thread lubricants have less effect on clamping force. Tightening by degrees is sometimes referred to as torque-to-angle or "TTA". If tightened to the yield point, it is referred as torque-to-yield or "TTY". TTY fasteners are generally replaced on assembly.

Consider something as simple as the lubrication of threads. When clean and dry, threads require 30 to 40 percent greater torque to obtain the same clamping force as obtained with engine oil on the thread (see Appendix A). Always clean and lubricate threads with engine oil unless manuals specifically state otherwise. Using other lubricants, especially some anti-seize compounds, can cause thread failure because friction drops so low that clamping force exceeds the strength of the fastener. Some compounds that do not change clamping force appreciably at specified torque are silicone sealers and anaerobic adhesives or sealers (Loctite). Silicone sealers prevent leakage around threads extending into oil passages or water jackets. Anaerobic adhesives on critical fasteners prevent loosening caused by vibration.

IDENTIFYING THREADS

To identify threads, measure the outside diameter and the pitch. Measure the outside diameter across the threads with a micrometer (see Fig.4-7). Measure-

ments of diameter will be less than the given thread size because of tolerances, the specified diameter being the maximum allowed. Identify the pitch by matching the thread profile on a thread pitch gauge to the threads on the fastener (see Fig.4-8).

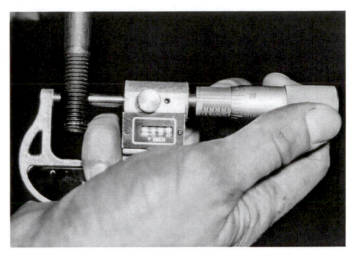

Fig.4-7 Measuring diameter across the thread

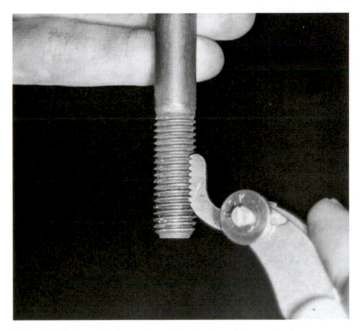

Fig.4-8 Identifying thread pitch with a pitch gauge

Calculate numbered machine screw sizes by multiplying the number of the screw by .013in (.33mm) and then adding .060in (1.5mm). There is no pitch specification except as threads per inch. Calculate the actual pitch by dividing one (1) by the number of threads per inch. A number 10-32 machine screw is identified as follows:

Measured diameter	= .188in
Calculated 10 x .013	= .130 + .060
	= .190in
Threads per inch	= 32 per gauge
Pitch	= 1/32
	= .031in

Threads manufactured in inches, 1/4in and larger, have diameters in 1/16in increments. Common diameters used in automotive engines are 1/4, 5/16, 3/8, 7/16 and 1/2 inch. As with machine screws, pitch is not specified except as the number of threads per inch and therefore must be calculated. A 3/8-24 capscrew or bolt might be identified as follows:

Measured diameter	= .370in
Nearest size	= 3/8in or .375in
Threads per inch	= 24 per gauge
Pitch 1/24	= .042in

Metric fastener diameters are in 1-millimeter increments. Unlike threads in inches, pitches for metric fasteners are in millimeters. Common pitches for automotive fasteners are .75, 1.0, 1.25, 1.50, and 1.75mm. An 8 x 1.25 fastener might be identified as follows:

Measured diameter	= 7.9mm
Nearest size	= 8mm
Pitch	= 1.25mm per gauge

The various combinations of diameters and pitches, or threads per inch, are organized into thread systems. Inch system threads are specified under the Unified system. Coarse threads are identified with the abbreviation of UNC or NC and fine threads as UNF or NF. The automotive industry has its own interchangeable equivalents identified as USS coarse threads and SAE fine threads. The International Standards Organization or Systeme Internationale (ISO or SI) systems have no coarse and fine distinctions. The most common combinations of diameter and threads per inch or pitch are:

| | Coarse | Fine |
Diameter	USS	SAE
1/4	20	28
5/16	18	24
3/8	16	24
7/16	14	20
1/2	13	20

Diameter	Pitch in Millimeters
6	1.00
8	1.25
10	1.50
12	1.75

USING PIPE THREADS AND FITTINGS

Tapered pipe threads serve a different purpose than the threaded fasteners discussed so far. In automotive applications, these threads seal against hydraulic leaks and do not clamp, adjust or perform any of those functions of straight threads. They seal because the threads taper at a rate of 3/4in per foot along their length (see Fig.4-9). This taper causes the threads to wedge in place and seal firmly. This wedging action accounts for the difficulty sometimes encountered in removing pipe plugs and other threaded fittings.

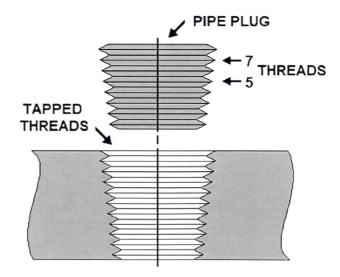

Fig.4-9   Pipe threads taper 3/4in per foot

Identifying pipe threads is deceptive because their nominal size is a rough approximation of the inside diameters for pipe. That is, a 1/4in threaded fitting has an approximate inside diameter of 1/4in but a maximum outside diameter of .540in making a visual judgment of size tricky. The table below contains basic dimensions to help with the identification of threads under the American Standard Taper Pipe Thread or SAE Short systems.

Inside Diameter	Outside Diameter	Threads Per Inch	Tap Drill
1/8	.405	27	11/32
1/4	.540	18	7/16
3/8	.675	18	19/32
1/2	.840	14	23/32
3/4	1.050	14	15/16

To tap threads for pipe plugs or fittings, follow just two basic steps. First, ream holes after tap drilling with a tapered pipe reamer as this makes cutting the threads and aligning the tap easier. Second, use cutting oil and tap into the work not more than five full turns of the tap and then check the thread engagement with the fitting. The fitting should engage five full turns when finger tight and up to seven turns when tightened with a wrench. If it does not engage five turns, tap the number of additional turns required to gain five full turns of engagement. Limiting the engagement of tapered pipe threads assures that fittings tighten into the threads and do not bottom out against the undersides of fittings prior to sealing (see Fig.4-10).

Fig.4-10　Check that fittings do not bottom out before the threads seal

Tapered fittings in cars manufactured under metric standards share the same taper and pitch (threads per inch) but there are differences in small and large end diameters. This is because the metric fittings use different points along the same taper for each nominal size. Fittings from each system will thread into the same tapped holes but with differences in the length of engagement.

Straight pipe threads are used in automotive production as well but primarily for tubing or uses unrelated to the engine. The nominal sizes and threads per inch are the same with slight differences in actual diameters.

REMOVING BROKEN FASTENERS

Occasionally, we find ourselves trying to remove a fastener that we have broken, or have found broken, in engine castings. Removing a broken fastener is not really difficult providing we are patient and follow a particular sequence of operations. If we rush into the job thoughtlessly, we are likely to find ourselves trying to Figure out what to do about broken tools, usually an easy-out, stuck inside the broken fastener. Proceed with caution.

First, place a center punch mark in the center of the fastener. Second, drill a pilot-hole approximately one half of the fastener diameter on center and in alignment with the centerline of the fastener. Drill all the way through the length of the fastener. It is the failure to do these first steps correctly that causes most difficulties with the removal of broken fasteners.

Pilot holes located off center or out of alignment cannot be corrected simply by redrilling because the second drill only follows the first hole. The only way to drill on a new centerline is drill through a drill bushing (see Fig.4-11). An option to drilling is boring a hole on center using a two-lip end mill, but this requires positioning the piece in a milling machine. A possible time saver is the use of left-handed drills for drilling pilot holes. Because left-handed drills run in the reverse direction, they frequently hook into the broken fastener and unscrew it. Left-handed drills are available from sales outlets for mechanics tools. Of course, left handed drills also require a reversible drill motor.

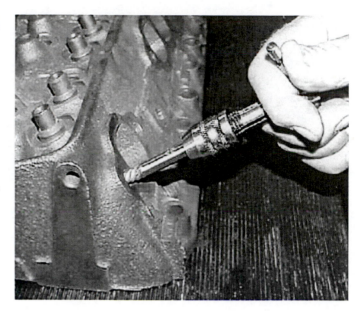

Fig.4-12   Using an easy-out to remove a fastener

Usually, the easy-outs are successful if used as directed. If not, drill to the tap drill size (see Appendix A) of the broken fastener and then retap the hole to the original thread size. The tap picks up the original thread and breaks loose the fragments of the fastener.

If retapping the hole does not work, install a Helicoil (see the next section). Again, if the steps given are followed carefully, chances of successfully removing the fastener are very good. The most difficult jobs are those done in the vehicle because of the difficulty in getting access to the fastener.

INSTALLING HELICOILS

A Helicoil is a stainless steel coil of thread used to replace damaged threads (see Fig.4-13). Helicoils work well for repairing threads in engine castings because they require minimum enlargement of the original tapped hole. Engine castings have limited wall thickness and it is desirable to keep walls as thick as possible.

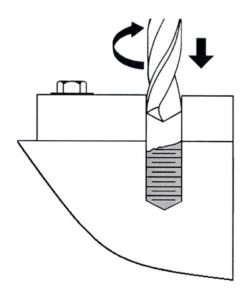

Fig.4-11   Using a drill bushing over the center to guide a drill

The next series of steps requires both patience and judgment. Try an easy-out in the pilot hole. Do not force it (see Fig.4-12). If the fastener does not break loose easily, stop, drill another 1/32 to 1/16in (1.0mm) oversize, and try the next larger easy-out. Again, do not force the easy-out. Instead, drill larger and try the next larger easy-out.

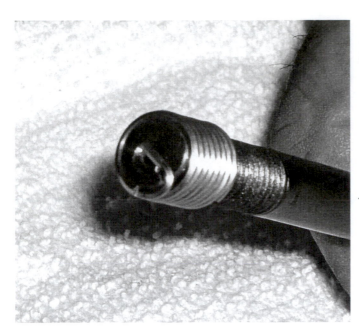

Fig.4-13   A Helicoil over an installing tool

Fig.4-14   Tapping for a Helicoil

Helicoils usually come in lengths equal to 1.5 times the thread diameter. That is, a 1/2in Helicoil would be 3/4in long installed. This length is necessary to obtain maximum thread strength. Because threads for Helicoils are drilled and tapped oversize, and because the Helicoils are made of steel, Helicoil strength is 50 percent greater than the original threads in castings.

The first step in installing a Helicoil is to tap drill oversize (see Appendix A). Drill slightly deeper in blind holes (holes closed on one end) to allow the Helicoil tap to cut to the full depth of 1.5 diameters. To be safe, first check the casting thickness to avoid drilling into water jackets or oil passages.

Next, tap the hole using the Helicoil tap of the specified size (see Fig.4-14). The Helicoil tap cuts the same thread pitch as the original thread, but the diameter is approximately 1/16in (1.5mm) larger. Use cutting oil or tapping fluid for all tapping.

When tapping, turn the tap clockwise one full turn and reverse a fraction of a turn to break off continuous chips. Be careful not to pack chips in the bottom of blind holes; they will cause the tap to bind. It may help to remove the tap halfway through the job and blow chips clear.

Next, place the Helicoil on the installing tool (see Fig.4-15). Blow chips out of the hole and screw the Helicoil into place. Install Helicoils one turn below the surface.   After the Helicoil is in place, break off the drive tang with a pin punch. The Helicoil can be kept from backing out of the hole by staking the exposed thread above the coil using a pin punch. The thread repair is now ready for the original size fastener.

Fig.4-15   Installing the Helicoil with the installing tool

REMOVING BROKEN TOOLS

Occasionally, taps and easy-outs break. A state of panic easily sets in at this point and efforts to remove the broken tool frequently add to the problem. First, do not attempt to drill out broken tools; they are equal in hardness to the drills and drilling simply will not work. Stand back and study the options.

Larger size taps are sometimes removed with torch and a punch. Success is more likely when lodged in iron castings, not aluminum, and for taps jammed in holes that go all the way through. First adjust the torch to a neutral or slightly oxidizing flame, heat the center of the tap red hot, take the torch away, and spray the tap with water. Then strike the center of the broken tap with a pin punch. Wear safety glasses because the tap is now brittle and fragments will break away. Repeat the process of heating and tapping with a punch until the tap falls through the hole. Then clean the hole with a wire bore brush and retap or repair threads with a Helicoil.

Another approach sometimes works with small diameter taps broken off in through holes. Simply take a pin punch and drive the broken tap through the holes (wear safety glasses). Brush the fragments of the tools out of the hole, tap drill for a Helicoil, tap and install a Helicoil.

Most difficult to repair are broken easy-outs, especially when pilot holes were not drilled all the way through the fastener. It may be necessary to sublet the repair to a machine shop equipped with an EDM (Electrical Discharge Machine) or a so called "metal disintegrator" (see Fig.4-16). The work is set up in the machine as if for drilling, but instead of drilling, a hollow electrode is fed through the work piece. An electric arc starts between the electrode and the work that puddles the broken tool, but since water runs through the hollow electrode, the molten metal chills and washes away before it can redeposit on the work. The surrounding metal is not damaged and the original thread can frequently be retapped and reused. Should the thread be damaged, a Helicoil can still be installed.

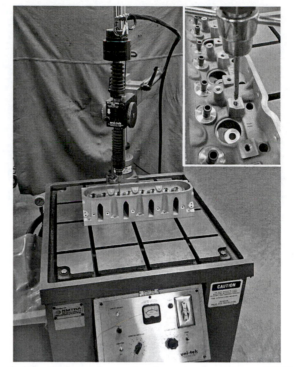

Fig.4-16 Removing a broken tool with an EDM

SUMMARY

Engine failures caused by improper clamping force or fastener failures are avoidable. While most failures are eliminated by correctly following torque specifications and recommended tightening methods, some situations call for more knowledge of fasteners. For example, technicians and machinists frequently need to first identify fasteners by diameter, pitch, and grade. In tightening these fasteners, they also need to understand the relationship between clamping force and turning torque or degrees of rotation.

Because thread repair is rarely a planned part of any job, the knowledge and skills required in removing broken fasteners and repairing threads is extremely useful. The ability to complete such repairs in the course of engine repair keeps jobs on schedule and makes these skills an important asset. Employers expect technicians to work effectively with fasteners and fittings of all types. Technicians and machinists must be familiar with threaded fasteners in Unified, SAE, and ISO metric fastener systems. In addition, they must also have knowledge of American Standard pipe threads and fittings. Such knowledge cannot be acquired except through study.

Chapter 4

THREADED FASTENERS

Review Questions

1. Machinist A says that fasteners stretch when tightened to specifications. Machinist B says that so long as specifications are not exceeded, stretch remains within elastic limits. Who is right?

   a. A only
   b. B only
   c. Both A and B
   d. Neither A or B

2. Machinist A says that the cross-sectional area of the minor diameter determines the strength of internal threads. Machinist B says that the cross-sectional area of the thread determines the strength of an external thread. Who is right?

   a. A only
   b. B only
   c. Both A and B
   d. Neither A or B

3. Machinist A says that the cross-sectional area of coarse internal threads is greater than that of fine threads. Machinist B says that the cross-sectional area of a fine threaded capscrew is greater than a coarse threaded one. Who is right?

   a. A only
   b. B only
   c. Both A and B
   d. Neither A or B

4. Machinist A says that thread pitch is the distance between threads. Machinist B says that it equals the distance a fastener travels per revolution. Who is right?

   a. A only
   b. B only
   c. Both A and B
   d. Neither A or B

5. Tapped head bolt holes in engine block and cylinder head castings use_____ threads.
   a. UNC or USS
   b. UNF
   c. UNEF
   d. SAE or UNF

6. Machinist A says that tightening fasteners by degrees of rotation gives more precise clamping force than tightening with a torque wrench. Machinist B says that this method of tightening increases clamping force. Who is right?

   a. A only
   b. B only
   c. Both A and B
   d. Neither A or B

7. Machinist A says that some anti-seize compounds decrease friction and thereby increase the clamping force at specified torque. Machinist B says that even with these compounds, tighten fasteners to specifications. Who is right?

   a. A only
   b. B only
   c. Both A and B
   d. Neither A or B

8. Machinist A says that some anaerobic compounds prevent loosening of threads. Machinist B says that some of these compounds seal threads. Who is right?

   a. A only
   b. B only
   c. Both A and B
   d. Neither A or B

9. Calculate the OD of number sized screws as follows
   a. N x .013in
   b. N x .060in
   c. N + .060in
   d. (N x .013in) + .060in

10. The outside diameter of a 10-32 thread is _____ in.
    a. .080
    b. .104
    c. .164
    d. .190

11. A fastener measures .370in. The nearest nominal size is
    a. 8mm
    b. 5/16in
    c. 3/8in
    d. 10mm

12. The pitch of a 10 x 1.25 metric thread is
    a. 1 divided by 1.25
    b. 1.25mm
    c. 1.25cm
    d. 10mm

13. The pitch of a 10-24 thread is
    a. 1 divided by 24
    b. .240in
    c. .024in
    d. .010in

14. Machinist A says that USS and SAE threads are automotive equivalents to the Unified coarse and fine thread series. Machinist B says that there are no coarse or fine series designations for metric fasteners under the ISO system. Who is right?

    a. A only   c. Both A and B
    b. B only   d. Neither A or B

15. Machinist A says that oil plugs, pressure fittings, and other pipe threads are tapered 3/4in per foot. Machinist B says that pipe thread sizes are equivalent to the nominal inside diameters of pipes. Who is right?

    a. A only   c. Both A and B
    b. B only   d. Neither A or B

16. Machinist A says that for pipe fittings, tap to five turns of the fitting. Machinist B says tap these threads to five turns of the tap. Who is right?

    a. A only   c. Both A and B
    b. B only   d. Neither A or B

17. Machinist A says accurately locate and center punch broken fasteners before drilling. Machinist B says drill in alignment with the centerline through the broken fastener. Who is right?

    a. A only   c. Both A and B
    b. B only   d. Neither A or B

18. Machinist A says that should the pilot-hole for an easy out be drilled off center or out of alignment, correct it by re-drilling on center. Machinist B says re-drill through a drill bushing or bore with end mill. Who is right?

    a. A only   c. Both A and B
    b. B only   d. Neither A or B

19. Machinist A says that Helicoil repairs require special oversize diameter taps. Machinist B says that Helicoils require special installing tools. Who is right?

    a. A only   c. Both A and B
    b. B only   d. Neither A or B

20. Machinist A says that to obtain full strength, fasteners require 1.0 diameter of engagement. Machinist B says that 1.5 diameters are required. Who is right?

    a. A only
    b. B only
    c. Both A and B
    d. Neither A or B

21. Machinist A says that each Helicoil tap has its own special pitch. Machinist B says that the pitch is the same as the thread being repaired. Who is right?

    a. A only
    b. B only
    c. Both A and B
    d. Neither A or B

22. Install Helicoils _____ the surface.
    a. flush with
    b. one-half turn below
    c. one full turn below
    d. one and a half turns below

23. Machinist A says that damaged spark plug threads require repair by means other than Helicoils. Machinist B says that Helicoil repairs require care to keep threads square with spark plug seats. Who is right?

    a. A only
    b. B only
    c. Both A and B
    d. Neither A or B

24. Machinist A says remove broken taps by drilling. Machinist B says see if it possible to drive small taps through holes or burn out larger taps with a torch. Who is right?

    a. A only
    b. B only
    c. Both A and B
    d. Neither A or B

25. Machinist A says that a "tap disintegrator" removes broken tools without damage to the surrounding metal. Machinist B says that this process is limited to broken tools jammed in through holes. Who is right?

    a. A only
    b. B only
    c. Both A and B
    d. Neither A or B

FOR ADDITIONAL STUDY

1. Select six metric and English capscrews at random. Measure their diameters with a micrometer and check their pitch or threads per inch with a thread pitch gauge. Also identify the fastener grade. Record findings below:

Capscrew	Measured Diameter	Nominal Size	Pitch-TPI	Grade
#1				
#2				
#3				
#4				
#5				
#6				

2. Select three pipe fittings or pipe plugs at random. Measure their diameter at the large end with a micrometer and check their pitch or threads per inch with a thread pitch gauge. Record findings below:

Fitting	Measured Diameter	Nominal Size	Pitch-TPI
#1			
#2			
#3			

3. 3.Describe the effect of lubricating threads on torque and clamping force.

4. 4.List in sequence the steps in removing a broken capscrew.

5. 5.List in sequence the steps in installing a Helicoil.

Chapter 5

ENGINE THEORY

Upon completion of this chapter, you will be able to:

- Explain each of the four engine cycles in a four-stroke engine.
- Describe the differences between gasoline and diesel engines and their combustion cycles.
- Relate valve opening and closing events to crankshaft rotation and piston travel.
- List the different valve train configurations used in production engines.
- Explain the operation of different zero-lash hydraulic valve trains.
- Describe different oil pump drive systems and oil circulation through engines.
- List oil additives and their purposes in engine protection.
- Calculate displacement, compression ratio, and clearance volume.
- Explain different fits and required clearances for engine parts in assembly.
- List cooling system components and explain their operation.
- Explain what occurs in gasoline engine combustion and list the by-products.

INTRODUCTION

The information covered in this chapter supports the technical content of this text. Theory is stated in brief terms and limited to essential points. Comprehension of this information aids in diagnosing engine malfunctions and in making critical judgments regarding engine service.

THE FOUR-STROKE CYCLE

Automotive engines, with few exceptions, operate on a four-stroke cycle. An air-fuel mixture enters the cylinder, is compressed, and then ignited. Upon ignition, gases expand and force the piston downward in the cylinder. Force and motion are transmitted from the piston through the connecting rod to the crankshaft. In this manner, reciprocating, or up-and-down, motion at the piston changes to rotary motion at the crankshaft. The relationship of valves, piston, and crankshaft is as follows:

1. On the intake stroke, the intake valve is open and the piston travels downward. The air-fuel mixture enters the cylinder because of low pressure in the cylinder and high atmospheric pressure outside the engine (see Fig.5-1).

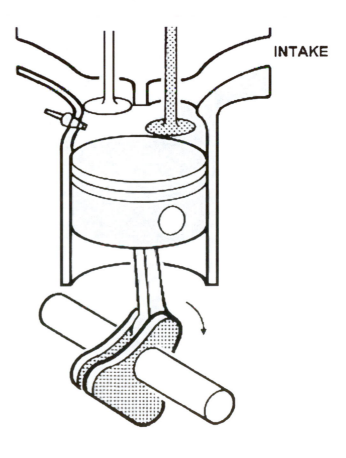

Fig.5-1 The intake stroke

2. On the compression stroke, both intake and exhaust valves are closed and the piston travels upward in the cylinder. The piston travel compresses the air fuel mixture and then ignition occurs (see Fig.5-2).

3. On the power stroke both intake and exhaust valves remain closed. Upon ignition of the air-fuel mixture, the expansion of burning gases forces the piston to travel downward in the cylinder (see Fig.5-3).

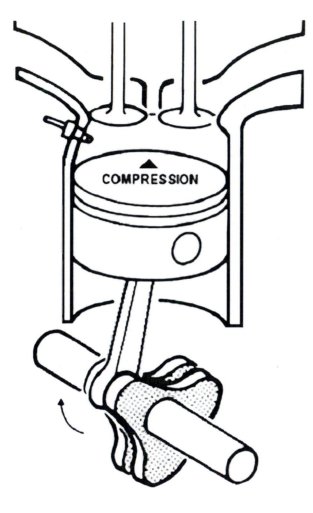

Fig.5-2   The compression stroke

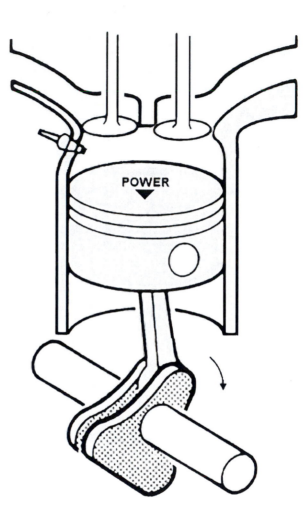

Fig.5-3   The power stroke

4. On the exhaust stroke, the exhaust valve is open and the piston travels upward in the cylinder. Burned gases are forced through the exhaust valve by residual pressure and the piston (see Fig.5-4).

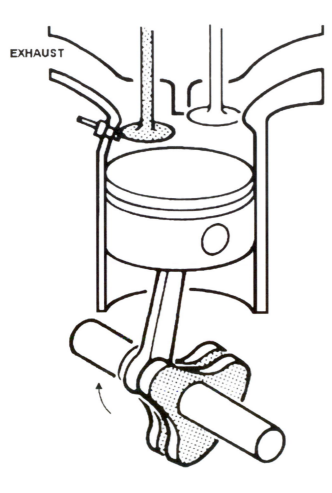

Fig.5-4   The exhaust stroke

Keep in mind that one stroke requires one half turn or 180 degrees of crankshaft rotation. Four strokes require two full turns or 720 degrees of crankshaft rotation. All cylinders complete the four-stroke cycle in two crankshaft revolutions.

COMPRESSION IGNITION ENGINES

Most diesel engines in passenger cars also operate on a four-stroke cycle. The difference between diesel and gas engines is that diesels have compression ignition and gas engines have spark ignition. Compression ignition is possible because of the somewhat low volatility of diesel fuel and the use of compression ratios of 20:1 or more. At such ratios, compression heats the air in the combustion chamber to approximately 1000 degrees F. Because diesel fuel ignites spontaneously at approximately 600 degrees F, combustion begins with injection of diesel fuel into the heated air.

The fuel delivery system injects fuel under very high pressure. For safety, injection pressures must exceed cylinder pressures and therefore the fuel injection nozzles open at pressures of 1,400 pounds per square inch (PSI) or more. Air only passes through the intake manifold into the cylinder on the intake stroke. There is no throttle valve to regulate airflow. Engine revolutions per minute (RPM) increase as fuel delivery to cylinders increases.

Diesel engines use two combustion chamber types and injection methods. Some inject fuel directly into the chamber and others inject into a pre-combustion chamber inserted into the cylinder head casting (see Fig.5-5). Fuel injection occurs at approximately 20 degrees before top-dead-center (TDC) on the compression stroke and continues past TDC on the power stroke. The fuel droplets first vaporize, begin burning, and then continue vaporizing and burning until consumed.

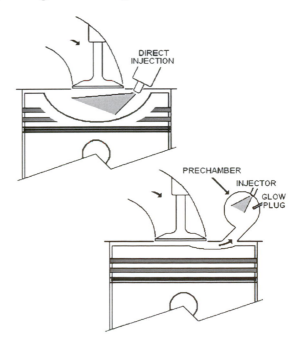

Fig.5-5   Diesel engine combustion chambers

Injection timing is critical in this process. If injection timing is late, combustion will be incomplete and result in power loss and considerable black smoke in the exhaust. If injection is early, detonation and damage to the engine results.

VALVE TIMING AND CAMSHAFTS

Because each valve must open and close every two-crankshaft revolutions, the camshaft operates at one-half crankshaft speed. Therefore, camshaft drive sprockets or drive gears have twice as many teeth as crankshaft sprockets or gears (see Figs. 5-6 and 7).

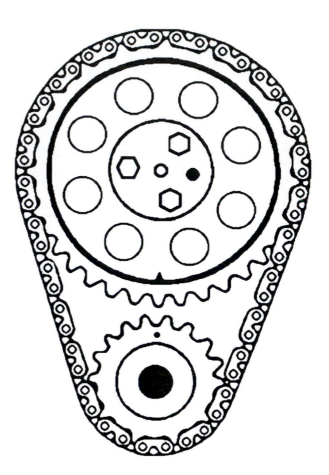

Fig.5-7   A chain driven camshaft

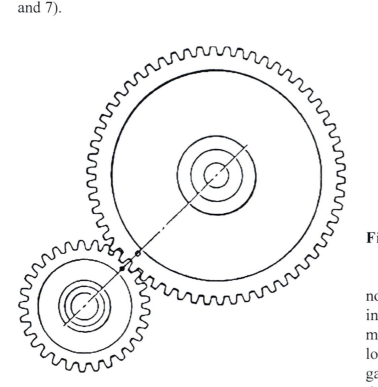

Fig.5-6   A gear driven camshaft

This description of the four-stroke cycle does not fully explain periods of valve opening or closing. Both intake and exhaust valves are open for more than one crankshaft stroke so that high and low pressure conditions can be used to promote gas flow in and out of the cylinder. For example, the intake valve is open for approximately 250 degrees of crankshaft rotation because cylinder pressure remains lower than atmospheric pressure. The exhaust valve is also open for approximately 250 degrees (see Fig.5-8). "Duration" is the period when each valve is open.

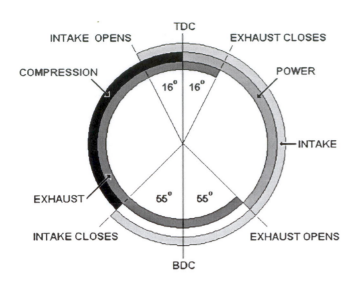

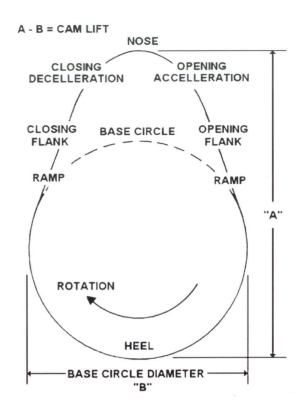

Fig.5-8 A valve timing diagram in crankshaft degrees

Shown in Figure 5-8 is a period when both intake and exhaust valves are open simultaneously. This is the "valve overlap" period between the end of the exhaust stroke and the start of the intake stroke. Valve overlap promotes gas flow because low pressure at the exhaust port draws in air and fuel, under higher pressure, through the intake port. As the exhaust valve closes, the piston continues downward, creating low pressure in the cylinder that continues to draw in air and fuel.

Another common term relating to valve trains is cam lift. Cam lift is the total range of travel for a valve lifter or cam follower from the valve-closed to the valve-open position. Lift is the difference between measurements from the nose of the cam to the heel and the base circle diameter (see Fig.5-9).

Fig.5-9 Cam lift measurements and terminology

Valve lift is usually greater than cam lift for engines using rocker arms. This is because rocker arms act as levers and open valves an amount greater than the cam lift (see Fig.5-10). Rocker arm ratios are commonly between 1.5:1 and 1.75:1.

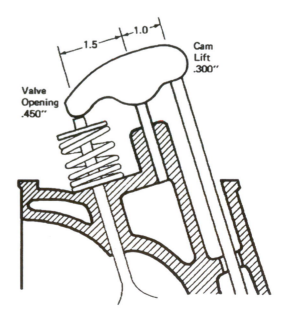

Fig.5-10 Rocker arm ratio

VARIABLE VALVE TIMING AND VALVE ACTION

Many manufacturers vary valve timing to broaden the power curve, improve fuel economy and reduce emissions. At low engine speeds, when volumetric efficiency is relatively low, advancing the intake cam centerline closes the intake valve earlier increasing cylinder pressure and torque. At high engine speeds, retarding the intake from the advance position and bringing the intake and exhaust center-lines together increases overlap, improves volumetric efficiency and increases power. Gradually making these adjustments over the full RPM range raises the average output without cylinder pressures rising to the level of potential detonation (see Fig.5-11). As for emissions, retarding the exhaust cam closes the exhaust valve later and promotes exhaust recirculation into the intake thereby reducing emissions. Some timing variations are shown in Figure 5-12. There is more to come on camshafts in Chapter 17.

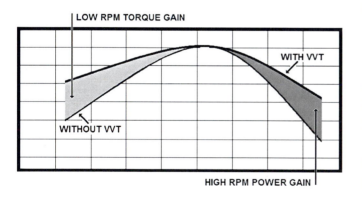

Fig.5-11 Gains with variable valve timing

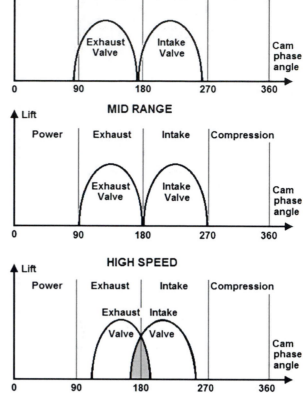

Fig.5-12 Cam centerlines optimized for performance

A timing actuator attaches to the front end of the camshaft to advance or retard valve timing several degrees in either direction (see Figure 5-13). Internally, these may be radial or linear type actuators but either type is hydraulically actuated by engine oil pressure directed through valves controlled by the engine control module (computer).

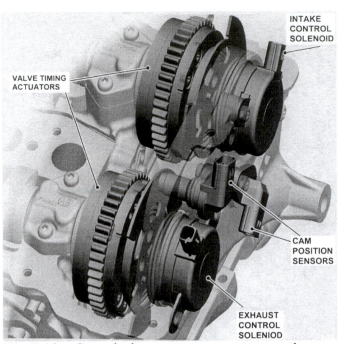

Fig.5-13 Cam timing actuators, sensors and control solenoids

Some manufacturers go further and vary valve lift and duration by shifting cam followers to select low or high performance cam lobes or by shifting between high and low rocker-follower ratios according to engine power demands. An example of a variable rocker-follower mechanism is shown in Figure 5-14.

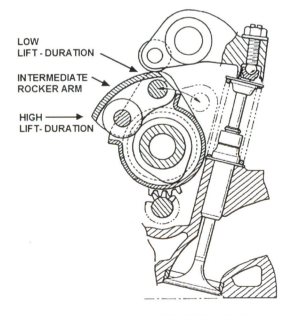

Fig.5-14 Example of variable lift and duration mechanism

To improve fuel economy, some manufacturers use the valve train to deactivate cylinders to reduce engine displacement at idle and while operating under light loads. Deactivation requires interrupting valve openings and shutting off fuel delivery to inactive cylinders. An example is shown in Figure 5-15.

Fig.5-15 Deactivating valves by controlling oil pressure to hydraulic valve lifters

All components of these systems require maintaining oil levels and scheduled oil changes using oil with the specified oil service rating to keep mechanisms clean and functioning as intended. Particle contamination leads to component failure and the specified oil viscosity and oil pressure are required for normal operation.

VALVE TRAIN CONFIGURATIONS

There are four basic valve train configurations that one can expect to find in passenger car engines. The earliest of these designs is still with us and is recognized as the pushrod operated valve system. In this system, the camshaft opens valves via a valve lifter, pushrod, and rocker arm (see Fig.5-16). In all valve trains, the valve is closed by the valve spring. The disadvantage in this design is the number of parts and the inertia that they add to valve operation. To overcome this inertia, greater valve spring pressures are required relative to the other designs. An advantage is found in the compact size of pushrod engines since going to overhead camshaft configurations typically adds to the overall height of the engine assembly.

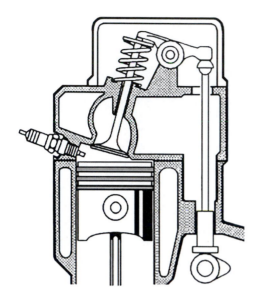

Fig.5-16   A complete pushrod valve train

Cam-in-head engines eliminate the need for pushrods and therefore some of the added inertia (see Fig.5-17). As with other overhead camshaft engines, a much longer cam drive belt or chain is required (see Figs, 5-18 and 19). Although simple in design, these engines still add height to the assembly.

Fig.5-17   A cam-in-head cylinder head

Fig.5-18   A belt driven overhead camshaft

Fig.5-19   A chain driven overhead camshaft with tensioners and guides

Overhead cam engines with rocker type followers are common and valve lifters and pushrods are eliminated. In the first type, the rocker arm acts as a cam follower. In the second type, the camshaft acts on the rocker arm from above and the rocker arm pivots at one end and opens the valve at the other (see Fig.5-20).

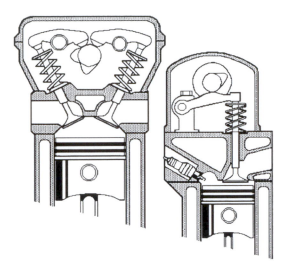

Fig.5-20   An overhead cam with a rocker follower (L) and the cam over a rocker arm (R)

The simplest of all designs is overhead camshaft with a bucket follower positioned directly between the camshaft and the valve (see Fig.5-21). All extra valve train parts are eliminated and, without the inertia, much less valve spring pressure is required.

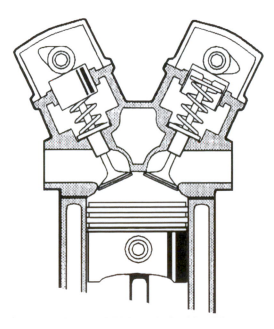

Fig.5-21 Overhead cams and bucket followers

As an enhancement to these overhead camshaft configurations, many engines now use four valves per cylinder (see Fig.5-22). While the number of parts increases, the size and weight of valves and springs are reduced and this further reduces inertia. The key advantage however is that two small valves on either the exhaust or intake side still have greater area than a single large valve and therefore increase flow rates and improve volumetric efficiency.

Fig.5-22   A four valve per cylinder head

VALVE LIFTERS AND LASH COMPENSATORS

Two types of valve lifters are used in pushrod operated overhead valve engines. Solid valve lifters have no internal parts and require clearance or lash in the valve train mechanism to ensure closing of the valve (see Fig.5-23). Hydraulic valve lifters are designed to maintain zero-lash in the valve train mechanism. Their advantages include quieter engine operation and the elimination of periodic lash adjustments required for solid valve lifters. Hydraulic valve lifters maintain constant pressure on the camshaft and therefore the anti-scuff properties of lubricating oils are critical. Hydraulic valve lifters maintain zero-lash as follows:

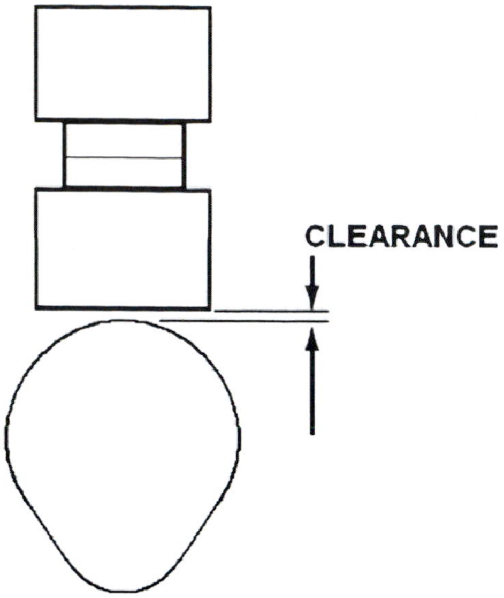

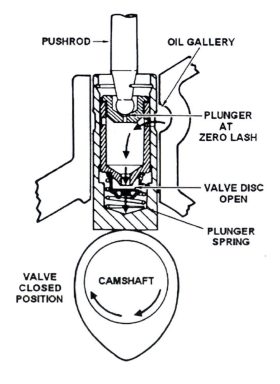

Fig.5-23 Valve lash, or clearance, with solid valve lifters

1. In the valve-closed position, oil flows through the lifter bore to the interior of the plunger and past the check valve (see Fig.5-24). The plunger return spring maintains zero-lash at this point.

Fig.5-24 Oil flow through a hydraulic valve lifter in the valve closed position

2. As the cam and lifter body rise to open the valve, the check-valve seats and trapped oil limits slippage of the plunger within the lifter body (see Fig.5-25). The valve opens just as it would with a solid valve lifter.

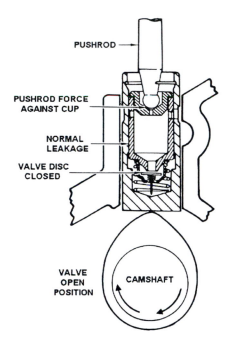

Fig.5-25 Blocked oil flow through a hydraulic valve lifter in the valve open position

3. As the camshaft returns to the valve-closed position, oil again enters lifter body and flows through the plunger past the open check valve. This oil replaces oil lost by leakage between the plunger and lifter body. This loss is referred to as "predetermined leakage" and is normal.

There also are metering devices under the pushrod seats that meter oil to the pushrods. They usually consist of a flat disc that limits oil flow through a hole in the pushrod seat (seat Fig.5-26).

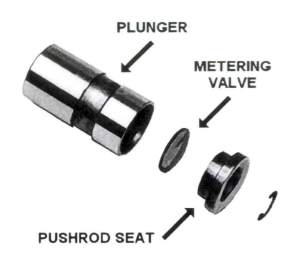

Fig.5-26 Note the valve metering oil flow through the pushrod seat

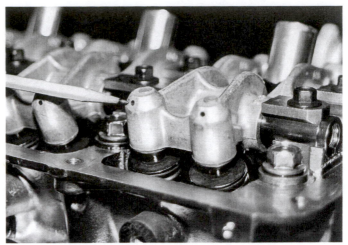

Fig.5-28 A hydraulic lash adjuster built into a rocker follower

Lash compensators are used to maintain zero-lash in overhead camshaft engines with rocker style followers (see Fig.5-27). Other hydraulic mechanisms are also found in the valve ends of rocker type followers and in bucket type followers (see Figs. 5-28 and 29). Internally, they operate similar to hydraulic valve lifters and are subject to the same lubrication and service requirements.

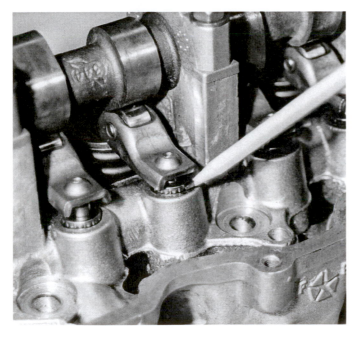

Fig.5-27 OHC cam follower with hydraulic lash compensator

Fig.5-29 The hydraulic mechanism built into a bucket follower

As stated, because zero lash lifters and followers are in constant contact with camshafts, it is more difficult to keep oil between surfaces and there is some tendency for scuffing. To overcome this, many pushrod engines now use roller lifters (see Fig.5-30). Rollers have also been added to rocker type followers in overhead camshaft engines for the same reasons (see Fig.5-31).

Fig.5-30 Roller hydraulic lifters for a pushrod engine

Fig.5-31 A roller rocker follower for an overhead cam engine (Chrysler)

ENGINE OILING

Lubrication by combinations of pressurized oil circulation, run off from bearings or other moving parts, and sometimes jets of oil directed at critical areas, prevents metal-to-metal contact between moving parts (see Figs. 5-32 and 5-33). Engine oil also acts to cool engine parts such as pistons and cylinders as it drains back into the oil pan. The pressurized oiling of engine bearings also aids in cooling as the oil absorbs heat from the crankshaft and bearing surfaces.

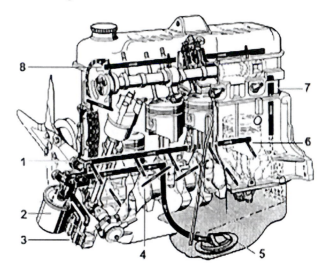

Fig.5-32 Oil system including pressure regulator (1) filter (2) pump (3) oil passage main to rod (4) pick up (5) main passage (6) return (7) cam bearing oil passage (8)

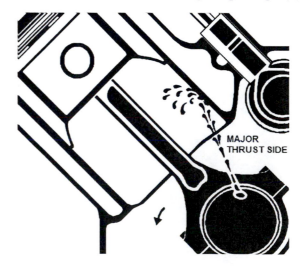

Fig.5-33 Oiling the major thrust side of a cylinder with oil from a rod bearing

The oil "sump" is the lowest point in the oil pan and contains the engine oil. Oil pans frequently use baffles to keep oil in the sump on hard braking, acceleration, or cornering (see Fig.5-34). A windage tray positioned between the crankshaft and the oil sump prevents turbulent air from the rotating crankshaft from aerating oil in the sump (see Fig.5-35).

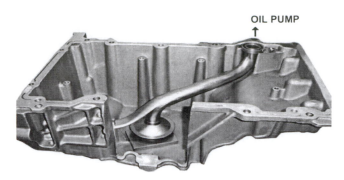

OIL PUMP

Fig.5-36 Oil pump pick up screen in the crankcase

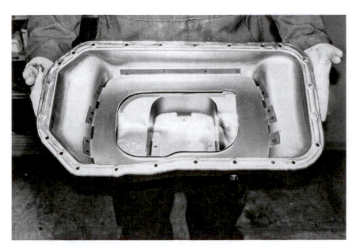

Fig.5-34 An oil pan with baffles

Oil pumps are driven by an extension of the distributor shaft, by direct engagement of a pump drive gear with a gear on the camshaft, or directly off of the crankshaft (see Figs. 5-37 and 38). Pumps driven directly or indirectly off the camshaft operate at one-half crankshaft RPM. Only crankshaft driven pumps operate at full crankshaft RPM.

Fig.5-35 A windage tray

Oil circulates from a pump mounted in the engine crankcase, the front timing cover, or outside the block. The pump draws oil from the sump of the oil pan through a tube extending from the pump inlet to the sump. A screen over the sump end of the pickup tube prevents large pieces of debris from entering the pump (see Fig.5-36).

DRIVE

PUMP

SCREEN

Fig.5-37 An oil pump driven by the distributor shaft (Pontiac)

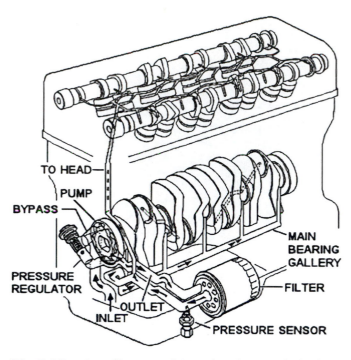

Fig.5-38   An oil pump driven by the crankshaft

Most oil pumps are gear or rotor types. Gear pumps use a pair of meshing gears in a closed housing. The pump drives the first gear, which in turn drives the second gear. A pressure relief valve regulates oil pressure (see Fig.5-39). In rotor pumps, a drive shaft drives the inner rotor, which in turn drives the outer rotor (see Fig.5-40). As with gear pumps, a pressure relief valve regulates oil pressure.

Fig.5-39   A gear-pump and pressure relief valve (Cadillac)

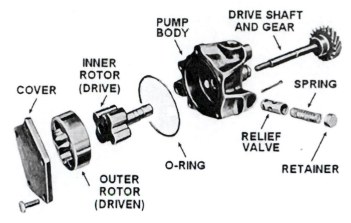

Fig.5-40   Rotor-pump parts including pressure relief valve (Chrysler)

Current engines use full-flow oil filtering systems. This means that oil pump output goes through a filter before circulating through the engine. Typical oil filter materials are resin treated paper, cotton, or other materials with very fine porosity.

Resin treatment prevents contaminants in the oil, such as water or acids, from attacking or restricting the filter. The porosity of the filter material must be fine enough to trap solids of any size but low in restriction so that flow is not inhibited.

There is a bypass valve in the filtering system that keeps dirty or restricted filters from limiting engine oiling. The bypass valve opens should the filter become plugged and oil pump output goes directly to the engine (see Fig.5-41).

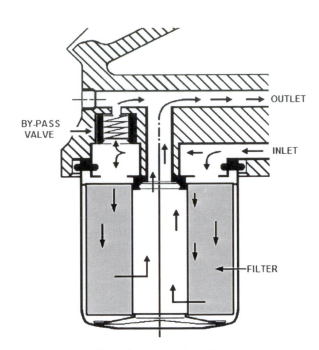

Fig.5-41 Oil filter bypass circuit

ENGINE OILS

Oil service classifications assist in linking the engine manufacturer's recommendations with oil marketing labels so that the proper oils may be selected. Service classifications are developed through the cooperative efforts of a number of professional organizations including:

SAE   Society of Automotive Engineers
API   American Petroleum Institute
ILSAC   International Lubricant Standardization Approval Committee

ASTM   American Society for Testing and Materials
AAM   Alliance of Automotive Manufacturers
EMA   Engine Manufacturers Association

API "S" classifications describe standards for gasoline "service type" engines and "C" classifications for "commercial type" diesel engines. The "S" or "C" designations are followed by other letters in alpha order to call out increasing service requirements. Service classifications SA through SH are now considered obsolete. SA oil was straight mineral oil without additives for the earliest engines while the oil classifications that follow have additives that address service requirements of each new generation of engines.

SAE ratings indicate that oils meet specified test standards for viscosity change over a temperature range. For example, an SAE 5W-30 oil has a flow rate equivalent to 5-weight oil at -22 degrees F (Cold Crank Simulator) and –31 degrees F (Mini Rotary Viscometer which simulates the oil pump) for "W" winter grades, and a flow rate equivalent to 30-weight at 212 degrees F. Oils that thin less with heat have a higher viscosity index or "VI".

For current emission controlled engines with catalysts and oxygen sensors, ILSAC-GF gasoline fuel minimum performance standards apply. Many passenger car manufacturers recommend GF "energy conserving" oil not exceeding 30-weight viscosity. Of these classifications, GF-1 through 4 are obsolete. GF oils have limits on phosphorous content and lower volatility to reduce oil burning and catalyst and oxygen sensor contamination. Fuel economy should also improve slightly (.5 percent minimum) when compared to other oils.

The concern for engine builders is that the phosphorous content is part of a zinc-phosphorous anti-scuff additive (ZDDP). For light duty service in late model passenger car engines with roller valve trains, this may not be a major concern but, in performance engines and early engines with flat tappets or followers, anti-scuff properties are important. Oils with viscosities exceeding 30-weight need not conform to phosphorous limits. For maximum anti-scuff protection, the author recommends investigating the ZDDP content in 10-40, 20-50, or synthetic oils. For non-converter or off-road applications, motorcycle oil, racing oil and CH-4 diesel oil introduced in 1998 oils typically provide protection. The later CI-4 and CJ-4 diesel oils have been reformulated for low emission diesel engines running on low sulfur fuel.

Oil additives are obviously very important in meeting service requirements and the various service ratings call for several. The selection of additives and their reasons for use are summarized as follows:

1. Viscosity index improvers to minimize viscosity change with heat and pour point depressants to maintain flow rate when cold. Compounds include metha-acrylate polymers, olefins, and styrenes.

2. Detergents to dissolve varnish and sludge. Compounds include barium, calcium, magnesium phenates and sulfonates.

3. Dispersants to keep sludge, carbon, and other materials from recombining and suspended in the oil so that they are removed with drain oil. Compounds include polymeric succinimides, succinate esters, akyl phenol amines and benzlyamides.

4. Scuff inhibitors, extreme pressure (EP) agents, and friction modifiers to reduce friction and wear. Compounds include zinc diothiophosphates (ZDDP) and organic compounds of phosphate, sulfide, and chlorine.

5. Antifoam and antioxidant agents and metal deactivators to prevent foaming and to slow oxidation of the oil. Compounds include silicone polymers, ZDDP, phenols, and amines.

6. Some of these additives also work in combination to prevent rust and corrosion. The alkaline additives neutralize acids that cause corrosion and other additives protect parts by forming a protective film over them.

The "C" classifications are slightly different in that each addresses the specific service requirements of different engines such as for turbo or non-turbo, 2 or 4-stroke, light or heavy duty, low or high speed, etc. These oils should be selected not by alpha order but instead by specific engine requirements. Classifications CA through CG-4 are obsolete.

<u>American Petroleum Institute Recommendations</u>

The current and previous API Service Categories are listed below. Vehicle owners should refer to their owner's manuals before consulting these charts. Oils may have more than one performance level. For automotive gasoline engines, the latest engine oil service category includes the performance properties of each earlier category. For diesel engines, the latest category usually, but not always, includes the performance properties of an earlier category.

ILSAC STANDARD FOR PASSENGER CAR ENGINE OILS

GF-5	Current	Introduced in October 2010, designed to provide improved high temperature deposit protection for pistons and turbochargers, more stringent sludge control, improved fuel economy, enhanced emission control system compatibility, seal compatibility, and protection of engines operating on ethanol-containing fuels up to E85.
GF-4	Obsolete	Use GF-5 where GF-4 is recommended.
GF-3	Obsolete	Use GF-5 where GF-4 is recommended.
GF-2	Obsolete	Use GF-5 where GF-4 is recommended.
GF-1	Obsolete	Use GF-5 where GF-4 is recommended.

GASOLINE ENGINES

SN	Current	Introduced in October 2010, designed to provide improved high temperature deposit protection for pistons, more stringent sludge control and seal compatibility. API SN with Resource Conserving matches ILSAC GF-5 by combining API SN performance with improved fuel economy, turbocharger protection, emission control system compatibility, and

protection of engines operating on ethanol-containing fuels up to E85.

SM	Current	For 2010 and older automotive engines.
SL	Current	For 2004 and older automotive engines.
SJ	Current	For 2001 and older automotive engines.
SH	Obsolete	OBSOLETE: For 1996 and older automotive engines.
SG	Obsolete	CAUTION: Not suitable for use in most gasoline-powered engines automotive built after 1993. May not provide adequate protection against engine build-up of engine sludge, oxidation or wear.
SF	Obsolete	CAUTION: Not suitable for use in most gasoline-powered automotive engines built after 1988. May not provide adequate protection against engine build-up of engine sludge.
SE	Obsolete	CAUTION: Not suitable for use in gasoline-powered engines built after 1979.
SD	Obsolete	CAUTION: Not suitable for use in gasoline-powered engines built after 1971. Use in modern engines may cause unsatisfactory performance or equipment harm.
SC	Obsolete	CAUTION: Not suitable for use in gasoline-powered engines built after 1967. Use in modern engines may cause unsatisfactory performance or equipment harm.
SB	Obsolete	CAUTION: Not suitable for use in gasoline-powered engines built after 1951. Use in modern engines may cause unsatisfactory performance or equipment harm.
SA	Obsolete	CAUTION: Contains no additives. Not suitable for use

in gasoline-powered engines built after 1930. Use in modern engines may cause unsatisfactory performance or equipment harm.

DIESEL ENGINES

CJ-4	Current	For high speed, four stroke engines designed to meet 2010 model year on-highway and Tier 4 non-road exhaust emission standards as well as previous model diesel engines. These oils are formulated for use in all applications with diesel fuels ranging in sulfur content up to 500ppm (.05% by weight). However, use of these oils with greater than 15 ppm (.0015% by weight) sulfur fuel may impact exhaust after-treatment system durability and/or drain interval. CJ-4 oils are especially effective at sustaining emission control system durability where particulate filters and other advanced after-treatment systems are used. Optimum protection is provided forcontrol of catalyst poisoning, particulate filter blocking, engine wear, piston deposits, low and high temperature stability, soot handling properties, oxidative thickening, foaming, and viscosity loss due to shear. API CJ-4 oils exceed the performance criteria of API CI-4 with CI-4 Plus, CI-4, CH-4, CG-4 and CF-4 and can effectively lubricate engines calling for those API Service Categories. When using CJ-4 oil with higher than 15 ppm sulfur fuel, consult the engine manufacturer for service interval.

CI-4 Current — Introduced in 2002. For high-speed, four-stroke enginesdesigned to meet 2004 exhaust emission standards implemented in 2002. CI-4 oils are formulated to sustain engine durability where exhaust gas recirculation (EGR) is used and are intended for use with diesel fuels ranging in sulfur content up to 0.5% weight. Can be used in place of CD, CE, CF-4, CG-4, and CH-4 oils. Some CI-4 oils may also qualify for the CI-4 PLUS designation.

CH-4 Current — Introduced in 1998. For high-speed, four-stroke engines designed to meet 1998 exhaust emission standards. CH-4 oils are specifically compounded for use with diesel fuels ranging in sulfur content up to 0.5% weight. Can be used in place of CD, CE, CF-4, and CG-4 oils.

CG-4 Obsolete — OBSOLETE: Introduced in 1995. For severe duty, high speed, four-stroke engines using fuel with less than 0.5% weight sulfur. CG-4 oils are required for engines meeting 1994 emission standards. Can be used in place of CD, CE, and CF-4 oils.

CF-4 Current — OBSOLETE: Introduced in 1990. For high-speed, four-stroke, naturally aspirated and turbocharged engines. Can be used in place of CD and CE oils.

CF-2 Current — OBSOLETE: Introduced in 1994. For severe duty, two-stroke cycle engines. Can be used in place of CD-II oils.

CF Current — OBSOLETE: Introduced in 1994. For off-road, indirect injected and other diesel engines including those using fuel with over 0.5% weight sulfur. Can be used in place of CD oils.

CE Obsolete — CAUTION: Not suitable for use in most diesel-powered automotive engines built after 1994.

CD-II Obsolete — CAUTION: Not suitable for use in most diesel-powered automotive engines built after 1994.

CD Obsolete — CAUTION: Not suitable for use in most diesel-powered automotive engines built after 1994.

CC Obsolete — CAUTION: Not suitable for use in diesel-powered engines built after 1990.

CB Obsolete — CAUTION: Not suitable for use in diesel-powered engines built after 1961.

CA Obsolete — CAUTION: Not suitable for use in diesel-powered engines built after 1959.

Not shown above are oils for a new generation of diesel engines expected in 2017. Proposed are new API CK-4 and FA-4 oils that also meet requirements for Tier 4 non-road exhaust emission standards and previous model year diesel engines. These oils are formulated for use in all applications with diesel fuels ranging in sulfur content up to 500 ppm although oils with greater than 15 ppm sulfur fuel may impact exhaust after-treatment system durability and oil drain intervals.

CK-4 oils are especially effective at sustaining emission control system durability where particulate filters and other advanced after-treatment systems are used. CK-4 oils are designed to provide enhanced protection against oil oxidation, viscosity loss due to shear, oil aeration as well as protection against catalyst poisoning, particulate filter blocking, engine wear, piston deposits, degradation of low and high-temperature properties and soot-related viscosity increase. These oils exceed the performance criteria of CJ-4, CI-4 with CI-4 PLUS, CI-4, and CH-4 and can effectively lubricate engines calling for those API Service Categories. When using CK-4 oil with higher than 15 ppm sulfur fuel, consult the engine manufacturer for service interval recommendations.

FA-4 XW-30 oils are specifically formulated for use in select high-speed four-stroke cycle diesel engines designed to meet 2017 model year on-highway greenhouse gas (GHG) emission standards. These oils are formulated for use in on-highway applications with diesel fuel sulfur content up to 15 ppm. API FA-4 oils are neither interchangeable nor backward compatible with API CK-4, CJ-4, CI-4 with CI-4 PLUS, CI-4, and CH-4 oils. Refer to engine manufacturer recommendations to determine if API FA-4 oils are suitable for use.

Oil containers should be marked with API-SAE service ratings and possibly a starburst for "Energy Conserving" or "Resource Conserving" oils. Note that the newest API SN oil service rating listed above also applies to ILSAC GF-5, the latest performance standard set by the International Lubricant Standardization and Approval Committee (ILSAC), a joint effort of U.S. and Japanese automobile manufacturers. Most automobile manufacturers are expected to recommend oils that meet ILSAC GF-5. Vehicle owners and operators should follow their vehicle manufacturer's recommendations on engine oil viscosity and performance standard. Be aware that engine oils may be rated for both spark and compression ignition engine service (see Fig.5-42).

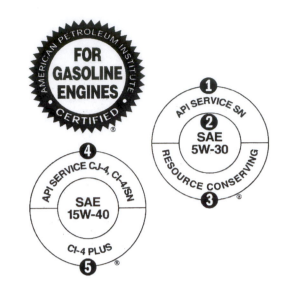

Fig.5-42 API gasoline engine Starburst and (1) Service Rating (2) Viscosity (3) Resource or Energy Conserving (4) Multiple diesel and gasoline service ratings (5) CI-4 high-level protection

Petroleum oils have limited ability to stand up under high temperature conditions. Under test conditions, oils reach their flash point at somewhere between 400 and 480 degrees F and internal engine parts exceed these temperatures at some points within the engine:

1. Crankshafts, rods, and bearings operate between 200 and 300 degrees F, well within acceptable limits for oils.

2. Pistons and related parts will operate at 400 degrees F and lower below the piston pin, but above the pin, piston crowns will see 700 degrees F and higher.

3. Upper piston rings see 300 to 600 degrees F.

4. Upper cylinder walls will see 300 to 500 degrees F.

5. Water jackets surrounding cylinders will range from 160 to 230 degrees F.

6. Exhaust valves will approach 1500 degrees F when the engine is operating under load. Exhaust gases run from 500 to 1400 degrees F.

7. Combustion chamber temperatures range from

Considering that aluminum melts at 1300 degrees F and iron at nearly double this temperature, these engine parts clearly do not reach combustion temperatures or the engine would destroy itself. High temperatures are not sustained but instead peak during combustion and cool on intake strokes, but the importance of oil's role in cooling parts cannot be under estimated. Engine oil carries away heat as it washes over surfaces without forming deposits on engine parts.

With the extreme performance requirements for engine oils, synthetics are growing in popularity with some manufacturers. Synthetic oils are generally derived from severely "hydrocracked" paraffinic base oil or from polyalphaolefin base stock. Generally, synthetics perform better over a wider range of temperatures; they flow better at low temperatures and maintain viscosity better at high temperatures. They also volatilize at higher temperatures and therefore create less emissions and less oil consumption. Like petroleum oils, they utilize additive packages to improve performance such as required for additional scuff protection.

The initial reluctance to adopt synthetics in wide application was largely to due to cost-benefit considerations. However, the refining of petroleum can be less than perfect and oils invariably contain minute quantities of sulfur and other impurities making them subject to earlier crankcase degradation.

A major factor in determining engine oil drains intervals is contamination. Water from combustion and from condensation, fuel dilution, and dirt ingestion limit usable oil life. Synthetics do not protect better than petroleum oils in respect to crankcase contamination. Oil drains remove damaging contaminants not trapped by the filter.

Operating parameters must be considered when selecting oils, service ratings, and oil drain intervals. The best possible oil and filter for the application should be used. In every case, the original equipment manufacturer's recommended drain intervals should be followed closely based on the operator's type of driving including severe conditions such as dusty environments, cold temperatures, short trips, heavy loads, etc.

ENGINE MEASUREMENTS

The first engine measurements to become familiar with are bore and stroke. Bore is the diameter of the cylinder, and stroke is the distance between TDC and bottom-dead-center (BDC) of piston travel (see Fig.5-43). In reading engine specifications, bore diameter is given first and stroke length second. The stroke offset of each crankpin of the crankshaft is half of specified stroke length.

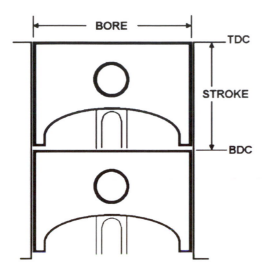

Fig.5-43   Bore and stroke measurements

Remember that a crankpin offset 2 inches from the center of rotation creates 4 inches of piston travel per stroke (see Fig.5-44). Displacement is a calculation of volume displaced by the piston travel in each cylinder. Total displacement is the sum of displacements for all cylinders in an engine and is calculated as follows:

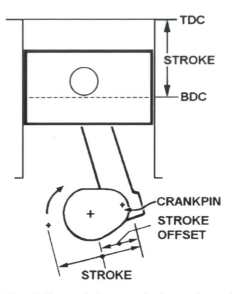

Fig.5-44   Offset of the crankpin and stroke length

Example: Calculate the cubic inch displacement (CID) of a four-cylinder engine with a 3.5 inch bore diameter and a 3.0 inch stroke length.

> N = Number of cylinders
> Pi = 3.1416
> D = Cylinder bore diameter
> R = Radius or half the bore diameter
> S = Stroke length

> N x π x R x R x S = Displacement
> 4 x 3.1416 x 1.75 x 1.75 x 3.0 = 115.4in3

Compression ratios (CR) are important because of their direct influence on engine efficiency. Calculate compression by dividing the total volume above the piston at BDC by the volume above the piston at TDC (see Fig.5-45). Note that the total volume above the piston at BDC includes displacement and that the volume above the piston at TDC includes the combustion chamber. The term "clearance volume" refers to the volume above the piston at TDC.

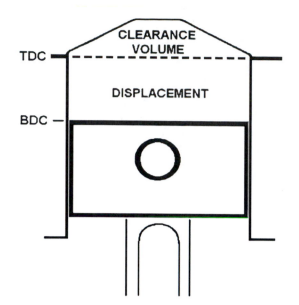

Fig.5-45 Compression ratio equals clearance volume plus displacement divided by clearance volume

Example: Calculate the compression ratio of an engine with 30 cubic inches and a 3.3 cubic inch clearance volume.

$$(CID + CV) \div CV = CR$$
$$(30 + 3.3) \div 3.3 = 10 \text{ (or 10:1CR)}$$

In precision or performance engine building, it is necessary to calculate the clearance volume (CV) because it is not available in specifications. Calculate the clearance volume by dividing cubic inch displacement by the compression ratio minus one. The compression ratio used as a base may be the specified ratio or one that the builder considers necessary for the particular engine project.

$$CID \div (CR - 1) = CV$$

Example: Calculate the clearance volume of a 40 cubic inch cylinder with a specified 9:1 compression ratio.

$$40 \div (9 - 1) = 40 \div 8 = 5 in3$$

As mentioned, the compression ratio directly affects engine efficiency. Engines with higher compression ratios compress the air-fuel mixture

more before ignition. This causes greater expansion after ignition and increases cylinder pressures. If combustion remains under control, increased power output and fuel economy is the result. However, high compression engines typically require high-octane fuel and the higher combustion pressures add to stresses on piston rings and other components. Currently, production engines limit compression ratios to help minimize the nitrogen oxide emissions associated with high combustion pressures and temperatures and to make possible the use of fuel with relatively low octane ratings in the range of 87 to 91.

"Deck clearance" and "deck height" are terms used when discussing compression ratios or changes in compression ratios due to machining. Deck clearance is the distance from the top of the piston to the block deck (see Fig.5-46). Clearances can be positive or negative depending on how they affect clearance volumes. Deck height is the distance from the main bearing centerline to the block deck (see Fig.5-47).

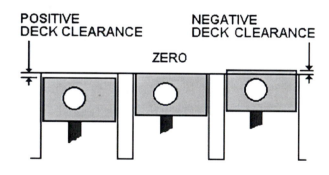

Fig.5-46 Positive and negative deck clearance relative to clearance volume

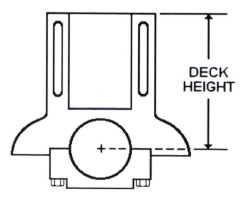

Fig.5-47 Deck height

FITS AND CLEARANCES

The term fit has specific meaning and importance to engine technicians and machinists. First, there are several types of fits in any mechanism. There are "running" and "interference" fits and each type requires carefully checking dimensions of component parts to ensure proper assembly and optimum performance.

Crankshaft bearings are an example of running fits. Clearance between the bearing surface and the shaft permits oil flow between the moving parts. This clearance generally increases as the shaft diameter increases to permit a proportionate increase in oil flow through the bearing. With too little clearance, there will be metal-to-metal contact and wear. With too much clearance, there will be too much movement of the shaft resulting in knocking and bearing failure.

Specified clearances are "diametral". This means that clearances are based on the diameter of the shaft as opposed to the radius. A specified clearance of .002in (.05mm) allows .001in clearance (.025mm) on each side of the shaft (see Fig.5-48).

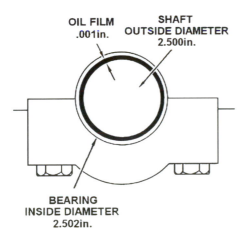

Fig.5-48   The oil film on one side of a shaft with .002in diametral clearance

In other places, there are interference or press fits. Such fits hold parts in assembly and there is no relative motion between the two assembled parts. The outside diameter of the press-fit part is larger than the inside diameter of the part it is forced into. For example, many piston pins are press fit through the small end of the connecting rod (see Fig.5-49).

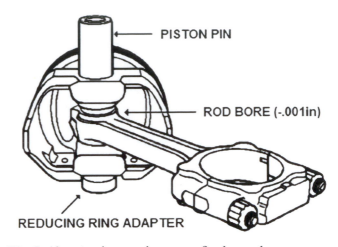

Fig.5-49   A piston pin press fit through a connecting rod

To ensure that fits of either type fall within specifications, each part must be carefully measured. Each part has a tolerance or an acceptable variation in size. With the exception of some "selectively fitted" parts, if parts are within tolerance, assemblies will be within specified limits for clearance or interference.

However, manufacturers frequently use selective fits and assemble by matching parts to each other to obtain optimum fits. This way, very precise assemblies are possible using more permissive levels of tolerance for individual parts. This reduces the cost of manufacturing for precision assemblies.

The major point of emphasis is the necessity to measure all engine parts so that fits are within specified limits. Precision measuring does as much to maintain a high standard of quality in engine service as any other single practice.

COOLING SYSTEM OPERATION

Internal combustion engines are less than 30 percent efficient. That is, of the heat energy released during combustion, less than one-third produces usable power. Approximately one-third escapes through the exhaust system and one-third radiates to the atmosphere through the cooling system.

The one-third handled by the cooling system alone is sufficient to melt pistons and destroy other engine parts if not controlled. Aside from melting pistons, operating temperatures must be lower still to prevent over expansion of moving parts and the resultant scuffing and scoring. Yet to maximize engine efficiency, operating temperatures must be high enough to keep fuels vaporized for complete combustion. High temperatures also vaporize crankcase blow-by gases so that they are scavenged by crankcase ventilation systems thus minimizing oil contamination and parts corrosion caused by acid buildup.

The basic components of a liquid cooling system include the engine water jackets, radiator, water pump, and hoses connecting the engine and radiator (see Fig.5-50). In the engine, heat energy transfers through the walls of the combustion chambers and cylinders into the coolant. The coolant circulates through the upper radiator hose into the radiator, where heat energy dissipates into the air. From the bottom of the radiator, coolant passes through the lower radiator hose to the water pump, is pumped through the engine, and passes through the upper radiator hose back to the radiator. Water pumps circulate as much as 4,000 gallons per hour to keep coolant temperatures down and to protect the engine.

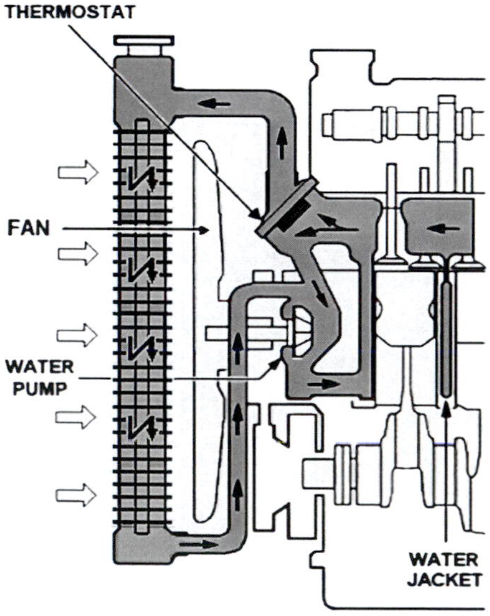

To improve efficiency, a thermostat regulates coolant flow and keeps the engine temperature in the optimum range. Thermostats regulate water flow by opening as coolant temperatures rise and by closing when coolant temperatures drop. A bypass hose or circuit limits coolant circulation to the engine water jackets when the thermostat is closed.

Fans increase the airflow through radiators at idle and low speeds so that there is sufficient heat transfer from the coolant to the air. Fans however require 6 to 8 horsepower (HP) and it is desirable to have them working only as needed for increased cooling. To limit the power drain, some fans are electric with thermostatic switches that permit fans to remain off when extra cooling is unnecessary (see Fig.5-51). Other engines use temperature controlled hydraulic fan clutches (see Fig.5-52). The hydraulic fluid is silicone and the clutch uses a hydraulic control valve operated by a bi-metal thermostat. The fan and fan clutch are in the radiator airflow and the clutch engages when the air temperature reaches approximately 160 degrees F.

Fig.5-51 An electric cooling fan and shroud

THERMOSTAT

FAN

WATER PUMP

WATER JACKET

Fig.5-50 Liquid cooling system including radiator, thermostat, pump, and water jackets

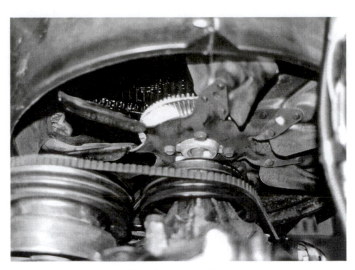

Fig.5-52 A fan clutch assembly

Coolant efficiency is improved by using a mixture of water and a coolant such as ethylene glycol and also by pressurizing the system. With 50 percent ethylene glycol, the boiling point of the coolant rises from 212 to 227 degrees F. By pressurizing the system to 15 PSI, the boiling point rises to 265 degrees F. It is essential that boiling be prevented because this is when coolant is lost.

Engine coolants also contain corrosion inhibitors that prolong the life of cooling system components and the engine. Coolant mixtures generally include 50 to 60 percent ethylene glycol, propylene glycol or Dexcool, each with a variety of additive packages. New coolant is on the basic side of the pH scale with a numeric value of 8.5 or more. Over time, coolant deteriorates and becomes more acidic making metallic engine components subject to corrosion therefore coolant is replaced when pH drops below 7.5.

COMBUSTION EFFICIENCY

Under ideal conditions, combustion would be complete and exhaust emissions limited to carbon dioxide and water vapor. Specifically, two molecules of iso-octane (pure laboratory gasoline) would combine with twenty-five molecules of oxygen. This combination would produce sixteen molecules of carbon dioxide and eighteen molecules of water vapor.

$$2\,C_8H_{18} + 25\,O_2 = 16\,CO_2 + 18\,H_2O$$

As mentioned, complete combustion only occurs with pure gasoline (iso-octane) and perfect fuel mixtures. Under real conditions, the mixture of fuel to air is frequently imperfect, and in addition, hydrocarbons cool as they contact the walls of the combustion chamber. The result is incomplete combustion and emissions of carbon monoxide and unburned hydrocarbons. Two iso-octane and eighteen oxygen molecules typically yield eleven carbon dioxide molecules, thirteen water molecules, and one molecule each of carbon monoxide and butane.

$$2\,C_8H_{18} + 18\,O_2 = 11\,CO_2 + 13\,H_2O + CO + C_4H_{10}$$

Under high engine loads, cylinder pressures and temperatures climb and another reaction occurs. Because air is 78 percent nitrogen, oxides of nitrogen also develop. Typically, this means two molecules of nitrogen and three of oxygen yields two molecules each of nitrogen monoxide and nitrogen dioxide. The different oxides of nitrogen are called "NOX."

$$2\,N_2 + 3\,O_2 = 2\,NO + 2\,NO_2$$

In today's world, it is mandatory that emissions be kept to the minimum. Simultaneously, no one wishes to sacrifice power output or fuel economy. To gain high efficiency and low emissions, precise control of spark timing, fuel mixture, and peak cylinder temperature is necessary.

The use of emission controls began in California in 1966 and federal requirements followed in 1968. These first controls were primitive by current standards. They reduced emissions but irritated drivers with drivability problems and many drivers complained that power and fuel economy suffered. Computers now integrate the control of fuel injection, electronic ignition, and spark timing. With catalytic converters, exhaust gas recirculation, and air injection into the exhaust, emissions are now reduced to a few percentage points of what non-controlled vehicles produce, while at the same time, fuel economy, power output, and drivability are all improved.

Air fuel ratio is the first factor to consider when comparing emissions to fuel consumption. Low emissions of carbon monoxide and unburned hydrocarbons, and low rates of fuel consumption, are possible at a 14.7:1 air-fuel ratio (see Fig.5-53). Oxygen sensors in the exhaust feedback information through on-board computers to maintain correct mixtures (see Fig.5-54).

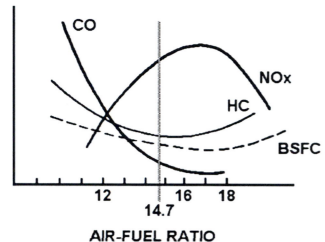

Fig.5-53   Emissions and fuel consumption

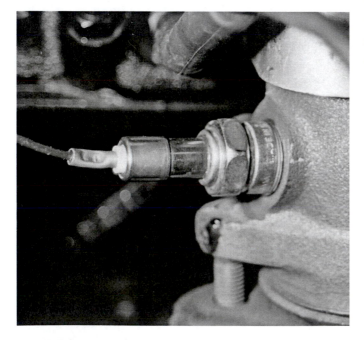

Fig.5-54   An "O2" sensor in the exhaust system

Note that oxides of nitrogen are high at this point. However, by controlling spark timing, NOX can be reduced without significantly increasing unburned hydrocarbons or fuel consumption (see Fig.5-55).

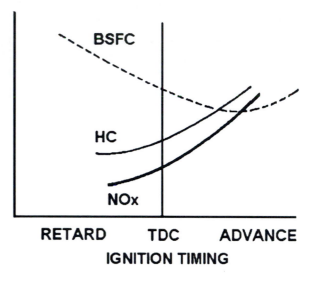
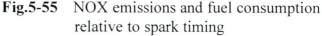

Fig.5-55   NOX emissions and fuel consumption relative to spark timing

The term "conversion efficiency" describes the rate that unburned hydrocarbons and carbon monoxide convert to carbon dioxide and water vapor within catalytic converters. Approximately 90 percent conversion occurs using 14.7:1 air-fuel ratios (see Fig.5-56). Note that this ratio is also the optimum for low fuel consumption.

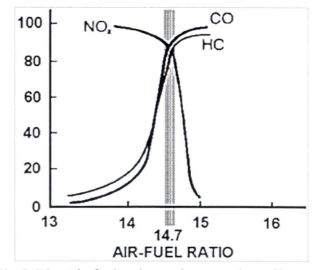

Fig.5-56   Air-fuel ratios and conversion efficiency of catalytic converters

Exhaust gas recirculation (EGR) into the inlet system reduces peak cylinder temperatures and oxides of nitrogen. EGR essentially dilutes the mixture in the cylinder with an inert gas since there is no remaining oxygen or fuel to support combustion. Note that at 5 to 10 percent EGR, hydrocarbon and NOX emissions are lowered without dramatic effect on fuel consumption (see Fig.5-57).

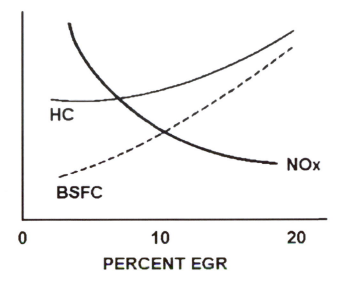

Fig.5-57   EGR and effect on emissions and fuel economy

SUMMARY

Technicians require a thorough understanding of engine theory to effectively diagnose and tune engines. The four-stroke cycle and valve timing are at the core of this understanding. Familiarity with valve train, pistons, crankshaft bearings, and engine oiling are also important especially in regard to diagnosing engine noise or oiling failures.

Knowledge of engine operating conditions, and engine oils, filters, and oiling systems, helps technicians understand requirements for engine protection and provide appropriate maintenance services to consumers. On occasion, technicians or service managers also need to explain maintenance requirements to consumers and a detailed knowledge of engine oils helps add clarity to explanations.

An understanding of engine measurements, fits, and clearances are necessary for proper engine assembly and machining or rebuilding. Even greater detail in this area is required for those involved in performance engine preparation. These technicians and machinists must predict changes in displacement and compression caused by changes in engine components or machining.

Engine cooling problems are a primary cause of engine failure. Engine builders must understand the function of components in system operation and the efficiency of coolants and coolant mixtures. The extended maintenance intervals in newer vehicles and self-service gasoline stations mean that fluid levels frequently go unchecked between services. This adds importance to the inspection of individual components in the system and to the maintenance of coolants and coolant concentrations.

The relationships of spark timing and fuel mixtures to combustion efficiency are critical elements in understanding engine management systems and emission control. Instead of "tune-ups", we now have long-term service intervals and continuous monitoring of engine management via computers with onboard diagnostic capabilities.

Unless the requirements for combustion efficiency are understood and maintained, performance and economy suffers and emissions increase.

Chapter 5

ENGINE THEORY

Review Questions

1. Technician A says that a four-cylinder engine fires every 360 degrees of rotation. Technician B says that all cylinders fire in 720 degrees of rotation. Who is right?

   a. A only   c. Both A and B
   b. B only   d. Neither A or B

2. Technician A says that the intake valve opens after TDC. Technician B says that the intake valve closes after BDC. Who is right?

   a. A only   c. Both A and B
   b. B only   d. Neither A or B

3. Technician A says that the exhaust valve closes after BDC. Technician B says the valve closes before BDC. Who is right?

   a. A only   c. Both A and B
   b. B only   d. Neither A or B

4. Valve lift in pushrod engines
   a. exceeds cam lift
   b. is equal to cam lift
   c. is less than cam lift
   d. is equal to the radius of the base circle

5. Technician A says that compression in a diesel heats air in the chamber to approximately 1000 degrees F. Technician B says that diesel fuel ignites at approximately 600 degrees F. Who is right?

   a. A only   c. Both A and B
   b. B only   d. Neither A or B

6. Technician A says that both valves are open at TDC on the exhaust stroke. Technician B says that both valves are closed at TDC compression stroke. Who is right?

   a. A only   c. Both A and B
   b. B only   d. Neither A or B

7. Technician A says that duration is the time both valves are open. Technician B says that it is the time that both valves are closed. Who is right?

   a. A only   c. Both A and B
   b. B only   d. Neither A or B

8. Cam lift is equal to the
   a. radius of the base circle
   b. diameter of the base circle
   c. maximum measurement of the cam lobe minus the radius of the base circle
   d. maximum measurement of the cam lobe minus the diameter of the base circle

9. Overlap is the period when
   a. both valves are closed
   b. the intake valve is closing
   c. the exhaust is opening
   d. both valves are open

10. Technician A says that an advantage to overhead cam valve trains is the reduced inertia of moving valve train parts. Technician B says that an advantage to pushrod operated overhead valve engine designs is the compact size. Who is right?

    a. A only   c. Both A and B
    b. B only   d. Neither A or B

11. Technician A says that the check valve in a hydraulic lifter opens when the valve opens. Technician B says that check valves open when valves are closed. Who is right?

    a. A only
    b. B only
    c. Both A and B
    d. Neither A or B

12. In a full-flow filtering system, a plugged oil filter causes oil to
    a. bypass the filter and return to the sump
    b. bypass the filter and lubricate the engine
    c. bypass the filter and return to the inlet side of the oil pump
    d. stop flowing

13. In some rods, there is an oil hole through the rod bearing. This oil hole
    a. permits oil to flow into the rod bearing
    b. lubricates the camshaft
    c. lubricates the cylinders
    d. prevents excess oil pressure in rod bearings

14. Oil baffles in the oil pan
    a. keep oil in the sump on braking, acceleration, or cornering
    b. minimize air turbulence in the sump
    c. allow oil to return more quickly to the sump
    d. prevent overfilling the sump

15. Maximum engine oil pressure is limited by the
    a. crankshaft bearing clearance
    b. oil service rating
    c. oil pump pressure relief valve
    d. engine RPM limit

16. The most current duty gasoline engine oil listed below is
    a. CD
    b. CE
    c. SL
    d. SAE 20-50

17. Technician A says change oil because the additives become depleted. Technician B says change the oil because it wears out. Who is right?

    a. A only
    b. B only
    c. Both A and B
    d. Neither A or B

18. Technician A says calculate the displacement of a cylinder by multiplying Pi times cylinder diameter times stroke length and then dividing by 2. Technician B says multiply Pi times the radius squared times the stroke length. Who is right?

    a. A only
    b. B only
    c. Both A and B
    d. Neither A or B

19. The crankpin in an engine with a 4-inch stroke length is offset from the center of rotation
    a. 2in
    b. 4in
    c. 8in
    d. 4in ÷ N

20. The formula for compression ratio is
    a. displacement divided by clearance volume
    b. clearance volume divided by displacement
    c. clearance volume plus displacement divided by displacement
    d. displacement plus clearance volume divided by clearance volume

21. Technician A says measure deck clearance from the crankshaft centerline to the block deck. Technician B says measure from the block deck to the piston at TDC. Who is right?

    a. A only
    b. B only
    c. Both A and B
    d. Neither A or B

22. Technician A says measure deck height from the crankshaft centerline to the block deck. Technician B says measure from the block deck to the piston at TDC. Who is right?

   a. A only
   b. B only
   c. Both A and B
   d. Neither A or B

23. A crankshaft bearing with .002in clearance has an oil film _____ inches thick.
   a. .0005
   b. .001
   c. .002
   d. .004

24. Precision assemblies are most economically manufactured by
   a. using very close tolerances
   b. hand fitting parts
   c. selectively fitting parts
   d. machining each individual part to fit

25. Technician A says that one-third of the heat of combustion goes out the exhaust. Technician B says that one-third goes into the cooling system. Who is right?

   a. A only
   b. B only
   c. Both A and B
   d. Neither A or B

26. Technician A says that pressurizing a coolant mixture of 50 percent ethylene glycol to 15 PSI raises the boiling point to 265 degrees F. Technician B says without pressure, this mixture raises the boiling point to 227 degrees F. Who is right?

   a. A only
   b. B only
   c. Both A and B
   d. Neither A or B

27. Technician A says that when the engine is cold, coolant bypasses the thermostat and recirculates within the engine. Technician B says that when the engine is cold, coolant recirculates within the pump. Who is right?

   a. A only
   b. B only
   c. Both A and B
   d. Neither A or B

28. Electric cooling fans are switched on by the
   a. computer
   b. coolant switch
   c. temperature gauge
   d. computer, AC controls, or coolant switch

29. Fan clutches engage by
   a. ambient heat
   b. a coolant sensor
   c. the computer
   d. a coolant sensor and computer

30. Technician A says introducing exhaust gas into the intake stream reduces unburned hydrocarbons. Technician B says that precise control of fuel mixtures reduces nitrogen oxides. Who is right?

   a. A only
   b. B only
   c. Both A and B
   d. Neither A or B

FOR ADDITIONAL STUDY

1. When does compression begin in the cylinder?

2. The time both intake and exhaust valves are open is called _____.

3. Describe diesel engine fuel injection, ignition, and combustion.

4. List the advantages to overhead camshaft and pushrod engines.

5. Explain the operation of a "zero-lash" valve train.

6. What is the most current gasoline engine oil service rating and what changes were made from the prior rating?

7. List six engine oil additive groups and their purposes?

8. Calculate the change in displacement when boring a 3.5 x 3.0 four-cylinder .030in oversize.

9. The engine above has a 9:1 compression ratio. What is the clearance volume for one cylinder?

10. A set of pistons has a minimum diameter of 3.4985in and a uni-lateral tolerance of -.0000 to +.0010 of an inch. What is the maximum diameter?

11. The piston above has a minimum clearance of .0015 of an inch. Based upon the minimum piston diameter, to what size should you hone the cylinder bore?

12. A cooling system is filled with 50 percent ethylene glycol and water and has a 15-PSI radiator cap. At what temperature will this coolant boil?

13. Why is the boiling of coolant to be avoided?

14. In perfect combustion, what gases are emitted?

15. At what mixture of air and fuel is conversion efficiency of the catalytic converter greatest?

Chapter 6

ENGINE DIAGNOSIS

Upon completion of this chapter, you will be able to:

- Recognize the signs and indicators of engine wear.
- List the checks for cracks in engine castings or head gasket failure.
- Compare power balance, compression, and cylinder leakage testing
- Interpret results of power balance, compression, and cylinder leakage testing.
- Explain methods of checking valve timing.
- List engine problems and how they appear on a vacuum gauge during testing.
- Explain methods of testing exhaust backpressure.
- List different noises, conditions when they occur, and their causes.
- Explain how to test engine oil pressure and list causes for incorrect readings.
- List the steps in cooling system pressure testing and what problems to watch for.

INTRODUCTION

Engine diagnosis must be as exact as possible so that satisfactory repairs can be made at minimum cost. Failure to perform complete tests and inspections generally leads to making unnecessary repairs and missing some that are necessary. Either situation leads to unsatisfactory or incomplete repairs.

With computerized engine controls comes the capability of on board diagnostics. Of particular value to the technician, is the ability of the computer to detect variations in crankshaft speed via crankshaft position sensors. With low compression, the crankshaft slows on the power stroke relative to the other cylinders and a diagnostic trouble code is set in the computer. However, it is important to note that the drivability technician that detects an engine problem using computer diagnostics is likely to forward the engine to a heavy-duty technician for in depth diagnosis and repair. The heavy-duty

technician will then test compression or perform other conventional tests to identify the precise problem before going ahead with repairs: leaking valve seals, worn piston rings, leaking valves, etc.

Most technicians consider diagnosis especially important when valve grinding is being considered because valve grinding increases vacuum on the intake stroke making the piston ring sealing critical. Worn rings not only allow blow-by past piston rings on the compression stroke, they also allow oil to bypass piston rings on the intake stroke. While the engine may run smoothly, the customer will be unhappy with the increased oil consumption. Avoid problems by testing engine condition before providing service. If ring sealing is poor, valve grinding will be unsatisfactory without also replacing piston rings.

Worn or heat-damaged oil seals also account for considerable oil consumption, especially through intake valve guides (see Figs. 6-1 and 2). If rings do pass testing, oil consumption is reduced after valve service because valve guides are resealed. There are many tests for engine condition and most repair facilities are equipped to perform the tests discussed here.

Fig.6-1   Valve stem seal hardened and cracked by heat

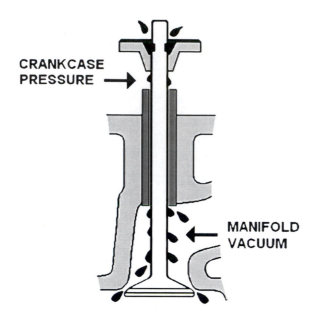

CRANKCASE PRESSURE →

← MANIFOLD VACUUM

Fig.6-2   Oil pullover through intake guides

LOOKING FOR SIGNS OF ENGINE WEAR

A worn engine gives outward signs of its condition. The most obvious of these is the blue gray exhaust smoke that accompanies oil burning. Still, do not confuse the exhaust smoke associated with oil burning with black smoke caused by an over rich fuel mixture or the condensation of water vapor in the exhaust. Oil burning caused by poor ring sealing is most evident under acceleration, especially after the engine has been running at idle. Oil burning caused by oil passing through valve guides is most evident on deceleration.

To understand how oil enters the combustion chamber, first consider that vacuum draws in the oil. Second, consider when and where vacuum is greatest. For example, when under load, piston speed is high, the throttle is open, and high vacuum in the cylinder draws oil past piston rings. When decelerating, piston speed is also high but the throttle is closed creating high vacuum in the intake ports, which draws oil through valve guides.

Of course, evidence of this can show up in exhaust smoke since there is insufficient oxygen to burn this oil in the combustion chamber and most of it is pumped through exhaust ports into

hot manifolds. Catalytic converters complicate the visual diagnosis of exhaust smoke because they so thoroughly convert hydrocarbons that evidence of smoke all but disappears. To diagnose visually, evaluate exhaust in the first few minutes after cold start up before the converter heats up and do not confuse steam from the condensation of water vapor with oil. Of course, if the engine is using oil, the converter will eventually become ineffective.

Mileage records of oil consumption should be checked to help evaluate the extent of engine wear. Although normal rates of oil consumption vary widely among engines, a badly worn engine can be expected to use 1 quart of oil per 1,000 miles or more. This is a lot of oil considering that manufacturers have been making every effort keep oil consumption low so as to not contaminate or degrade emission controls such as exhaust gas recirculation systems or catalytic converters.

There is also a definite power loss that accompanies engine wear although the loss may go unnoticed until the engine is run under load. Often the wear is so gradual that the power loss is not appreciated until the engine is reconditioned and producing full power again.

Check for water in the oil or oil in the water. These conditions are proof of gasket failures or cracks in cylinder heads or blocks.

CHECKING THE BLOCK ASSEMBLY

There is a test for combustion gases in the cooling system called a block check. The presence of gases is caused by cylinder head gasket leaks or cracks in engine castings. To perform a block check, draw vapors from the top of the radiator tank through the chemical test solution (see Fig.6-3). If combustion gases are present, there is a change in solution color from blue to yellow.

Fig.6-3   Testing for the presence of combustion gas in the cooling system

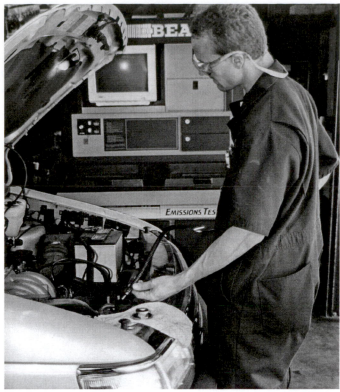

Fig.6-4   Checking for unburned hydrocarbons at the radiator with an exhaust analyzer

Another method of testing involves the use of an exhaust gas analyzer to detect blown head gaskets or cracks by placing the tester probe over the radiator filler neck and checking for the presence of unburned hydrocarbons (see Fig.6-4). Remove the radiator cap only when cold and avoid placing the tester probe directly into the coolant.

One possible test requires no equipment. By removing the water pump drive belt, upper radiator hose and thermostat, combustion gases leaking into the system force coolant through the thermostat opening. For this test, run the engine at a high idle for a minute and check for combustion gases escaping from the thermostat location.

TESTING POWER BALANCE

In power balance testing, we look for the RPM drop when a cylinder is disabled. Little or no RPM drop indicates a weak cylinder typically due to low compression or erratic combustion caused by fuel or ignition problems. On engines with breaker point ignition systems, connect a tachometer and read the RPM drop as one spark plug wire at a time is disconnected and grounded (see Fig.6-5). With electronic ignitions, not computer controlled, cylinder-to-cylinder power balance can be checked by grounding one ignition lead at a time lead with the engine off and restarting the engine to look for changes in engine RPM at idle. Pay attention to changes in engine sound or roughness and then check cylinder-to-cylinder differences. In an engine with normal compression and ignition, the RPM drop is 10 percent or more; the exact amount depending upon the number of cylinders. RPM drop for a cylinder with low compression is less than 10 percent, possibly even zero. Be sure to check if spark plugs have been firing normally before condemning the cylinder.

For computer controlled engines, diagnostics are best done using scan tools because the management system will maintain engine RPM even with a cylinder disabled. Although this feature can be circumvented for testing, very specific product knowledge is required.

Fig.6-5   Insulated pliers for spark plug boots

The test is convenient for identifying the cylinder or cylinders with problems. To be specific about the cause of low compression, it will be necessary to run a compression or cylinder leakage test.

TESTING COMPRESSION

The compression test is probably the most widely used test of engine condition (see Fig.6-6). Most repair shops perform compression tests and testing can usually be completed within an hour. The test measures the pressure produced in individual cylinders at cranking speed. From this, it can be determined if the leakage is occurring past the piston rings or past the valves. Keep in mind however that only the compression rings buildup cylinder pressure and that defective oil control rings can be present even with acceptable test results.

Fig.6-6   Testing cylinder compression at
cranking speed with throttle open

Be sure to run the test properly for accurate results. Run the engine a few minutes so that a normal amount of oil is present at the piston rings for sealing. Open the throttle so that air is available to the cylinders and to prevent siphoning fuel into cylinders from low-speed circuits in carburetors. Remove all spark plugs so that the cranking speed is maintained and ground the coil output lead for safety.

When removing spark plugs for compression testing, keep them in order and visually check deposits for clues to engine condition. For example, a heavy buildup of deposits on one side of electrodes suggests poor valve guide sealing (see Fig.6-7). A closer check often proves that the deposits are building up on the side facing the intake valve. Heavy deposits all around the spark plug suggest poor piston ring oil control.

To separate piston ring leakage from valve leakage, run the test run dry and then wet. If dry testing shows proper compression pressures in each cylinder, further testing is not required. If compression is low or varies beyond specifications, a wet test is in order. Retest cylinders with low compression after injecting approximately one tablespoon of oil through the spark plug hole. If wet testing produces little or no change in results, the low compression is due to leaking valves. If wet testing raises test results by 10 percent or more, the compression loss is occurring past the piston rings. Low compression in adjacent cylinders suggests a blown head gasket.

Fig.6-7   Spark plug deposits caused by leakage through valve guides

Methods of interpreting compression tests vary. Some references specify that compression tests should be within a 20-PSI range; others specify that test results be within a 10 percent range. Some manufacturers specify that the low cylinder should be 75 percent or more of the high cylinder reading. Follow the specifications of each individual manufacturer.

Compression testing cannot readily distinguish compression loss between exhaust or intake valves. Other tests such as cylinder leakage or manifold vacuum testing further isolate the source of leakage. As with other tests, compression test-

ing is limited. For example, "worn rings" may turn out to be a scored cylinder wall, a damaged piston, or broken rings. It is important to check the valve adjustment if test results are uneven or suggest valve sealing problems.

TESTING CYLINDER LEAKAGE

Cylinder leakage testing provides a more detailed analysis than compression testing. Perform this test by pressurizing each cylinder with compressed air; reading the percentage of total leakage on a gauge, and observing the source of escaping air (see Fig.6-8).

Fig.6-8   A Snap-on cylinder leakage tester with line pressure left and cylinder leakage right

Run the engine a few minutes, remove the air cleaner and oil cap, and open the throttle for the test. Find leaking piston rings by listening for the escape of air from the crankcase oil-fill tube. Find a leaking exhaust valve by listening for escaping air at the exhaust pipe. Escaping air at the air intake suggests a leaking intake valve. Air leaking from the cylinder next to the one tested suggests a blown head gasket.

With tests only, determining necessary engine repairs is not always easy. For one thing, normal leakage increases with the circumference of the piston rings therefore expect small diameter cylinders to test with less than 10 percent leakage and larger cylinders to test higher than 10 percent. If testing suggests leaking valves, as with compression testing, check valve adjustment. Check oil consumption rates, exhaust smoke, and for oil fouled spark plugs before passing or condemning an engine that tests in the marginal zone.

Conditions for testing also account for marginal test results in engines that run well. For example, for minimum leakage, the piston must be at TDC. Sometimes this is difficult to do for all cylinders. Unless the engine has been run, there may not be sufficient oil on the cylinders for normal ring sealing. Also, the tester requires at least 100-PSI line air pressure.

On occasion, engines pass tests for rings and valves but still run rough. Unless a fuel system or ignition problem is apparent, check for proper valve action. Remove a valve cover and look for variations in valve lift as a worn camshaft causes these symptoms. Camshaft wear sufficient to cause rough engine operation will be visible, and measuring valve lift will not likely be required. Of course, valve noise is the first clue to a camshaft wear problem.

CHECKING VALVE TIMING

There are several ways of checking valve timing but first consider the common causes of incorrect timing. One is improper cam drive assembly, and others are the failure of cam drive sprockets, chains, or gears (see Fig.6-9).

Fig.6-9 A camshaft sprocket badly worn by the timing chain

The obvious signs of incorrect valve timing include firing through the intake or exhaust system. Compression testing also results in low readings. Engines with a timing chain that has "jumped a tooth" sometimes run smoothly but the power loss is extreme. Tests for valve timing generally involve checking the valve position relative to crankshaft position. For example, when the timing marks are at TDC on the compression stroke and the ignition rotor points to the cylinder number 1 position, both valves should be closed. Valves that are open, just closing or just opening show incorrect valve timing. It should be mentioned here that, if the error in valve timing is due to jumping a tooth, and if the distributor is cam driven, ignition timing would also be incorrect.

A more exact test is possible at TDC on the exhaust stroke. Because this is the valve overlap position, both valves are normally open slightly. The exhaust valve should be just closing, and the intake valve should be just opening. Both valves should move simultaneously when the crankshaft is turned slightly forward and backward.

Since excessive backlash is common with timing chains, check cam drive backlash whenever considering a valve grind or overhaul. Because valve timing retards, a high-backlash condition lowers cylinder pressure and does not allow an engine to run with normal power. More importantly, the timing chain and sprockets commonly fail, causing erratic engine operation and sometimes engine failures that include bent valves. When backlash exceeds 5 degrees, include replacement of the cam drive sprockets and camshaft timing chain in any repair estimate for valve grinding. In engines with high mileage, it is not uncommon to find over 10 degrees backlash. Unless hydraulic chain tensioners are used, excessive backlash may be confirmed statically by testing as follows:

1. Turn the engine forward until the timing marks are at TDC (see Fig.6-10)

Fig.6-10   Turning the engine forward to TDC

2. Mark the position of the rotor. Turn the engine backward by hand until the rotor just begins to move (see Fig.6-11).

Fig.6-11   Watching for motion at the distributor rotor or reluctor

3. Read the amount of backlash at the timing marks.

Since overhead cam engines use hydraulic chain tensioners, they cannot be tested in the same way. Instead, look for signs of excessive slack with an ignition timing light. With excessive slack, ignition timing jumps back and forth especially on deceleration. With direct ignition and no distributor, timing is picked up from the crankshaft assembly and there is no simple test short of disassembly and inspection.

TESTING MANIFOLD VACUUM

Manifold vacuum is a good indicator of engine tune and condition. When compression testing is unclear, follow-up with a vacuum test. First, it is important that "normal" vacuum gauge readings be understood. Normal vacuum in the manifold varies with altitude and engine design. Gauge readings must be comparable to other engines of the same type in good condition but be aware that vacuum drops 1 inch for every 1,000-foot rise above sea level.

Connect the vacuum gauge with as short a hose as possible so that the needle responds quickly to changes in pressure. Some gauge readings and their associated valve train defects include the following:

1. Low and steady readings caused by manifold leaks, late ignition timing, or possibly by worn piston rings (see Fig.6-12).

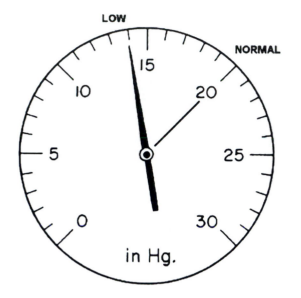

Fig.6-12   A low and steady vacuum reading

2. Intermittently dropping readings caused by sticking valves (see Fig.6-13).

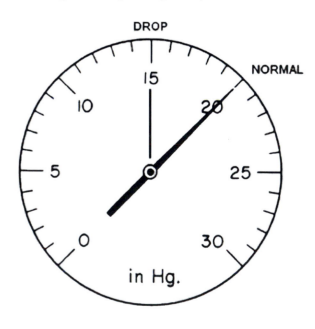

Fig.6-13   An intermittent drop in the vacuum reading

3. Rapid needle oscillation at idle caused by leaking valves (see Fig.6-14).

4. Rapid needle oscillation on acceleration, similar to that seen in Figure 6-14, caused by weak valve springs.

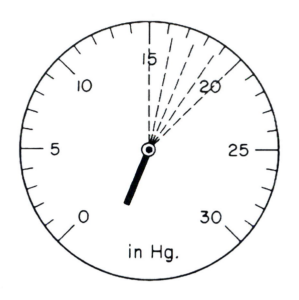

Fig.6-14   Rapid needle oscillation on a vacuum gauge

Test for ring leakage and exhaust restriction as follows:

1. 1.Connect the gauge and observe the normal reading at idle (see Fig.6-15).

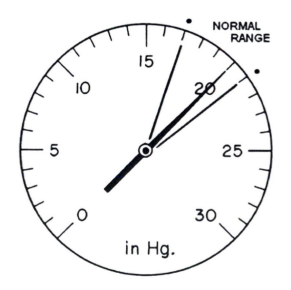

Fig.6-15   Normal vacuum reading of 17 to 21 inches of mercury

2. Increase the engine speed to approximately 2,000-RPM.

3. Close the throttle rapidly and observe the gauge reading. A rebound of less than 3 inches of mercury (in/Hg) over the reading at idle suggests piston ring leakage (see Fig.6-16). If the needle returns slowly to normal, or there is no rebound, there is an exhaust restriction.

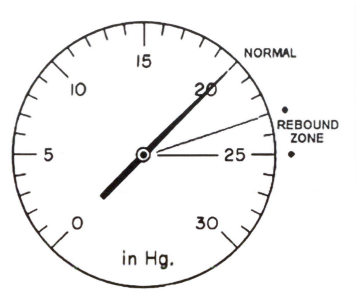

Fig.6-16 A normal increase in vacuum, called "rebound," when the throttle is closed

4. 4.Verify exhaust restriction by reading vacuum at 3,000-RPM. There is a restriction if the reading drops below idle vacuum.

Low but steady readings suggest late timing, vacuum leaks, or worn piston rings. Also, unsteady readings at constant RPM suggest valve problems. Of course, needle action on the vacuum gauge is more obvious at idle or low speeds. The need for tuning, especially ignition timing and valve adjustment, and manifold vacuum leaks must be eliminated as the causes of abnormal vacuum gauge readings.

TESTING EXHAUST BACK PRESSURE

Exhaust restrictions cause a noticeable loss of power. Restrictions result from collapsed pipes and restricted flow through mufflers and catalytic converters (see Fig.6-17). In early stages, probably only the power loss is noticeable, but in later stages, a distinct change in the sound of the exhaust is detectable.

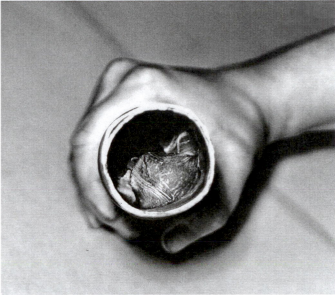

Fig.6-17 Restriction caused by a collapsed inner wall of a double walled pipe

While a lack of rebound on a vacuum gauge suggests exhaust restriction, this test is not always sufficiently sensitive, especially in a no-load condition. An alternative test is to directly measure exhaust system backpressure. Connect a pressure gauge to the exhaust tube feeding the EGR valve, an oxygen sensor port or remove the exhaust pressure check valve from an air injection manifold as shown in Fig.6-18. Under a no-load condition, exhaust pressure should not exceed 1.5 PSI.

Fig.6-18 Remove air injection check valve and connect gauge

Some models have provisions for tapping into the exhaust system ahead of the converter. For engines without fittings in the exhaust, kits are available that provide for drilling a hole in the exhaust and plugging it after testing (see Fig.6-19).

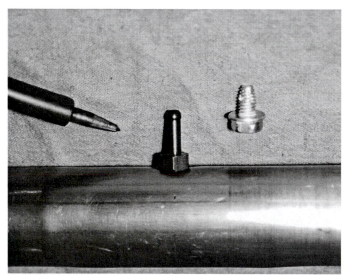

Fig.6-19 Punch, pressure fitting, and self-tapping plug to plug hole after testing

TESTING WITH A SCAN TOOL; AN INTRODUCTION

A scan tool is an electronic diagnostic tool that communicates with the on-board computers to gather codes, view the data stream, and perform functional tests (see Figs.6-20 and 21). Scan tools range in functionality from basic code readers to manufacturer specific features for the various systems like engine, transmission, air bag, ABS brakes, and body controls. Some scan tool companies support many automotive manufacturers' systems in one tool.

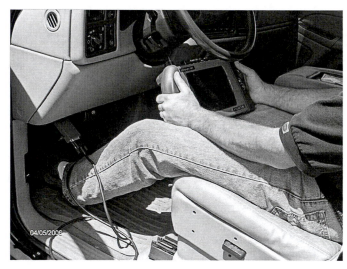

Fig.6-20 A Snap-On Modis connected via a 16 pin connector

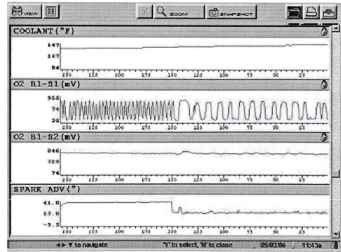

Fig.6-21 Graphing data stream parameters

For engine diagnosis, the ability to read and clear codes is essential in a scan tool (see Fig. 6-22). With the implementation of OBDII (on board diagnostics) in the mid-90s, code definitions were becoming more standardized. The five character codes are broken down as follows:

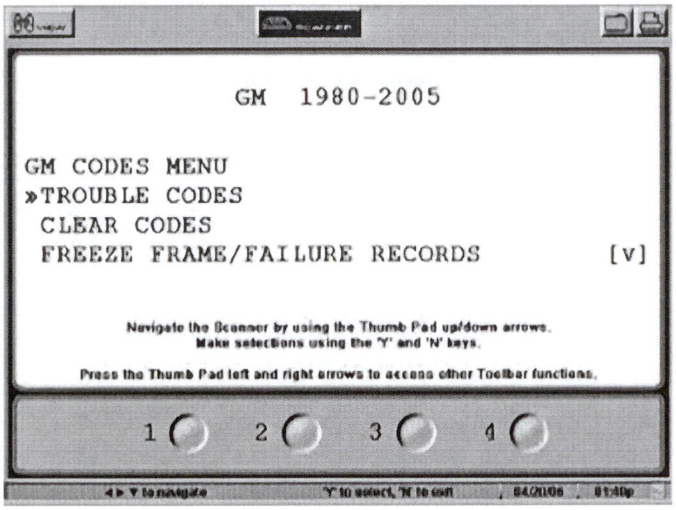

GM 1980-2005

GM CODES MENU
»TROUBLE CODES
 CLEAR CODES
 FREEZE FRAME/FAILURE RECORDS   [v]

Navigate the Scanner by using the Thumb Pad up/down arrows.
Make selections using the 'Y' and 'N' keys.

Press the Thumb Pad left and right arrows to access other Toolbar functions.

1 ○   2 ○   3 ○   4 ○

Fig.6-22 Snap-On Modis screen with codes menu displayed

The 1st character
    B = Body
    C = Chassis
    P = Powertrain
    U = Network

The 2nd character
    0 = Generic code
    1 = Manufacturer specific code

The 3rd character
    1 = Fuel & Air Metering
    2 = Fuel & Air Metering (Injectors)
    3 = Engine Misfire
    4 = Auxiliary Emissions
    5 = Vehicle Speed & Idle Control
    6 = Computer Output Circuits
    7 = Transmission
    8 = Transmission
    9 = Transmission

The 4th and 5th two-digit number identifies the fault.

The designation for a code P0303 looks like this:
    P = Powertrain
    0 = Generic code
    3 = Misfire code
    03 = Cylinder number 3

Functional or ATM (asynchronous transfer mode) tests can also be helpful in diagnosing engine related problems. An injector test can disable one cylinder at a time and act as a power balance test to locate a cylinder that is contributing less than the others (see Fig. 6-23). An EGR (exhaust gas recirculation) test can operate the EGR control to determine if the passages are restricted. An ECT (engine coolant temperature) test can simulate varying engine temperatures so a technician can monitor at what temperature the cooling fan turns on and what changes in fuel mixture are occurring. These are just a few of the many tests that scan tools can perform.

GM 1980-2005

SELECT TEST MODE.   PRESS Y TO CONTINUE.
 IAC CONTROL   [^]
 INJECTOR #1 DISABLE (YES/NO)
»INJECTOR #2 DISABLE (YES/NO)   [v]

Navigate the Scanner by using the Thumb Pad up/down arrows.
Make selections using the 'Y' and 'N' keys.

Press the Thumb Pad left and right arrows to access other Toolbar functions.

1 ○   2 ○   3 ○   4 ○

Fig.6-23 Disabling each fuel injector to check power balance using

Besides engine diagnosis, scan tools can also be used after engine installation. As careful as we are during installation, a malfunction indicator light (MIL) or Check Engine light sometimes turn on after engine start-up (see Fig. 6-24). Some faults may not turn on the MIL until two drive-cycles have completed, which could cause a problem if a vehicle is returned to a customer too soon. A drive-cycle is a specific sequence of engine temperatures, vehicles speeds, engine loads, and acceleration/deceleration rates that may take days to complete.

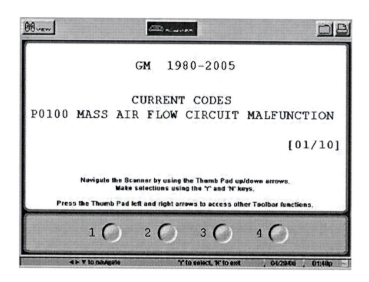

Fig.6-24 An OBDII code list showing a MAF (mass air flow) sensor code displayed using a Snap-On Modis

DIAGNOSING ENGINE NOISES

On occasion, engine noise alone suggests the need for engine repair. However, some engine noises may be characteristic of a particular engine and not necessarily an indication of pending failure. Evaluating the seriousness of noises requires the experience of a skilled technician (see Fig.6-25). Some common engine noises and conditions include the following:

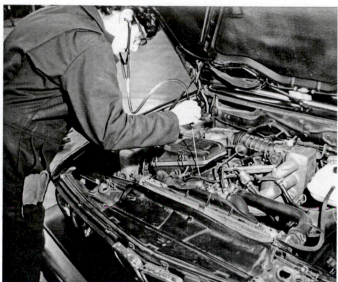

Fig.6-25 A stethoscope used to locate an engine noise

1. Hydraulic valve lifters are a common source of noise complaints. The noise is a knock occurring at camshaft speed or half engine RPM. The faulty lifter is isolated by listening at each rocker pivot point with a stethoscope while the engine is idling. Check also for a damaged valve spring, pushrod, rocker arm, or excess valve lash because any one of these cause noises similar to defective lifters. Even mechanical fuel pumps sometimes make noise at valve train speed.

2. Excessive main bearing clearance causes a deep metallic knock. The noise is most audible when the engine is under load, accelerating, or just started. Low oil pressure accompanies excess main bearing clearance.

3. Crankshaft endplay beyond acceptable limits is heard as a sharp metallic rap when releasing or engaging manual transmission clutches. Test for this condition by placing a dial indicator at one end of the crankshaft and prying the crankshaft to its limits of travel (see Fig.6-26).

Fig.6-26 A dial indicator used to check crankshaft endplay

1. Connecting rod noises caused by excessive clearance make a metallic rap when the engine is running under a light load. The noise becomes louder and increases in frequency as engine speed increases. Isolate the defective connecting rod bearings by grounding one spark plug at a time. The engine noise noticeably decreases when grounding the spark plug wire to the defective cylinder.

2. Piston slap is a noise associated with excess piston-to-cylinder wall clearance. Piston slap is a dull metallic rattle heard at idle and under light engine load. Many of these noises disappear when the piston expands during warm-up and do not necessarily affect engine reliability. When the slap remains after warm-up, piston or ring failures are possible. Grounding the spark plug does not affect the noise.

3. The noise caused by excessive piston pin clearance is a light metallic rap at idle and at low speeds. Noisy piston pins are also isolated by grounding spark plugs but the affect is different. With the spark plug wire grounded, noise from a loose piston pin increases in frequency. Pin noise after replacing piston rings are replaced is common because of

increased drag and loose pin fits. The noise usually diminishes and disappears as friction between the piston rings and cylinders decreases during break-in.

Judging from these examples, the technician must note the noise frequency and the influence of oil pressure or engine temperature on the noise. Frequency separates crankshaft and piston assembly noises from valve train noises. Oil pressure influences bearing and hydraulic lifter noises, and temperature influences piston noise.

Keep in mind that some noises come from sources other than internal engine parts. Drive belts, alternators, compressors, air pumps, and fuel pumps all make noise. It is sometimes a good idea to run the engine after disconnecting all drive belts or removing the mechanical fuel pump from the engine. The noise in question just might disappear. To prevent overheating, be sure to run the engine for only a few minutes.

Check also for damaged flywheel covers because they may rub against rotating parts. Even flywheels knock if loose or, as in the case of automatic transmission flexplates, broken (see Fig.6-27).

Fig.6-27 A broken flexplate

TESTING ENGINE OIL PRESSURE

Of course, serious oil pressure problems are not dealt with as complaints but are seen as complete engine failures. There is however, a range of problems that lead to complaints, but with luck, there will be no major failures. To understand these problems, it is necessary to clearly understand normal oil pressure characteristics.

1. Engines frequently idle with less than half the oil pressure seen at highway speeds.

2. Oil pressure drops by half once the engine fully warms up.

3. Some engines normally run oil pressures less than 15 PSI hot at idle. Idle RPM alone effects whether or not there will be a complaint.

4. Original equipment gauges and warning lights are often of little value in comparing actual pressure to specifications.

Common complaints include warning lights coming on and low readings on gauges at idle or high speeds. Before attempting to find the cause of such complaints, go through the following basic checks:

1. Check the engine oil level and properties of the oil. While the implications of low levels are obvious, contaminated oil will also cause low pressure. If contaminated with water, the oil will appear something like an emulsion and the engine potentially has major problems. If diluted with fuel, it will smell like it. When in doubt, drain the oil, replace the filter, and refill with the correct quantity and quality of oil.

2. An engine low on oil may have normal pressure at idle but pressure drops or surges when turning, stopping, starting out, or running at high speeds. Oil washing away from the oil pump inlet allows air to enter the pump and this causes erratic pressure.

3. Engines with high oil levels are occasionally found. Such engines could have been overfilled or be contaminated with water or fuel. When

overfilled with oil, the engine will probably leak but not necessarily have low pressure. If diluted with fuel, oil pressure will be reduced.

To properly test oil pressure requires accurate readings taken at idle and approximately 2,000 RPM. Because many engines are equipped with warning lights or inadequate gauges, it is sometimes best to remove the pressure sending unit and attach a pressure gauge directly to the engine (see Fig.6-28). With a gauge attached, watch for the following patterns:

Fig.6-28 An oil pressure gauge installed in place of a pressure sending unit

1. Low pressure at idle RPM and near normal pressure at higher speeds. This pattern indicates high oil flow rates caused by excessive main bearing clearance or a worn oil pump.

2. Low oil pressure at all speeds also indicates high flow or possible pump wear. Do not replace the oil pump without first checking main bearing clearance. Of course, it is possible for a missing oil passage plug to prevent the build-up of normal oil pressure.

3. Normal pressure at idle and low pressure at higher speeds indicates a restriction at the oil pump inlet. The pump receives sufficient volume at idle RPM but starves at higher speeds. Remove the oil pan and check for sludge or some other restriction at the pump pick-up screen.

4. Should pressures be acceptable at idle and high speeds, replacing the pressure-sending unit will frequently eliminate the complaint.

TESTING COOLING SYSTEMS

Warped and cracked cylinder heads, scuffed cylinders, pistons, and other major engine damage are evidence of the destructiveness of overheating. The first step in diagnosis is to check coolant level in the cold engine and, depending upon the level, proceed in one of two directions. If the level is normal, the engine is running hot but has not yet boiled off coolant. Check ignition timing, fuel mixture, and the intake system for vacuum leaks first.

If the coolant level is low, pressure test the system to locate leaks (see Fig.6-29). Check carefully around hose connections, radiator seams, heater cores and hoses, water pump, and core plugs for leaks and repair as needed. Then refill the system with water, allow the engine to warm up and check coolant circulation. Thermostats fail open or closed. If stuck open, the engine is slow warming up, but if stuck closed, there is no circulation and the engine overheats. Although dropping it into hot water can test it, it is more cost effective to simply replace it.

Check the radiator for both circulation and cooling efficiency. First check lower radiator hoses for rigidity because if too soft, or if a coil spring inside is missing or damaged, suction from the water pump will collapse the hose and stop coolant circulation (see Fig.6-30). As for cooling efficiency, there are a number of conditions that can reduce heat transfer to the atmosphere. Most common of these is restricted flow through the core caused by deposits. To test, warm the engine and compare inlet and outlet coolant temperatures and look for at least a 20 degrees F difference. If the temperature difference is less than this, clean or replace the radiator.

Fig.6-30 A coil reinforcing a lower radiator hose

Also check fan operation. Check clutch fans by shutting off the warm engine and turning the fan by hand (see Fig.6-31). If normal, the clutch engages and rotation is stiff. Electric fans switch on and off through coolant sensors and air conditioning controls or computers. To check electric fans, warm up the engine see if it switches on. If it fails to switch on, bypass controls and sensors and jumper power directly to the fan. If the fan runs, the sensor or controls are at fault. If it does not run, replace the fan motor.

Fig.6-29 Pressure testing a cooling system

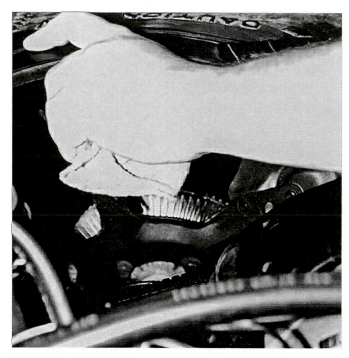

Fig.6-31 Checking a clutch fan for drag

SUMMARY

While onboard computer diagnostics warn of engine problems and yield trouble codes indicating the nature of the problem, the level of diagnosis is not sufficiently detailed to go ahead with engine repair. The technician must know the precise problem and assess how far to go to into the engine to correct the problem. To do this, the technician must be familiar with test procedures such as cooling system pressure testing, compression or cylinder leakage testing, and other tests such as backpressure and manifold vacuum.

There are of course indications of engine wear such as reduced power, increased oil consumption, and possibly misfiring cause by oil fouled spark plugs. These are likely to be the complaints made by drivers and technicians should listen carefully to drivers prior to diagnosis.

When the complaint is overheating, technicians must be especially careful in their diagnosis. They must not only check the components of the cooling system, they should also check for engine damage caused by overheating such as ring wear, burned valves, valve seal failure, and piston or bearing noises. Camshaft timing belt or chain

failures are common. Technicians must be knowledgeable in diagnosing valve-timing errors and detecting possible valve damage caused by such failures. They must also be capable of replacing timing belts or chains as part of repairs or preventative maintenance.

As mentioned, there is a range of other tests that technicians need to perform. Because manifold vacuum changes with compression, valve timing, valve leakage, or intake manifold leakage, testing manifold vacuum helps verify a number of problems. Exhaust backpressure caused by failures in converters, mufflers, and exhaust pipes reduces power and elevates temperature in exhaust ports leading to valve train failures. An understanding of how engine bearings, pistons, and the valve train are affected by oil pressure, temperature, and engine speed or load is necessary in diagnosing engine noises.

Chapter 6

ENGINE DIAGNOSIS

Review Questions

1. Technician A says that blue exhaust smokes indicates oil burning. Technician B says that black exhaust smokes indicates rich air-fuel mixtures. Who is right?

   a. A only
   b. B only
   c. Both A and B
   d. Neither A or B

2. Technician A says that blue exhaust smokes on acceleration indicates poor guide sealing. Technician B says that blue exhaust smokes on deceleration indicates poor ring sealing. Who is right?

   a. A only
   b. B only
   c. Both A and B
   d. Neither A or B

3. Technician A says that catalytic converters make diagnosis by exhaust smoke difficult. Technician B says check exhaust smoke in the first few minutes of engine operation. Who is right?

   a. A only
   b. B only
   c. Both A and B
   d. Neither A or B

4. Steam in the exhaust of a cold engine is
   a. normal
   b. caused by a blown head gasket
   c. caused by a cracked cylinder head
   d. not normal

5. Technician A says to detect combustion gases in the cooling system, draw coolant vapors through the chemical solution. Technician B says detect these gases by drawing coolant through an exhaust gas analyzer. Who is right?

   a. A only
   b. B only
   c. Both A and B
   d. Neither A or B

6. Technician A says that because piston speed is high and the throttle is closed on deceleration, oil pulls past rings. Technician B says that because piston speed is high and the throttle is open on acceleration, oil pulls through guides. Who is right?

   a. A only
   b. B only
   c. Both A and B
   d. Neither A or B

7. Technician A says that power balance testing separates valve from ring problems. Technician B says this test shows if a cylinder has reduced output. Who is right?

   a. A only
   b. B only
   c. Both A and B
   d. Neither A or B

8. Technician A says that compression testing does not show which valve leaks. Technician B says that wet compression testing separates valve leakage from ring leakage. Who is right?

   a. A only
   b. B only
   c. Both A and B
   d. Neither A or B

9. Poor valve guide sealing is indicated by spark plug deposits
   a. on one side of electrodes
   b. all around the plug
   c. which are oily
   d. which are wet with fuel

10. Technician A says that cylinder leakage testing shows which valve leaks and ring sealing problems. Technician B says that a flat camshaft will not show up on this test. Who is right?

    a. A only
    b. B only
    c. Both A and B
    d. Neither A or B

11. Technician A says that incorrect valve lash affects compression testing. Technician B says that providing there is lash, adjustments do not affect cylinder leakage tests. Who is right?

    a. A only
    b. B only
    c. Both A and B
    d. Neither A or B

12. Technician A says that flat camshafts show up in compression tests but not in cylinder leakage tests. Technician B says that neither test finds problems with oil control rings. Who is right?

    a. A only
    b. B only
    c. Both A and B
    d. Neither A or B

13. Technician A says that when an engine fires through the intake under load, an intake valve is leaking. Technician B says that an exhaust valve not opening causes this. Who is right?

    a. A only
    b. B only
    c. Both A and B
    d. Neither A or B

14. Technician A says that timing chain stretch causes valve timing to retard. Technician B says that it causes a loss in low-end power. Who is right?

    a. A only
    b. B only
    c. Both A and B
    d. Neither A or B

15. Valve timing is most accurately checked at
    a. TDC compression
    b. TDC exhaust
    c. the crankshaft
    d. the crankshaft with a degree wheel

16. Technician A says, in pushrod engines with distributors, measure timing chain backlash by turning the engine forward to TDC, backward until the rotor moves, and then read backlash at the timing indicator. Technician B says that backlash is apparent when ignition timing advances and retards on deceleration. Who is right?

    a. A only
    b. B only
    c. Both A and B
    d. Neither A or B

17. Technician A says that any rapid needle oscillation in vacuum testing indicates valve problems. Technician B says that this could be a ring problem. Who is right?

    a. A only
    b. B only
    c. Both A and B
    d. Neither A or B

18. For engine diagnosis, connect a vacuum gauge to a _____ vacuum source.
    a. ported
    b. venturi
    c. PCV
    d. manifold

19. Altitude affects vacuum gauge readings as follows
    a. not at all
    b. readings go down 1 in/Hg per 1000ft. increase
    c. readings go down 1 in/Hg per 1000ft. decrease
    d. readings go up 1 in/Hg per 1000ft. increase

20. Technician A says that vacuum leaks or worn rings cause low and steady vacuum readings. Technician B says that advanced ignition timing cause low readings. Who is right?

    a. A only   c. Both A and B
    b. B only   d. Neither A or B

21. A technician determines valve problems with a vacuum gage by checking
    a. vacuum at idle
    b. rebound on deceleration
    c. vacuum drop on acceleration
    d. needle action at idle

22. Technician A says that restricted exhaust shows as a slow rebound on a vacuum gauge. Technician B says that exhaust pressure exceeding 1.5 PSI at idle indicates a restriction. Who is right?

    a. A only   c. Both A and B
    b. B only   d. Neither A or B

23. Technician A says test exhaust backpressure in the pipe ahead of the catalytic converter. Technician B says that a good place to test is at air injection manifolds. Who is right?

    a. A only   c. Both A and B
    b. B only   d. Neither A or B

24. Technician A says that to determine the cause of an engine knock, listen for the frequency. Technician B says observe the effects of temperature and oil pressure. Who is right?

    a. A only   c. Both A and B
    b. B only   d. Neither A or B

25. When grounding a spark plug wire decreases an engine knock, the problem on that cylinder is
    a. excess main bearing clearance
    b. excess rod-bearing clearance
    c. excess piston pin clearance
    d. a defective hydraulic valve lifter

26. When an engine knock diminishes as the operating temperature reaches normal, the noise when cold is caused by
    a. excess piston clearance
    b. excess rod-bearing clearance
    c. excess piston pin clearance
    d. a hydraulic lifter that bled down

27. Technician A says that normal oil pressure at idle and low pressure at high speed indicates an oil inlet restriction. Technician B says that low oil pressure at idle and normal pressure at high speed indicates high main bearing clearance. Who is right?

    a. A only   c. Both A and B
    b. B only   d. Neither A or B

28. Technician A says that a thermostat that fails to open prevents the engine from reaching normal operating temperature. Technician B says that a thermostat that is stuck open causes engine overheating. Who is right?

    a. A only   c. Both A and B
    b. B only   d. Neither A or B

29. Technician A says that coolant circulation is restricted by blockage in the radiator. Technician B says that a collapsed lower radiator hose causes such restrictions. Who is right?

    a. A only   c. Both A and B
    b. B only   d. Neither A or B

30. Technician A says that electric fans switch on through coolant sensors or air conditioning and computer controls. Technician B says that should the fan fail to switch on, replace the fan motor. Who is right?

   a. A only
   b. B only
   c. Both A and B
   d. Neither A or B

FOR ADDITIONAL STUDY

1. List signs of engine wear that drivers notice.

2. List at least three ways to find a failed cylinder head gasket.

3. Compare differences in test results between power balance, compression, and cylinder leakage tests.

4. At what point in engine rotation should you check valve timing? What are you looking for at this point?

5. List the causes for needle oscillation on a vacuum gauge.

6. List the causes for low readings on a vacuum gauge.

7. What causes low oil-pressure at idle speed only? What causes low oil-pressure at running speed only?

8. For purposes of diagnosing engine noises, list those engine components effected temperature and those effected by oil pressure.

9. In diagnosing engine engines, how are crankshaft bearing or piston noises separated from valve train noises?

10. What kinds of failures cause high exhaust backpressure?

Chapter 7

ENGINE DISASSEMBLY

Upon completion of this chapter, you will be able to:

- Describe the extent of engine disassembly typically required for a valve grind and a minor engine overhaul.
- List the steps in cylinder head disassembly.
- List the steps in short block disassembly and precautions to be taken at each step.
- List the reasons for numbering piston and connecting assemblies on disassembly.
- Explain how to protect the crankshaft during removal of piston and connecting rod assemblies.
- Explain the need to keep camshafts and followers in order.
- Describe the disassembly procedures and precautions required for protection of lifter bores during camshaft removal or replacement.
- Select camshaft-bearing drivers and explain the procedure for removing bearings.
- Explain how to use an oxy-acetylene torch for removing pipe plugs in oil passages.
- Explain how to remove water jacket core plugs without damaging the bores.

INTRODUCTION

Procedures for the disassembly of engines vary from model to model. For this reason, check service references prior to beginning disassembly. However, those procedures common to a majority of engines are reviewed in this chapter.

Economics determine to some extent the repairs to be made as well as the procedures to be followed. For example, perhaps it is decided that engine block repairs are to be made after the cylinder heads have been removed for valve service. In such a situation, replacing piston rings and crankshaft bearings with the engine block mounted in the chassis would be a serious consideration. As long as the required engine service requirements are not too extensive, such procedures save several hours in labor charges. On the other hand, if the repair called for is a major engine overhaul or engine rebuilding, removing the engine from the chassis and more complete service procedures are required.

Handle parts and store them so that they will not be lost. Consider that head bolts, for example, are special high strength fasteners that are not necessarily easy to replace if lost or misplaced. Careful handling pays; if serviceable parts are damaged, additional machining, or parts replacement will be necessary. It is not unusual to have to resurface cylinder heads because of gouges or scratches picked up in handling, even though surfaces are within limits for flatness.

In general, it is best to follow good mechanical practice. Work according to correct procedures, keep engine parts clean and organized, and handle parts carefully. These basic work habits are a part of the craftsmanship essential to quality engine work.

HINTS FOR DISASSEMBLY IN THE CHASSIS

As mentioned, engine repairs are frequently performed with the engine mounted in the chassis to reduce labor costs. For example, valve service is typically done with the block in the car. Other examples include the replacement of a single defective piston or the replacement of piston rings in one or more cylinders. Simple ring and valve jobs are also often done in this manner. The valves are ground, piston rings replaced, and connecting rod bearings replaced.

Disassembly for valve service or other repairs generally begins by removing cylinder heads from engines. This also involves removing intake manifolds and either removing exhaust manifolds or disconnecting exhaust pipes from the manifolds. An air-conditioning compressor can sometimes be tied against a fender well with wire so that it will not have to be disconnected from the system. Remember that disconnecting the compressor requires evacuating the system and capturing the refrigerant.

Be aware that some engines have a loosening sequence for head bolts, usually the reverse of the sequence for tightening, and that failure to use the sequence causes warpage. It is recommended that all cylinder heads be removed cold to prevent warpage.

Oil pan removal frequently requires disconnecting and dropping steering linkages so that the oil pan may be lowered from the engine block. Sometimes oil pan removal requires unbolting engine mounts and raising the engine block a few inches. Where clearance is really tight, loosening the pan, unbolting the oil pump, and rotating the counterweights upward is necessary before the oil pan can be removed. Once the cylinder heads and oil pan are removed, it is possible to begin removing piston and connecting rod assemblies.

DISASSEMBLING CYLINDER HEADS

Cylinder head disassembly is done with the aid of spring compressors for pushrod and overhead cam cylinder heads (see Fig.7-1 and 2). This is routine work but it is wise to observe a few special precautions. First, valve spring retainers sometimes lock up or seize on the valve keepers. This is caused by the wedging action of their matching tapers. When locked, the valve springs cannot be compressed to remove the keepers. Loosen them by placing a socket or an old piston pin against the retainer and rapping it with a soft-faced hammer. Be careful to strike straight down the center of the valve stem or the valve will bend (see Fig.7-3). Also be sure that the valves do not hit anything as they pop open or this too will bend them.

Fig.7-1 Using an air operated valve spring compressor

Fig.7-2 Tools for compressing valve springs in bucket follower bores of OHC heads

Fig.7-3 Striking the retainer with a piston pin to free it from keepers

Second, valve tips occasionally mushroom and prevent the valve from slipping out of the valve guide. Clean up the tip of the valve stem with a file or air grinder (see Fig.7-4). Don't rush and attempt to drive the valve through the valve guide. The result will be a ruined valve guide and additional time lost in repairing it.

Fig.7-4 Deburring a valve stem tip on disassembly

Third, measure the valve stem height while valves are still in order of assembly (see Fig.7-5). This is critical on engines with nonadjustable rocker arms because valve stem length changes the position of the plunger in hydraulic valve lifters or lash compensators. After reassembly, the length must be the same as before disassembly or valves may be held open. Specifications for valve stem height are not always available because the manufacturers frequently use their own gauges (see Fig.7-6). Unfortunately, these fixed gauges are not on hand in most independent shops. It is best to record measurements taken on teardown for use later in assembly.

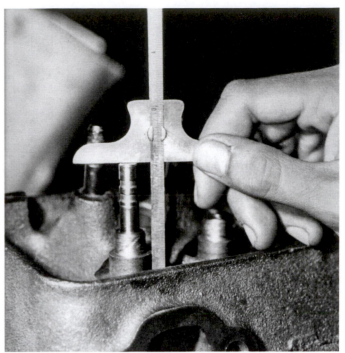

Fig.7-5 Measuring from the tip of the valve stem to the spring seat

Fig.7-6 Gauge for checking installed stem height (BHJ)

There are additional procedures recommended for overhead camshaft cylinder head tear down. First, keep cam followers in order as they have wear patterns matched to each cam lobe and must be reassembled in original order or wear rates are accelerated. Second, because there are surfacing limits for these heads, and because original thickness specifications are not always available, measure the head thickness on tear down.

Over the years, it has also become standard procedure to keep valves in order for reassembly. Unless oversized stems are used, there is minimal value in this practice since all valves and valve guides must be measured for wear and checked against specifications. Keeping valves in order during disassembly might be useful only so that mating valve guides or rocker arms can be checked for damage. Clearances when reassembled will be changed very little because only parts measuring within service limits will be reused.

NUMBERING CONNECTING RODS

There are checks and procedures to follow prior to unbolting connecting rods. For example, check to see that the connecting rods are numbered according to their cylinder position, as they are sometimes found not numbered. A number corresponding to the cylinder should be stamped on each half of the connecting rod (see Fig.7-7). The numbers are usually stamped on the same side of in-line engines or on the pan rail side on V-block engines (see Fig.7-8). Check carefully because the rod numbering, if present, could be incorrect. If the rods are not numbered, steel rods may be stamped while still on the crankshaft. It is however always safe to engrave rod numbers for rods of any material whether in or out of the engine.

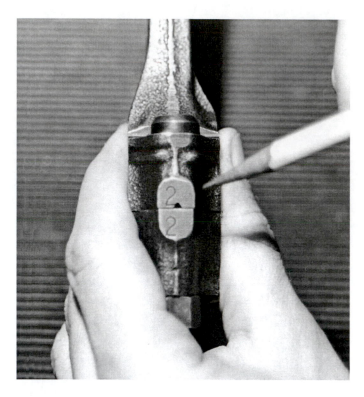

Fig.7-7   Cylinder number stamped on both sides of connecting rod parting lines

Fig.7-8   Engraving rod numbers

Numbering them in this manner does several things. First, stamping both the connecting rod and the cap aids the technician in replacing the cap on the rod in the original position. If the cap should is replaced on the wrong rod or in the wrong direction, the housing bore of the connecting rod will be out of round beyond acceptable limits.

Second, stamping connecting rods on the same side of in-line engines or on the pan rail side on V-blocks engines aids the technician in replacing piston and connecting rod assemblies in the cylinder in the proper direction. If the piston is replaced in the cylinder in the wrong direction, valve interference may occur (see Fig.7-9) or pistons may make noise due to the incorrect offset of piston pins (see Fig.7-10). If the connecting rod direction is reversed, oil spurt holes will point the wrong way, and rod bearings may be positioned on fillets of crankpins.

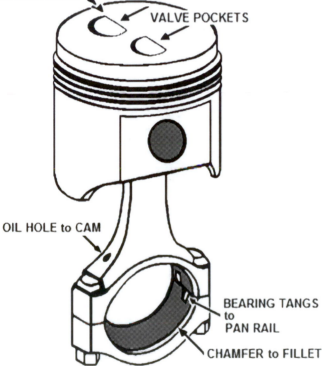

Fig.7-9   Points to recall for reassembly of connecting rods and pistons

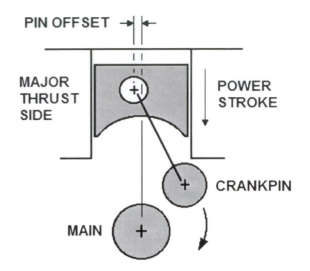

Fig.7-10   The piston pin offset to minimize rock at the top of the stroke

Third, stamping the connecting rods while they are assembled to the crankshaft minimizes any distortion that may occur as a result of the stamping. Instead of stamping, it is possible to number the connecting rods with an engraver. Whichever the method of numbering, be sure to number connecting rods before disassembly to eliminate confusion as to position of assembly.

RIDGE REAMING

Cylinders wear primarily in the top inch of piston ring travel and a ridge forms where the piston reverses direction. Unless removed, damage to ring lands occurs as pistons are pushed out of the cylinder and piston rings catch on the ring ridge (see Fig.7-11). Tools called "ridge reamers" remove the ring ridge (see Fig.7-12). Take care in using these tools to see that ridge reaming does not extend into the area of ring travel. Because of the irregular wear patterns in the top of the cylinder, it is common to find that ring ridges cannot be fully removed without reaming into the piston ring travel. If reamed into the ring travel, rings will not seal and the cylinder will require boring oversize. However, even partially removing the ridge makes piston and connecting rod removal easier and minimizes damage. In every case, use minimum force in removing rod and piston assemblies.

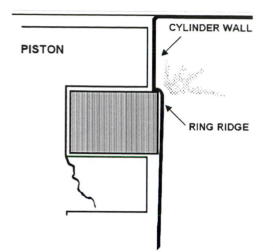

Fig.7-11   A piston ring striking a ring ridge

Fig.7-12   Using a ridge reamer

Problems with over-cutting with ridge ream-ers are such that many technicians simply push pistons past the ridge, at most removing the carbon first, and consider any engine where this is not pos-sible a candidate for reboring. In newer engines with moly or low-tension rings, there is rarely any amount of ring ridge and ridge reaming is typically unnecessary.

REMOVING PISTON AND ROD ASSEMBLIES

It is all too common to find that otherwise service-able crankshafts are damaged during the removal of piston and connecting rod assemblies (see Fig.7-13). The cause of the damage is failure to keep connecting rod bolts from contacting crankpins when connecting rod bolts slide past the crankshaft. Prevent this damage by placing a length of rubber tubing over each connecting rod bolt during disas-sembly (see Fig.7-14).

Fig.7-13   Crankpin damage from rod bolts

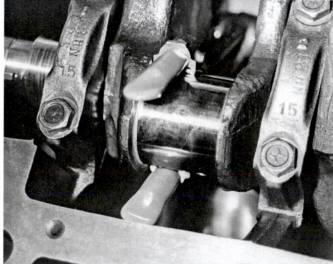

Fig.7-14   Tubing on rod bolts protects the crankshaft

Remove one rod cap at a time and place tubing over the rod bolts and then push or drive each rod and piston assembly out through the top of the block. Use a hardwood dowel or hammer handle as a driver so that the rods or pistons are not damaged. Do not attempt to remove assemblies through the bottom side; they wedge against the main bearing webs and attempting to move them in either direction results in broken pistons.

Use a variation in this procedure for diesel engines. Because compression ratios in diesels frequently exceed 20:1, and relatively minor changes in deck clearance after machining cause major changes in compression ratios, measure and record deck clearance during disassembly so that it may be rechecked on reassembly (see Fig.7-15).

wear limit generally accepted for backlash between timing gears is .006in (.15mm) (see Fig.7-17).

Fig.7-16   Checking timing chain slack

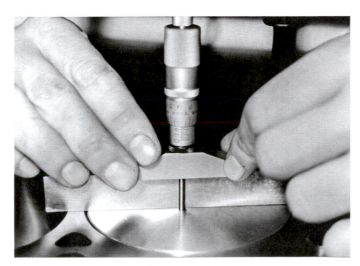

Fig.7-15 Measuring deck clearance with a depth mike

Fig.7-17   Checking timing gear backlash with a feeler gauge

REMOVING THE TIMING CHAIN AND SPROCKETS

Check timing chain slack or timing gear backlash in pushrod engines prior to removing the camshaft or crankshaft. When timing chain slack is less than 1/2in, the timing chain and sprockets are considered by many technicians to be serviceable (see Fig.7-16). It is also true that many technicians automatically change sprockets and chains during overhauls. If slack is excessive, replace both the sprockets and the chain. Use care not to twist a serviceable timing chain when removing it. The

In chain driven overhead cam engines, chain tensioners are spring loaded and sometimes eject themselves from their bores as soon as the cam sprocket is removed. Unless planning to remove the timing cover, place a wedge in between the two sides of the chain to hold the tensioner back in the assembly (see Fig.7-18).

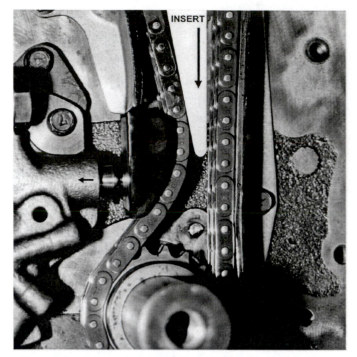

Fig.7-18   A wedge holds the chain in place while removing the cam sprocket

REMOVING THE CRANKSHAFT

As with connecting rods, the main bearing caps are not always marked for position or direction. Mark the caps with numbers on one side and corresponding points on the main bearing webs (see Fig.7-19). Main bearing housing bores will be out of round if main bearing caps are repositioned in the wrong location or in the wrong direction.

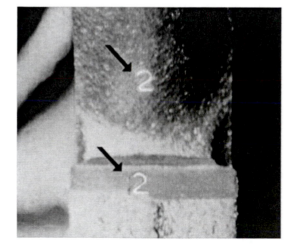

Fig.7-19   Marking main caps

Once the main caps are marked, remove them and set them aside. With the timing chain removed, lift the crankshaft out of the crankcase and stored it on end to prevent the distortion that occurs if stored lying flat. Then remove the main bearings and keep them in order for later inspection.

Replace the main bearing caps on the block after removing the crankshaft and bearings. This prevents misplacing the bearing caps or capscrews. It is also suggested that block machining operations be performed with the main bearing caps tightened in place so that engine block distortion caused by tightening the capscrews is the same as in the assembled engine. Observing this practice ensures that machining accuracy is best in the assembled engine rather than in the bare block.

REMOVING THE CAMS AND LIFTERS FROM PUSHROD ENGINES

In engine overhauls, reusing serviceable parts makes the repair more affordable. Therefore, it is important that valve lifters be kept in order so that they may be replaced in their original locations. Store the lifters in a box, or possibly in a wooden block with holes drilled in it, and mark their positions. If lifters are replaced in incorrect positions, camshaft failure is almost certain. Remember to keep rocker arms or cam followers from overhead camshaft engines in order for the same reasons.

Valve lifters are sometimes difficult to remove because of varnish buildup on the portion of the valve lifter extending below the lifter bores. Small tools are available for pulling out valve lifters. It is sometimes possible to remove the camshaft by standing the block on end and turning the camshaft one full turn to move valve lifters out of the way. With the camshaft removed, push lifters out the bottom side of the lifter bores using a pushrod. Never push valve lifters from the bottom because damage to the lifter bores results.

Check to see if there is a camshaft thrust plate used in the engine to limit camshaft end-play (see Fig.7-20). If present, loosen the thrust plate before attempting to remove the camshaft.

Remove the retaining screws through holes in the timing gear or sprocket.

Fig.7-20   A thrust plate in relative position with camshaft components

REMOVING CAMSHAFT BEARINGS

Various manufacturers make tool sets for removing and installing cam bearings. It is important to select drivers properly or to adjust them so that bearing bores in the engine block are not scored during the process of removing bearings (see Figs. 7-21 and 22). Surface damage to these bores, if not corrected, causes the new cam bearings to distort or misalign. Extra care taken at this point will make engine reassembly easier. In fact, efforts to remove cam bearings with improper tools such as chisels or punches can ruin the block.

Fig.7-21   An adjustable cam bearing driver

Fig.7-22   Driving out a cam bearing

Remember that in many engines, cam bearings must be removed to thoroughly clean oil passages. In addition, some caustics used in degreasing attack the bearing material. Once removed, it is necessary to install a new cam bearing set.

REMOVING OIL PLUGS AND CORE PLUGS

Oil passages in engine blocks are plugged at least at one end with pipe plugs (see Fig.7-23). It is necessary to remove these plugs to scrub oil passages clean. Keep in mind that hot tank cleaning of the engine block loosens deposits in passages and severe scoring of the new engine bearings results unless passages are thoroughly cleaned. In fact, solid contaminants left in oil passages are the major cause of bearing damage.

Fig.7-23　Pipe plugs at outer ends of oil passages

Pipe plugs in iron blocks are most easily removed by first heating them red hot with an oxy-acetylene torch and then lubricating threads with paraffin (see Fig.7-24 and 25). The plug is then easily removed using a socket wrench or oil drain plug wrench

Fig.7-24　Heating the oil plug and applying paraffin

Fig.7-25　Removing plug after applying paraffin

Do not try this method on pipe plugs in aluminum castings as the aluminum melts at the same temperature that the plugs turn red. It sometimes helps to heat the plug and the surrounding area using a propane torch. Some machinists report success in removing plugs in aluminum blocks by quickly heating the plugs with a TIG torch. In attempting this, use care not to heat the plug red as aluminum melts at this temperature.

One end of each oil passage is usually plugged with small diameter expansion or core plugs (see Fig.7-26). These core plugs are easily driven out with a 1/4in (6mm) diameter rod inserted from the opposite end of the passage once the pipe plugs are removed.

Fig.7-26　Core plugs in oil passages

Also remove water jacket core plugs for hot tank cleaning. Remove them even if new because caustics remove the surface plating that protects them against corrosion. One quick way of removing them is to use a punch to turn them sideways in the core plug hole and then a pry bar to pull them back out (see Figs. 7-27 and 28).

Fig.7-27 Turning a core plug to the inside before removing it

Fig.7-28 Using a rolling head pry bar to remove the plug.

Do not leave old core plugs in the water jackets as they disrupt the flow of coolant. As with cam bearing bores, take care not to damage the bore or the replacement plugs will not seal.

SUMMARY

Disassembly is an important stage in engine repair or rebuilding. It is an opportunity to view internal damage and be sure that all required repairs are included. However, care and organization are necessary to prevent damage in handling and to prevent the loss of small parts.

Careful diagnosis and job analysis help reduce labor by limiting teardown to the level required for repair. There is a tremendous difference in levels of disassembly required for engine overhaul, valve grinding, or leak repair. Technicians unfamiliar with a particular vehicle should learn to use information such as found in factory manuals, labor guides, and computer service references.

For thoroughness and for productivity, technicians and machinists should learn to record measurements on disassembly that will be necessary for reassembly. Some measurements, cylinder head thickness, for example, also tell us if parts are already outside of limits. Certain tricks of the trade are helpful in getting repairs done in a timely manner. Methods of removing stuck valve lifters, oil plugs, and core plugs as covered in this chapter are examples of labor saving tricks.

Chapter 7

ENGINE DISASSEMBLY

Review Questions

1. Free valve spring retainers from valve keepers by
   a. using penetrating oil
   b. striking the tip of the valve stem with a hammer
   c. striking the spring retainer with a hammer
   d. driving straight down on the spring retainer using a piston pin or similar driver

2. Remove valves with mushroomed tips from valve guides by
   a. filing or grinding away the mushroom
   b. driving out the valve guide with the valve
   c. driving the valve through the valve guide
   d. soaking the valve guide and valve stem with penetrating oil

3. Machinist A says that installed valve stem height is a critical dimension on assembly. Machinist B says measure and record this dimension on tear down because specifications are not always available. Who is right?

   a. A only   c. Both A and B
   b. B only   d. Neither A or B

4. If reusing the camshaft and rocker arms in an OHC engine,
   a. regrind them
   b. keep them in order
   c. polish them
   d. clean them in a hot tank

5. Because installed stem height specifications are sometimes unavailable, measure this dimension
   a. only on reassembly
   b. before removing valves
   c. to select spring shims for reassembly
   d. from the retainer to the spring seat

6. To protect the crankshaft while removing or installing piston and rod assemblies,
   a. place bearings on crankpins
   b. place bearings in the rod
   c. remove or replace rods with the crank removed
   d. place rubber tubing on rod bolts

7. Machinist A says that before removing piston and rod assemblies, check rod numbering and mark as necessary. Machinist B says that measure and record deck clearance if needed. Who is right?

   a. A only   c. Both A and B
   b. B only   d. Neither A or B

8. The major reason for removing the ring ridge before taking pistons out is to keep from
   a. damaging the piston pin
   b. breaking the piston ring land
   c. scratching the cylinder wall
   d. breaking the connecting rod

9. Machinist A says mark main caps for position and direction on tear down. Machinist B says replace caps in assembly after tear down for later machining operations. Who is right?

   a. A only   c. Both A and B
   b. B only   d. Neither A or B

10. If reusing the camshaft and valve lifters in pushrod engines,
    a. regrind them
    b. keep them in order
    c. polish them
    d. clean them in a hot tank

11. Machinist A says remove stuck lifters in pushrod engines by pushing from the top after removing the camshaft. Machinist B says remove them by driving from the bottom side with a punch. Who is right?

    a. A only   c. Both A and B
    b. B only   d. Neither A or B

12. Take care removing cam bearings not to damage
    a. the bearings
    b. bearing housing bores
    c. journals
    d. bearing drivers

13. Cam bearings are destroyed if left in the block during
    a. cleaning in caustics
    b. steam cleaning
    c. washing in solvent
    d. machining

14. Remove threaded oil plugs by
    a. using an impact driver
    b. using a breaker bar
    c. heating around each plug before removal
    d. heating each plug before removal

15. Remove core plugs in water jackets
    a. by heating them first
    b. by unscrewing them at room temperature
    c. by turning the plug to the inside and prying it out
    d. only if leaking

FOR ADDITIONAL STUDY

1. How are spring retainers freed from valve keepers? What precautions are necessary when freeing them?

2. What measurements should be recorded when disassembling cylinder heads?

3. Which valve train components must be kept in order?

4. List the reasons for numbering connecting rods? When should they be numbered?

5. How is the crankshaft protected when removing piston and connecting rod assemblies?

6. How should main bearing caps be numbered?

7. How are core plugs removed?

8. How are oil plugs removed?

Chapter 8

CLEANING ENGINE PARTS

Upon completion of this chapter, you will be able to:

- Describe differences in cleaning solutions used in hot tanks and spray cabinets.
- Explain the operation and temperature settings for degreasing in a cleaning oven.
- Compare airless blasting, bead blasting, and parts tumbling processes.
- List precautions in shot blasting or bead blasting castings.
- List places where cleaning by hand is necessary.
- List and explain processes for removing rust and scale from water jackets.
- Explain waste handling measures relative to parts cleaning necessary for hazardous waste compliance.

INTRODUCTION

The thorough cleaning of engine parts is essential to quality engine service but it is time-consuming and there is a temptation to shortcut steps. Carefully select procedures so that cleaning is completed as thoroughly and as quickly as possible.

A major cause of poor performance in reconditioned engines is the presence of solid contaminants in valve guides and engine bearings. Deposits collect in engines for 100,000 miles or more without causing problems, but they loosen in cleaning and, if not removed, they circulate throughout the engine after servicing. In fact, particle contamination is the leading cause of bearing failure.

A common source of hidden dirt is under the floor of V-block intake manifolds for carbureted or throttle body engines (see Fig.8-1). With these "wet" manifolds, exhaust heat is channeled through this area to heat the floor of the manifold and choke mechanisms and carbon builds up both within the passage and on the underside. To clean the underside, remove the pins holding the heat shield in place by wedging a chisel under the head of each pin. Replace the shield by reinstalling the pins or by drilling and tapping for screws. Baking in a cleaning oven, after removing sensors and vacuum switches, thoroughly cleans the crossover passage inside the manifold.

Fig.8-1   Carbon deposits under a V-block manifold after hot tank cleaning

Another commonly overlooked source of dirt is the oil pump pickup screen. It may be soaked, scrubbed, or flushed, but very probably it will remain dirty unless taken apart for cleaning (see Fig.8-2). If the pickup cannot be taken apart, replace it.

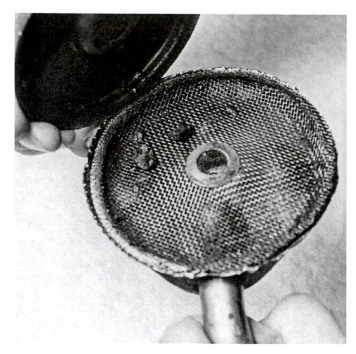

Fig.8-2 Opening an oil pump screen to expose dirt

USING SOLVENT AND COLD SOLUTIONS

The most common cleaning solution made available to mechanics is cleaning solvent. Solvent is suitable for small degreasing jobs but requires scrubbing and scraping by hand and such labor must be kept to a minimum. Solvent tanks are available with pump systems for improved efficiency (see Fig.8-3). Cleaning solvent is safe for all metallic materials used in automotive engines.

Fig.8-3 Cleaning parts in a solvent tank with pump

Some shops subscribe to solvent and tank maintenance services that periodically replace and recycle the solvent. This is an important consideration in today's world where we must look for ways of conserving resources and reducing the environmental impact of our businesses.

Cold soak solutions are available for cleaning of engine parts. These are usually general-purpose cleaners for "ferrous" iron and steel and "nonferrous" aluminum, brass or bronze. Parts soak in these solutions until clean and are then rinsed in water. Cold solutions are not particularly desirable for shops specialized in engine building because the cleaning action is so much slower than with hot solutions.

CLEANING IN HOT TANKS

Hot solutions are considerably faster than cold solutions. The tanks operate at temperatures of 170 degrees F or more. As with cold solutions, chemicals may be selected for ferrous or nonferrous metals or for both. These solutions are basic, or alkaline, and maintained at approximately pH 12 and care must be taken to place only the correct materials in these tanks. For example, cam bearings will be ruined by most ferrous solutions. If in doubt, separate ferrous and nonferrous metals with a magnet and select chemicals accordingly.

Common selections for degreasing are caustics for iron and steel and detergents for other metals including aluminum. A shop with only one tank may safely clean all parts in detergent, but with decreased efficiency.

Consider engine materials carefully when selecting solutions. For example, if working with aluminum-magnesium alloy parts, production common to the aviation industry may be better suited than automotive products. Chemical manufacturers offer recommendations for special job requirements.

Be aware that placing working assemblies such as pistons and connecting rods in these solutions may cause moving parts to bind because residues collect internally in the assemblies. Also, parts are so thoroughly degreased that corrosion readily takes place after removing them from the hot tank. Rinse parts in hot water, blown them dry, and spray them with corrosion inhibitor as quickly as possible. Rubber parts such as motor mounts or harmonic balancers should never be cleaned in caustics.

Equipment is available with variations in tank size and agitation devices. For example, many small non-production shops use hot soak tanks without parts agitation or spray jets. Production shops however seek the speed and efficiency gained by the addition of parts agitation or spray jet systems (see Fig.8-4). To conserve energy, it is recommended that hot tanks be double walled for insulation and heating systems equipped with electric ignition and time clocks.

Fig.8-4   Loading parts into a spray-jet hot tank

As mentioned, parts are subject to corrosion after degreasing and it is best practice to rinse them in hot water, blow them dry, and treat them with rust inhibitor immediately upon removal from hot tanks. Rust inhibitors range from light oils sprayed onto parts or water based phosphate dips. It is also recommended that clean parts be stored in a dry area removed from cleaning activities.

DEGREASING IN OVENS

The use of oven cleaning has expanded greatly in automotive machine shops. Interest in this process grew out of the search for processes that are more efficient and more environmentally acceptable. With hot tanks there is the problem of disposing of caustics and accumulated sludge and, even in small shops, there are easily 300 gallons or more of waste to dispose of each year.

The oven cleaning process requires specially constructed ovens with a primary chamber for parts cleaning and an exhaust afterburner to control emissions (see Fig.8-5). The parts heat rapidly to approximately 375 degrees F to initiate the vaporization of the more volatile vapors. Prior to this point, the exhaust afterburner comes on and remains on to oxidize vapors in the stack at approximately 1,300 degrees F. The primary chamber temperature continues rising to a pre-selected limit, 600 to 700 degrees F for iron and steel, and the burner cycles on and off to maintain this setting. In some ovens, should the volatile vapors catch fire and cause a sudden temperature rise, water injection is introduced to lower the temperature. After the cleaning cycle, three to four hours including cooling time, the sludge is reduced to a very small quantity of loose ash that can be collected for disposal or included with hot tank sludge.

Fig.8-5 Loading parts into a cleaning oven

With iron castings, because the amount oxidation increases with temperature, the low side of the recommended temperature range minimizes the work necessary to strip oxides following heat cleaning. All castings continue to go through hot tank cleaning to remove oxides and debris.

Only aluminum castings require precautionary adjustments to the process. A primary chamber temperature of not more than 500 degrees F reduces the possibility of loosening valve guides and seats in cylinder heads. With precipitation hardening alloys, slow cooling allows time at temperatures for precipitation. For all metals, slow cooling prevents distortion caused by unequal cooling rates within complex castings. The availability of an oven also facilitates other shop operations. For example, ovens are used to straighten heads, help in removing and installing valve guides, install cylinder sleeves and for pre and post heating when welding.

USING AIRLESS SHOT BLASTERS

Another cleaning system that greatly increases cleaning efficiency is airless shot blasting (see Fig.8-6). The process is called "airless" because a high-speed impeller, not compressed air, propels shot against slowly rotating parts. The shot is .030in. (.75mm) or larger in diameter and is usually steel although it is possible to use zinc or other media to lessen the severity of the impact. The automatic cleaning cycle requires minimum labor and approximately 10 minutes plus a few minutes of tumbling to remove shot. Blasting aluminum with steel shot is not recommended.

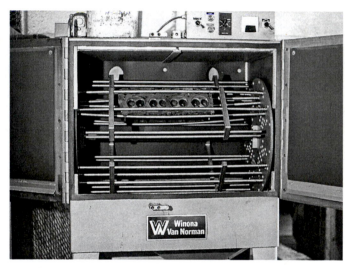

Fig.8-6 An airless shot blaster

Degrease and dry parts for efficiency and to minimize shot retention in castings. airless shot blasting ideally follows oven cleaning since parts are thoroughly degreased and dry. The efficiency of oven cleaning in combination with airless shot blasting reduces final cleaning requirements to a level easily handled by detergents in a spray washer. Again, this is an important environmental consideration since this reduces disposal requirements for caustics and large volumes of sludge.

BEAD BLASTING

Carbon removal on many engine parts is a particular problem. Because solutions used for cleaning are

primarily degreasers, they remove only the carbon held in oily deposits. Carbon collected on those engine parts directly exposed to hot exhaust or combustion gases are difficult to remove. Blasting with glass beads is a commonly used process for this purpose (see Fig.8-7).

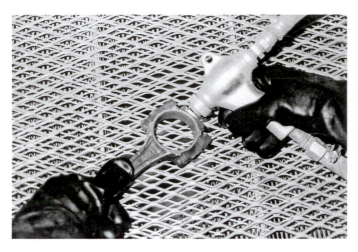

Fig.8-7   Bead blasting with glass beads

As with airless blasting, parts must be degreased and dry or beads embed in the grease. Any oil deposits reduce cleaning efficiency and increase shot retention. For cylinder heads, limit bead blasting to combustion chambers, ports, and valves to minimize chances for shot retention in threaded holes, oil passages, and water jackets. Bead blasting has further application in removing rust or gasket material from other engine parts thereby saving considerable time otherwise spent hand scraping or wire buffing.

Keep beads out of threaded holes by reinstalling fasteners and blocking other holes with heavy duty duct tape. After blasting, tumble castings and check for beads by blowing through holes with compressed air (see Fig.8-8). Remove beads while dry; flushing water jackets with wet solutions will not remove them. Keep in mind that failure to remove beads causes problems with fasteners, engine bearings, and water pump bearings and seals.

Fig.8-8 Tumbler to remove glass shot and debris

BLASTING WITH BAKING SODA

The problem with most blast processes is shot retention. Blasting with baking soda solves this problem since the baking soda is water soluble. Air pressure is the force behind the baking soda media and blasting is done in a closed cabinet (see Fig.8-9). For efficiency, degrease parts before blasting. After blasting, flush with water and the media will disappear in solution with the water. While the baking soda is non-toxic and can go into sewers, users must consider the nature of contaminants picked up in blasting prior to deciding upon the means of disposal.

Fig.8-9   Blasting cylinder heads with baking soda

BLASTING WITH HIGH PRESSURE WATER

The industry is increasingly under pressure to reduce the use of hydrocarbon cleaning solvents for degreasing. Because of this, blasting with water and aqueous cleaner at 600 PSI is gaining favor. Parts are placed in a closed cabinet and blasted with a high pressure stream of water (see Fig.8-10). There are two nozzle options to choose from, jet and wand. The force of this blast enhances the ability to clean hard to reach areas such as oil passages.

Fig.8-10   High pressure blast cabinet and nozzle

TUMBLING SMALL PARTS

For small parts, tumbling is an excellent alternative to bead blasting. Labor is reduced because after valves, valve springs, spring retainers and other small parts are loaded into the basket with media, and a timer is set to run the tumbler automatically through an approximate 20-minute cycle. The basket is square which adds to the agitation as the basket rotates. The peening action produced by rotating of the parts in the media removes carbon and other deposits. The corners of the basket also dip into solvent as the basket rotates washing debris from both the media and parts.

With steel media approximating 5/16in across, problems with media retention are minimal (see Fig.8-11). While it is common to protect valve stems from the severity of the action with tubing, it is the author's experience that adding extra media lessons the impact sufficiently to eliminate the need for protection.

Fig.8-11   A small parts tumbler and media

USING HAND AND POWER TOOLS

Cleaning machines do not entirely eliminate cleaning with hand and small power tools. For example, with caution, gasket scrapers remove tightly bonded gasket sealer and material. Although it is possible to use a broken ring segment to scrape ring grooves, there are ring groove scraping tools that fit the various standard ring groove widths and depths and reduce the risk of cutting into base metal (see Fig.8-12).

Fig.8-13　Scrubbing an oil passage clean with a bore brush

Fig.8-12　Scraping piston ring grooves

After hot tank cleaning, scrub oil passages with bore brushes. Select brushes with diameters and lengths suitable for oil passages, valve lifter bores, and crankshaft oil passages. Scrub by hand or use drill motors to drive the brushes (see Fig.8-13).

A wire buffer is a convenient tool for the last minute removal of gasket material or sealer. The buffer provides an alternative to hand scraping or bead blasting (see Fig.8-14). Wear safety glasses and avoid buffing near edges or holes in sheet metal as the buffer can grab the work and pull it from your hands.

Fig.8-14　Using a wire buffer to remove carbon from a piston head

An air motor and flare brush combination was the forerunner of bead blasting for cleaning combustion chambers and ports (see Fig.8-15). Although not as fast or thorough as bead blasting, this is the best alternative if blast equipment is not available. Use care however in cleaning aluminum castings, as the action is severe and can cut into the base metal.

Fig.8-15 Using a flare brush to remove carbon from a combustion chamber

REMOVING RUST AND SCALE

Occasionally, water jackets in cast iron cylinder heads and engine blocks plug up with rust and scale, especially in marine engines. Such deposits inhibit the transfer of heat from engine parts to the coolant. Unless removed, these deposits cause overheating and engine damage.

Cold acid descalers remove rust from iron castings but check with local regulators before setting up for descaling. For safety and to remain out of the defined corrosive range, keep descalers low in concentration and above pH 2.5. To descale, first degrease and thoroughly rinse parts, as any carry-over of the caustic degreaser into the acid descaler tends to neutralize it (a change to water and salt). Of course, use care when working with these solutions to prevent splashing the solution onto skin or clothing. Face shields and chemically resistant gloves and apron are recommended in working around any chemicals.

Should local regulations pertaining to descaling be prohibitive, run cooling system cleaner through the engine before removing it from the chassis. This at least loosens deposits to a point where they readily wash out of castings during routine cleaning. Collect and recycle any coolant and cleaner drained from the engine.

WORKING UNDER REGULATIONS

Since 1986, most automotive repair and machine shops are subject to the requirements of the Resource Conservation and Recovery Act. State and local agencies typically enforce provisions of this federal law for purposes of reducing the generation of hazardous wastes and to see that such wastes are properly handled. State and local governments also have the option of raising compliance standards as they often do depending on critical regional issues.

Any business that generates 220 pounds of waste per year is considered a hazardous waste generator. Once classified as such, the business must obtain an Environmental Protection Agency identification number and comply with regulations for storage, handling and shipping of wastes. A product, or by-product, becomes a waste when the business no longer uses it and must store or dispose of it. In most cases, such wastes must be properly contained, labeled, and kept in storage for no more than 180 days. For shipping, the waste hauler must be licensed and complete a manifest documenting the nature of the waste, the generator, and the destination. Wastes can only be shipped to licensed disposal facilities.

Wastes are hazardous if they fail tests for ignitability, corrosivity, reactivity, or toxicity. In brief terms, the standards for testing are as follows:

1. Ignitability testing requires that the waste not have a flash point below 140 degrees F meaning that it will not catch fire below this temperature.

2. Corrosivity testing requires that the waste not have a pH value below 2 or above 12.5. The lower numbers mean that the material is acidic and higher numbers that the product is basic or alkaline. For comparison, water has a pH of 7 and is neutral.

3. Reactive materials are those that become violent when exposed to water or other materials or release cyanide gas, hydrogen sulfide, or similar gases when exposed to acidic solutions.

4. Toxicity under these rules applies to heavy metals that "leach" into water during testing. Tests for leachability essentially find if the metals are likely to find their way into the ground water. Heavy metals typical of machine shops potentially include cadmium, copper, and lead.

Most common in machine shops are cleaning solvents and caustics or detergents for degreasing. These cleaners are contaminated with heavy metals such as copper and lead from bearings and cadmium from plated fasteners. Most of these materials are considered hazardous waste once removed from use and waste handling regulations go into effect immediately. Some materials such as solvent are recycled, but the shop does not escape responsibility for proper handling, storage, and transportation. Like cleaning materials, assume that grinding coolants are contaminated.

Generally, the best approach is to eliminate hazardous products where possible, reduce their use, or recycle. Some examples of these measures include the following:

1. Degreasing in an oven equipped with functional emission controls reduces the generation of hot tank sludge. Remember that the residual ash from ovens, although small in volume, is still a hazardous waste and requires handling as such.

2. Separate sludge and oil from hot tank solutions and add chemical to the solution rather than replace it. Once separated, the sludge can sometimes be dried to reduce the volume and weight of waste (check with local regulators). An alternative is to have the hot tanks pumped and the contents taken away by a waste hauler.

3. Separate oil and sludge from grinding coolant and reduce the number of times it is replaced. To do this however, may require treating the coolant to prevent bacterial growth. For some processes, surfacing for example, consider milling instead of grinding.

4. Again, there are services available to maintain hot tanks and to recycle solvents. While they may add to overhead cost, depending upon local regulators, they might be the best alternative.

Most important is that both shop management and personnel take the intent of regulation seriously and comply in every way possible. While regulatory bureaucracies can be frustrating, your families and those of your customers, live in the area, breathe the air, and drink the water.

SUMMARY

Cleaning is one of the most troublesome parts of engine repair and rebuilding. While it must be thorough, it is labor intensive and time consuming, especially in non-production shops. It is relatively easy to clean parts externally, but it is difficult to thoroughly clean oil passages and other internal areas where dirt, rust, and scale can hide. Should contaminants not be removed, be assured that they will cause engine damage.

A variety of products are used including solvents, detergents, and caustics. An equally wide variety of equipment is used including solvent tanks, spray cabinets or hot tanks, cleaning ovens, tumblers and blast equipment. Those responsible for cleaning must learn when to use particular products and equipment, when not to use particular products or equipment, and how to protect engine parts while cleaning.

Cleaning equipment should be selected to reduce labor but there are situations when cleaning by hand is the only practical way. Hand scraping, wire buffing, and bead blasting require labor or that someone stand in front of equipment. Keep this work to a minimum.

Cleaning processes also generate hazardous material and waste handling problems. Employees must be well informed as to safety and labeling, storage, and waste handling requirements.

Chapter 8

CLEANING ENGINE PARTS

Review Questions

1. The primary cause of bearing damage is
   a. lack of oil
   b. dirt or particle contamination
   c. improper assembly
   d. misalignment

2. Machinist A says strip intake manifolds of sensors and vacuum switches prior to oven cleaning. Machinist B says remove any heat shields on the underside to reach deposits. Who is right?

   a. A only   c. Both A and B
   b. B only   d. Neither A or B

3. Before assembling to the pump, _____ oil pump pickups.
   a. check for position
   b. degrease
   c. degrease and bead blast
   d. clean or replace

4. Machinist A says that detergent degreasers damage parts made of aluminum. Machinist B says that detergents are safe for cleaning iron, steel, or aluminum. Who is right?

   a. A only   c. Both A and B
   b. B only   d. Neither A or B

5. Cam bearings left in blocks or heads are damaged if cleaned in
   a. ovens
   b. detergents
   c. solvent
   d. degreasers

6. The time required for degreasing is reduced by
   a. agitating the parts
   b. heating the solutions
   c. spraying the solution onto parts
   d. combinations of these

7. Machinist A says disassemble working assemblies prior to hot tank degreasing. Machinist B says that assemblies such as pistons and rods sometimes lock up after immersion. Who is right?

   a. A only   c. Both A and B
   b. B only   d. Neither A or B

8. Machinist A says that surface rust forms especially fast on castings fresh out of hot tanks. Machinist B says that to prevent corrosion, rinse castings immediately, treat with rust inhibitor, and store in a dry place. Who is right?

   a. A only   c. Both A and B
   b. B only   d. Neither A or B

9. Phosphate dips after degreasing
   a. remove rust and scale
   b. prevent rust
   c. remove paint
   d. prepare for painting

10. Machinist A says that the primary chamber temperature in a cleaning oven operates at approximately 1300 degrees F. Machinist B says that the afterburner in a cleaning oven operates at approximately 500-700 degrees F. Who is right?

    a. A only   c. Both A and B
    b. B only   d. Neither A or B

11. Machinist A says that the secondary burner in a cleaning oven comes on before 400 degrees F and remains on. Machinist B says that the primary chamber burner cycles on and off to maintain the preset temperature. Who is right?

    a. A only   c. Both A and B
    b. B only   d. Neither A or B

12. Oils volatilize by approximately _____ degrees F.
    a. 250-350
    b. 350-450
    c. 500-700
    d. 1100-1400

13. Airless shot blasters generally use
    a. metal shot
    b. glass shot
    c. glass beads
    d. silicon carbide abrasive

14. Machinist A says that blasting with steel shot is faster and less labor intensive than bead blasting. Machinist B says blasting aluminum with steel shot is not recommended. Who is right?

    a. A only   c. Both A and B
    b. B only   d. Neither A or B

15. Machinist A says that bead blasting is safe for cleaning combustion chambers and ports. Machinist B says take precautions to prevent beads from contaminating water jackets and oil passages. Who is right?

    a. A only   c. Both A and B
    b. B only   d. Neither A or B

16. Prior to bead blasting, _____ parts.
    a. de-carbonize
    b. steam clean
    c. de-scale
    d. degrease and dry

17. The bead blaster uses _____ for cleaning.
    a. glass
    b. sand
    c. glass or sand
    d. silicon carbide

18. Machinist A says that beads in cooling systems plug thermostats. Machinist B says that the beads damage water pump bearings and seals. Who is right?

    a. A only   c. Both A and B
    b. B only   d. Neither A or B

19. When tumbling small parts, protect
    a. spring retainers
    b. valve faces
    c. valve springs
    d. valve stems

20. Machinist A says remove deposits from ring grooves with a ring groove scraper. Machinist B says scrape ring grooves with segments of the old rings. Who is right?

    a. A only   c. Both A and B
    b. B only   d. Neither A or B

21. Machinist A says that wire flare brushes are used to remove deposits from combustion chambers. Machinist B says that caution is required when using flare brushes on aluminum heads. Who is right?

    a. A only   c. Both A and B
    b. B only   d. Neither A or B

22. Machinist A says that immersion in alkaline solutions removes rust in water jackets. Machinist B says that immersion in acid solutions remove rust. Who is right?

    a. A only   c. Both A and B
    b. B only   d. Neither A or B

23. Machinist A says that immersion in acids removes grease from castings. Machinist B says that immersion in caustics removes rust and scale but not the grease. Who is right?

    a. A only
    b. B only
    c. Both A and B
    d. Neither A or B

24. A shop becomes a "hazardous waste generator", by RCRA standards, when waste exceeds
    a. 2,200 Pd per year
    b. 2,200 Pd per month
    c. 220 Pd per year
    d. 220 Pd per month

25. Machinist A says that sludge in hot tanks is considered hazardous waste. Machinist B says that the sludge only becomes a waste once removed from the hot tank. Who is right?

    a. A only
    b. B only
    c. Both A and B
    d. Neither A or B

FOR ADDITIONAL STUDY

1. What spray cabinet or hot tank cleaner is most effective in cleaning iron and steel? What about cleaning aluminum or iron and steel? What happens if the materials are placed into the wrong solution?

2. How are cylinder heads and blocks prepared for cleaning?

3. How are parts handled after degreasing?

4. At what temperature are iron castings degreased in cleaning ovens? What about Aluminum castings?

5. What media is used in airless blasters on iron castings?

6. How are glass beads kept out of cylinder head water jackets, bolt holes, and oil passages when bead blasting?

7. What parts are cleaned by tumbling?

8. List the hazardous chemicals and wastes typically found in automotive machine shops?

9. What are the advantages of baking soda as a blast media?

10. What are the advantages of water based cleaning? What advantages are there in using these cleaners at high pressure?

Chapter 9

INSPECTING VALVE TRAIN COMPONENTS

Upon completion of this chapter, you will be able to:

- Compare methods of measuring valve guide wear.
- List measurements and points of inspection for valves.
- List points of inspection for valve springs.
- Explain test procedures for valve spring pressure.
- Visually inspect rocker arms, rocker arm studs, and pushrods.
- Identify wear problems in camshafts and followers or valve lifters.
- Compare methods of checking timing slack and timing gear backlash.
- List points of inspection for cylinder head castings.
- Explain how to check flatness of cylinder head deck surfaces.
- Compare methods of checking camshaft bore wear and bearing clearance in overhead camshaft cylinder heads.
- Compare methods of checking camshaft bore alignment in overhead camshaft cylinder heads.

INTRODUCTION

Valve grinding is often viewed as simple and routine and changes in engines that complicate the job are often not considered. Newer engines may have one, two, or four camshafts and two, three, four, or even five valves per cylinder. Head temperatures cycle up to 250 degrees F and exhaust valves operate at 1500 degrees F. Valve performance in these engines simply cannot be ensured without detailed inspection and machining during valve service. The simple valve job that was adequate in years past is not satisfactory given the internal stresses in current power plants.

DETERMINING VALVE GUIDE WEAR

Check valve guide wear or "bell-mouthing" that occurs at each end of valve guides. The wear is greatest at the port end of the valve guide because of the extreme heat and deposits carried into the valve guide on the valve stem. Measure valve guide wear by taking measurements of the valve guide in the least worn part of the guide, the middle, and at the port end where it is most worn (see Fig.9-1). The difference in the two measurements is the amount of wear.

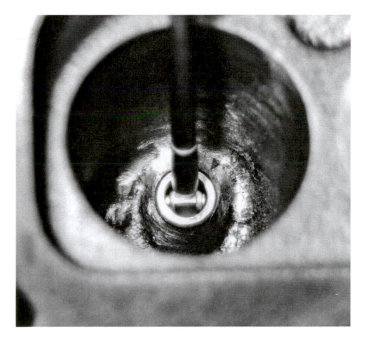

Fig.9-1 Measuring valve guide wear

Valve guide wear is extremely important in deciding upon what procedures to use in restoring specified valve stem-to-guide clearance. Many manufacturers do not give specifications for valve guide diameter but instead give the valve stem diameter and the range of valve stem-to-guide clearance. The valve stem must first be measured and compared to specifications.

Valve guides can be checked more quickly using go-no-go methods. First adjust a micrometer to the valve stem diameter plus maximum valve stem-to-guide clearance. Next, set a telescoping gauge to the micrometer setting and then check if it enters the valve guide. If it enters, the guide is worn beyond limits. If it does not enter, clearance is within service limits.

Be aware that oversized valve stems are used to correct for wear in integral valve guides. Should this be the case, compare valve guide diameters to measured valve stem diameters to find the clearance or wear.

Some manufacturers recommend checking guides by measuring "valve rock." To do this, open the valve slightly and rock the valve margin against a dial indicator (see Fig.9-2). Because the valve extends out of the guide, valve rock readings exceed specified stem-to-guide clearance.

Be sure not to confuse specifications for stem-to-guide clearance with valve rock when checking valve guides by this method. Check valve guide condition carefully during routine valve service procedures. Most high mileage cylinder heads, especially from pushrod engines, have stem-to-guide clearance exceeding specified limits. Correct sealing at the valve face is impossible to maintain with excessive valve guide clearance because the valve closes at an angle to the seat (see Fig.9-3).

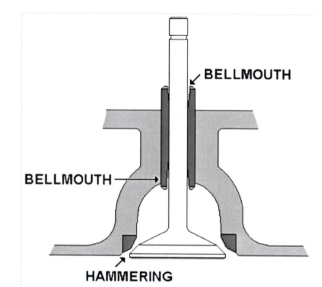

Fig.9-3   Cocking of a valve on the seat caused by valve guide "bell-mouthing"

CHECKING VALVES

It saves time to inspect valves before regrinding the faces. Set aside valves that show damage wear beyond service limits before cleaning if possible. Certainly do not reface such valves.

Check the thickness of valve margins first (see Fig.9-4). The thickness of the margin after grinding should be no less than 1/32-in (.8mm) or half of the new thickness, whichever is greater. For passenger car engines, replace valves thinner than 1/32in or they will burn. For heavy-duty applications, replace valves that are thinner than one-half of new thickness. If in doubt, check the manufacturer's recommendations regarding minimum specifications.

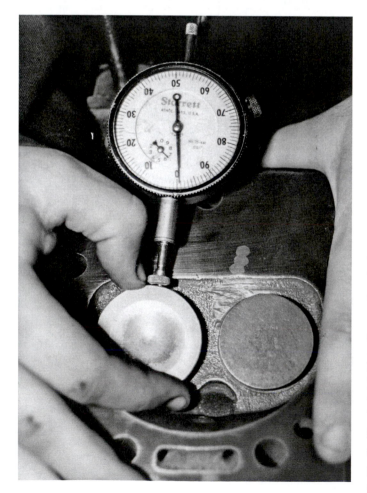

Fig.9-2   Measuring valve rock

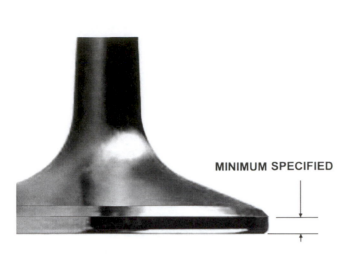

Fig.9-4   The thickness of valve margins

Fig 9-5   Heat damaged valves

Measure valve stem wear with a micrometer. Service limits allow .001 to .0015in (.03-.04mm) taper or variation in the diameter of the valve stem along its length. Keep in mind that manufacturers do specify stem diameters and that valve stems worn under minimum specified diameters will not assemble within specified limits for clearance. Be sure to check specifications because some valve stems use a taper to compensate for uneven heat expansion along the stem.

Also inspect the heads of valves visually for damage such as warpage, burning, or cupping (see Fig.9-5). Keep in mind that valve burning follows a sequence beginning with unequal cooling, warpage, and then burning. Cupping occurs when the valve overheats all around the head. Look also for a necking down of the valve stem beneath the head that suggests stretching (see Fig.9-6). Replace valves with such defects as they indicate exposure to extreme engine heat and the kinds of damage that cause failure. For some engines, it is safer to replace exhaust valves, not regrind them, during valve service because of the frequency of heat damage and the potential for failure.

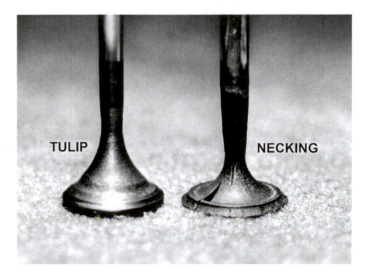

Fig.9-6   Cup (tulip) shaped, necked stem

Check for grooves worn in the faces of valve stems caused by defective "valve rotators" (see Fig.9-7). Valve rotators cause valves to rotate each time they open and prevent warpage by maintaining uniform heat distribution around the valve head. Although rotators sometimes sit under valve springs, they most often serve as valve spring retainers. While rotators were common for many years in heavy-duty truck engines, they are now common in passenger car engines, especially for exhaust valves. Replace rotators when grooves appear in the faces of the valve stems that show signs of non-rotation.

Fig.9-7　　Non-rotation pattern

Check keeper grooves for wear or damage. Worn keeper grooves prevent the keepers from properly locking to the valve stem. Sometimes the grooves wear so evenly that they appear as if they were made that way. If in doubt, compare worn parts to replacement parts and wear will be apparent (see Fig.9-8).

Fig.9-8　　Stages of keeper groove wear

The importance of maintaining correct guide clearance and carefully checking valves cannot be over emphasized. Considerable heat transfers through valves, valve seats, and guides into the

cooling system, therefore, valve seats and guides not kept within service limits cause valves to overheat and fail (see Fig.9-9).

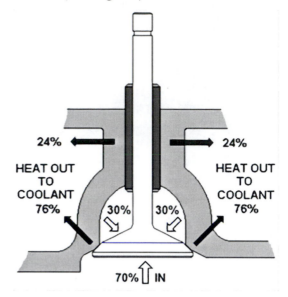

Fig.9-9　　Heat transfer through a valve

CHECKING NATURAL GAS VALVE TRAINS

Natural gas fuels are commonly known as compressed natural gas or CNG and liquid petroleum gas or LPG. Engines running on natural gas, or alcohol fuels such as ethanol and methanol, have particular considerations for valves and valve seats. Unlike gasoline, these fuels have no lubricating value for valve faces and valve seats.

Many engine builders eliminate valve rotators if so equipped. Although they keep valve temperatures uniform they also impart a rotating action to valves each time they close that can cause recession into the seats (see Figs. 9-10 and 11). Because gasoline engines are sometimes converted to alternative fuels without internal changes, check for the presence of rotators. If severe recession is present on teardown, consider replacing them with conventional spring retainers.

Fig.9-10 Valve face wear in a gasoline engine run on natural gas

Fig.9-11 Wear on the matching valve seat

To properly prepare engines for these fuels, it is important to select the right valve and seat materials. Valves and seats must retain hardness at very high temperatures and resist wear in the absence of lubrication from the fuel. For valves, "Inconel" alloy valves with hard chrome stems are recommended. The recommended valve seat material is a cobalt, non-magnetic, alloy capable of performing to 1600 degrees F. On inspection, if seats are magnetic, consider replacing them with the recommended material. Grinding seat widths

to maximum specifications also helps lower valve temperatures.

Valve guide sealing can also be a problem. Unless guides receive lubrication, guide wear can be extreme. Seals materials such as Viton are necessary to withstand the heat but care should be taken to allow some oil to reach valve stems and guides. Because of oiling requirements, changing from umbrellas to positive seals is not recommended.

TESTING VALVE SPRINGS

Test valve springs for pressure and warpage. Weak springs will not close valves and may break. A broken spring leads to severe engine damage by allowing the valve to drop into the cylinder (see Fig.9-12).

Fig.9-12 Piston damage from a "dropped" valve

Serviceable springs are within 10 percent of specified pressure at the given test lengths (see Fig.9-13). There are two sets of specifications, one for testing with the spring compressed to the valve closed length and another for testing with the spring compressed to the valve-open length.

Fig.9-13 A Rimac spring tester

Maximum allowable warpage for valve springs is 1/16in (1.5mm) for every 2in (51mm) of spring free length (see Fig.9-14). Replace weak or warped springs as they are showing the first signs of failure.

Fig.9-14 Checking spring distortion

INSPECTING CAMSHAFTS, LIFTERS, AND FOLLOWERS

Do not overlook hydraulic valve lifters, lash compensators, or camshafts in valve service. Even with correctly ground valves and seats, if noise or other problems remain, the customer will complain and the work will have to be redone. Thorough inspection procedures ensure complete repairs, good valve train operation, and customer satisfaction.

For pushrod engines, first visually check valve lifter and cam lobe wear patterns. To find how wear patterns develop, note that flat tappet type lifters are actually crowned and cam lobes tapered to promote rotation and reduce wear (see Fig.9-15). Occasionally, valve lifters stick in their bores and fail to rotate or rotate only intermittently. Varnish, dirt, or damage to lifter bores cause this and it leads to extremely rapid wear of both lifter bases and the camshaft (see Fig.9-16). Replace both the camshaft and lifters when non-rotation or intermittent rotation is apparent (see Fig.9-17).

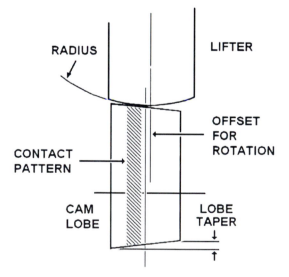

Fig.9-15 The radius or "crown" on lifter bases and lobe taper

Fig.9-16   A badly worn cam lobe

Fig.9-18 Badly worn lifter bases

Fig.9-17   Patterns of non-rotation (L) and intermit
tent rotation (R)

Fig.9-19   Edge-wear on a cam lobe

Experienced technicians hesitate to guarantee engines when lifter bases begin to wear in a concave pattern (see Fig.9-18). Sometimes, technicians replace single defective lifters to keep repair costs down but the repair cannot be guaranteed. Watch also for severe pitting and edge wear patterns on cam lobes that suggest approaching failure (see Fig.9-19). Again, replace camshafts and lifters in sets.

Roller lifters and cam lobes are less subject to wear (see 9-20). The camshafts are typically made of steel instead of nodular iron and the rollers have much less friction. Providing the hydraulic lifter operation is satisfactory and the roller bearings and camshaft pass inspection, both the lifters and camshafts are often reusable. Find the cam lobe wear by measuring the lift with a micrometer (see

Figs. 9-21 and 22). Variations exceeding .005in (.13mm) between exhaust cam lobes or between intake cam lobes are unacceptable. Check cam lobe base circles first for variations; should base circles vary, it will be necessary to measure the lift of each cam lobe and then make comparisons.

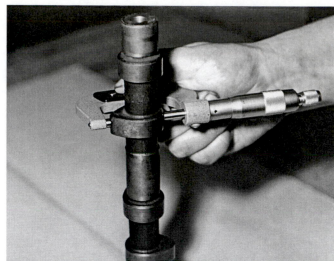

Fig.9-22 Measuring to determine cam lift

Measure camshaft journal diameters with a micrometer and check against specifications (see Fig.9-23). In aluminum heads, also measure bearing inside diameters and check for excessive journal clearance (see Fig.9-24). Unlike iron heads or blocks fitted with replaceable bearings, correcting clearance in aluminum heads requires operations including journal repair and line boring bearing housings.

Fig.9-20 Visually check wear on roller lifters or followers

Fig.9-23 Checking cam journal wear

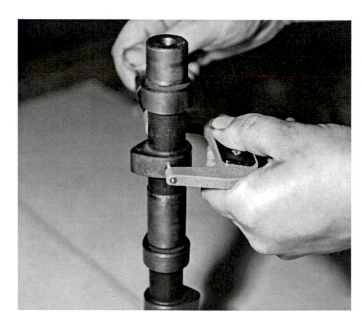

Fig.9-21 Measuring base circle diameter

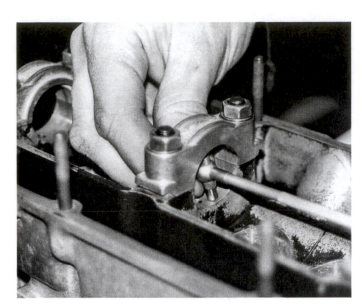

Fig.9-24   Measuring diameters and checking clearance in cam bearing bores

There is a hydraulic valve lifter leak down test that pinpoints defective lifters (see Fig.9-25). If body-to-plunger clearance is excessive or check valves malfunction, the lifter collapses almost immediately when loaded (see Fig.9-26). This test may also be done in a small arbor press as the difference in leak down rates between good and bad valve lifters is obvious, even by feel. Disassembly, cleaning, and retesting lifters typically does not pay and does not always cure the problem.

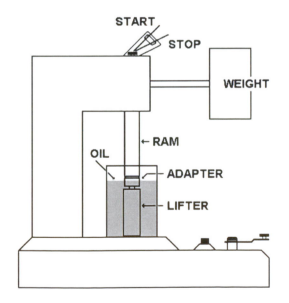

Fig.9-25   Testing for excessive leak down rates in hydraulic lifters

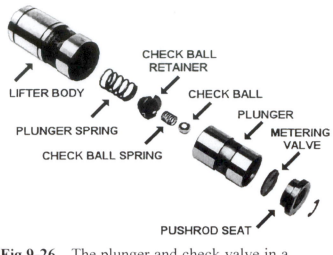

Fig.9-26   The plunger and check valve in a hydraulic lifter

Leak down rates are checked for "lash compensators" used in many overhead cam engines in the same way (see Fig.9-27). While they do not follow the cam or rotate, their operation is hydraulically the same as lifters in pushrod engines and are therefore subject to the same hydraulic failures.

Fig.9-27   Lash compensators are subject to the same problems as hydraulic lifters

Some overhead cam engines use "bucket" type cam followers that operate in precisely machined bores (see Fig.9-28). Both followers and bores are subject to wear and subsequent engine noise. Measure follower diameters and bores and check their sizes against specifications. If possible, correct clearances by fitting new or slightly larger diameter followers in worn bores.

Fig.9-28   Check cam follower wear and clearance in bores

Followers may be replaced but repairing bores may involve boring them oversize for sleeves. Other overhead cam heads use rocker arms that are subject to wear. Give particular attention to wear or scoring on surfaces that contact the camshaft (see Fig.9-29). Such wear eventually flattens the camshaft and both the camshaft and the rocker arms must be replaced. Some manufacturers now use a roller design for these followers and, as in pushrod valve trains; wear is greatly reduced (see Fig.9-30).

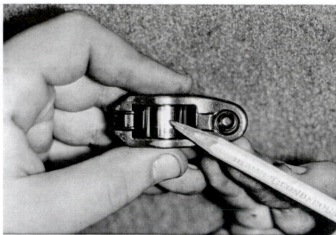

Fig.9-30   Checking rocker arm type cam followers for wear

INSPECTING ROCKER ARMS AND PUSHRODS

The condition of rocker arm faces in pushrod engines is often overlooked in inspection. Wear on the faces of rocker arms makes accurate adjustments of valve lash impossible because the feeler gauge will not slip fit precisely across the wear pocket (see Fig.9-31). While lash adjustments are unnecessary in hydraulic valve trains, the poorly matched surfaces of a worn rocker arm in assembly with a new or refaced valve stem accelerates wear.

Fig.9-29   Checking a roller in a rocker arm type cam follower

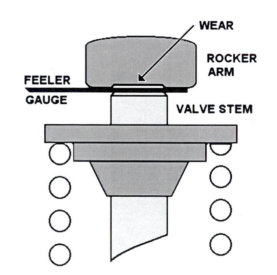

Fig.9-31   Lash adjustments with worn rocker arm faces

Inspect stamped steel rocker arms carefully including both the pushrod socket and the face (see Figs. 9-32 and 33). If wear is detectable in the face or in the pushrod socket, replace the rocker arm as it cannot be refaced.

When worn, stud mounted rocker arms sometimes wobble and cut into the sides of the studs (see Fig.9-34). Also check for variations in height, because press fit studs occasionally pull out of the cylinder head (see Fig.9-35). Find and correct the cause of studs pulling out such as incorrect press fit or piston-to-valve interference caused by timing chain failures.

Fig.9-32 Inspecting the pushrod socket of a stamped rocker arm

Fig.9-34 Check for cuts on the sides of rocker studs

Fig.9-33 Inspecting the face of a stamped rocker arm

Fig.9-35 Comparing rocker arm stud heights with a straight edge

When the pushrod socket wears, it affects the oiling of the rocker arm assemblies. In a new assembly, there is a reservoir of oil in the pushrod socket above the pushrod but, in a worn assembly, pushrod and socket wear eliminates the oil reservoir. Without the reservoir, oil flow to rocker assemblies is intermittent (see Fig.9-36). The solution is to replace the rocker arms and possibly the pushrods as well.

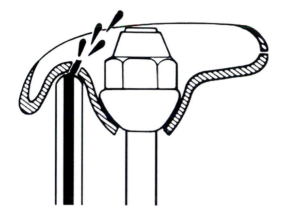

Fig.9-36   Intermittent oiling caused by pushrod and socket wear

The inspections for pushrods are simple. Inspect the ends for any visible wear patterns other than a smooth, round surface (see Fig.9-37). Compare them as a set for straightness and for variations in length. Except for dirty oil passages, which can be cleaned, any other conditions call for replacing the worn part

Fig.9-37   Check for worn pushrod ends

CHECKING TIMING CHAINS AND GEARS

To ensure reliability, check the condition of cam drive components. Checking for cam drive backlash is often the only realistic test because engine disassembly is not required. If valve service is part of an engine overhaul, it is possible to check further during tear down. The more thorough tests are those for chain slack and those for timing gear backlash (see Figs. 9-38 and 39). The maximum service limit for chain slack is 1/2in (13mm) although most technicians prefer much less than this. For gear tooth backlash, .006in (.15mm) is the limit. If wear is over service limits, all cam drive components should be replaced. When replaced, watch for valve timing marks on the gears or sprockets to ensure correct assembly.

Fig.9-38   Checking timing chain slack

Fig.9-39   Checking gear tooth wear or backlash

Note that discussion of timing chain slack has so far been limited to pushrod engines. This is because failures in these engines are common and occur without warning. Timing chains in overhead cam engines also stretch but usually make rattling noises before failure occurs. The noise develops when timing chain tensioners, usually actuated by engine oil pressure, wear or run out of travel (see Fig.9-40). Because these tensioners require engine oil pressure, there is no accurate method of testing chain slack in an engine at rest. Inspect the timing chain and tensioner in any overhead cam engine anytime that noise comes from the chain.

Fig.9-40   Hydraulic chain tensioner in an OHC engine

For reliability, replace overhead cam timing belts at regular intervals not exceeding 50,000 miles. Because these belts fail without warning and result in bent valves, it is best practice to replace these belts during any major engine repair. Also check water pumps and belt tensioners for bearing problems as they often cause timing belts to fail.

CHECKING CYLINDER HEAD CASTINGS

A good head gasket seal requires a flat, clean gasket surface on both the cylinder head and engine block. Overheating, gasket failure, or careless handling damages surfaces.

The first check is usually for flatness using a precision straight edge and feeler gauge (see Fig.9-41). A common limit for warpage is 004in (.10mm). That is, when a .004in (.10mm) feeler gauge passes under the straight edge, the head requires resurfacing. Be sure to check in several directions across the surface as cylinder heads warp and twist along their length.

Fig.9-41   Checking cylinder head flatness

Overhead cam cylinder heads that warp along their length cause camshafts to bind in their support bearings. Immediately on tear down, check for warpage through camshaft bearings (see Fig.9-42). When a straight edge will not go through camshaft bores or when bore diameters are "stepped" or have different diameters, use the camshaft for a straight edge (see Fig.9-43).

Fig.9-42   Using a round straight edge to check warpage through OHC bores

Fig.9-43 Checking for warpage by rocking the camshaft in its bores

Check also for damage caused by gasket failure. For this purpose, it is a good idea to keep the original head gaskets on hand and check them for evidence of leakage (see Fig.9-44). If leakage is apparent at the gasket, check the corresponding point on head and block surfaces. Depending upon the location of the leak, a "blown" head gasket sometimes permits combustion gases to burn or erode these surfaces (see Fig.9-45). Be sure to find the cause to prevent repeat failures.

Fig.9-44 Ahead gasket burned through by combustion gases

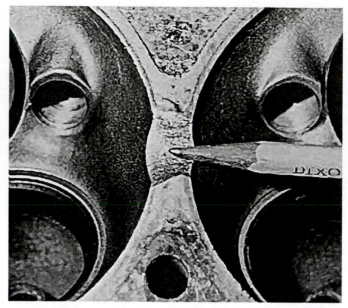

Fig.9-45 The burned surface of a cylinder head from head gasket failure

Corrosion damage around water circulation passages in aluminum cylinder heads is occasionally severe enough to prevent a head gasket seal. Sometimes corrosion damage extends across oil passages in the head and permits oil to enter the coolant and coolant to enter the oil (see Fig.9-46). Be sure to check visually for corrosion because such problems are often the reason for removing the head. If severe, resurfacing will not correct the condition without first welding the damaged areas (see Fig.9-47).

Fig.9-46 Severe corrosion on an aluminum cylinder head

Fig.9-47 Areas built up by welding prior to resurfacing

As mentioned, just careless handling damages surfaces. Look for nicks, dents, or scratches across gasket surfaces. Such defects cause gasket failures just as surely as severe operating conditions.

Crack detection is also very important but involves other technologies and specialized equipment. Chapter 11 covers crack detection and repair in detail.

SUMMARY

A thorough inspection of cylinder heads and small parts is necessary to determine the extent of repairs required and the small parts needing replacement. In newer overhead camshaft cylinder heads, this could be four valves, valve seats, springs, and valve guides per cylinder. There could be two camshafts, up to ten cam bearing bores, and cam followers to inspect. In addition to warpage along the deck surface, there could be alignment problems through cam bearing bores. Inspection of such cylinder heads is not a minor task.

In pushrod cylinder heads, there are fewer parts but the deck surfaces, valves, valve seats, and springs still require inspection. Because valve spring pressures are so much higher in pushrod engines, it is necessary to carefully inspect the camshaft, valve lifters, rocker arms and pushrods for wear.

Skill in the use of basic measuring tools is essential. Knowledge of potential problems is also required to perform visual inspections of castings and small parts.

Chapter 9
INSPECTING VALVE TRAIN COMPONENTS
Review Questions

1. Machinist A says that guides wear most in the center. Machinist B says that they wear most near the port end. Who is right?

    a. A only
    b. B only
    c. Both A and B
    d. Neither A or B

2. Machinist A says that measurements of valve rock equal stem-to-guide clearance. Machinist B says that valve rock exceeds stem-to-guide clearance. Who is right?

    a. A only
    b. B only
    c. Both A and B
    d. Neither A or B

3. Machinist A says that valve burning begins with valve warpage. Machinist B says that valve warpage begins with unequal cooling around the valve face. Who is right?

    a. A only
    b. B only
    c. Both A and B
    d. Neither A or B

4. The maximum allowable valve stem wear is _____ in.
    a. .0005-.0010
    b. .0010-.0015
    c. .0015-.0020
    d. .0020-.0025

5. Machinist A says inspect valves visually for burning. Machinist B says check keeper groove wear and look for signs of necking at the stems. Who is right?

    a. A only
    b. B only
    c. Both A and B
    d. Neither A or B

6. Approximately _____ percent of heat transfers through valve faces and seats.
    a. 100
    b. 75
    c. 50
    d. 25

7. Machinist A says that good valve springs have pressure within 10 percent of specifications. Machinist B says that good valve springs are square within 1/16in. Who is right?

    a. A only
    b. B only
    c. Both A and B
    d. Neither A or B

8. Machinist A says that flat tappet lifter bases are crowned and cam lobes tapered. Machinist B says that roller lifter cam lobes are not tapered. Who is right?

    a. A only
    b. B only
    c. Both A and B
    d. Neither A or B

9. Machinist A says that valve lifter non-rotation appears as a channel or groove worn into the lifter base. Machinist B says that non-rotation flattens cam lobes. Who is right?

    a. A only
    b. B only
    c. Both A and B
    d. Neither A or B

10. Malfunctioning hydraulic lifters are found by
    a. testing predetermined leakdown rates
    b. visually inspecting check valves
    c. measuring plunger body clearance
    d. visually inspecting metering valves

11. A camshaft is unacceptable for reuse when variations between exhaust lobes or intake lobes exceeds _____ in.
    a. 005
    b. .010
    c. .025
    d. .050

12. Predetermined leakdown is excessive when
    a. check valves malfunction
    b. valve stem heights are too short
    c. lifter bore clearances are high
    d. pushrods are too short

13. Machinist A says that to reuse cams and lifters in pushrod engines, install lifters in the same locations. Machinist B says that in overhead cam engines, keep cam followers in the same locations. Who is right?

    a. A only   c. Both A and B
    b. B only   d. Neither A or B

14. For valve grinds or overhauls, _____ stamped steel rocker arms.
    a. refaced
    b. reconditioned
    c. replaced as a set
    d. replaced as required

15. Inspect rocker arm studs for
    a. cuts on the sides and height above the head
    b. diameter
    c. taper
    d. diameter and height above the head

16. Machinist A says inspect pushrods for length, straightness, and end wear. Machinist B says check for clear oil passages. Who is right?

    a. A only   c. Both A and B
    b. B only   d. Neither A or B

17. Machinist A says that the maximum allowable timing chain slack in a pushrod engine is .006in. Machinist B says that maximum timing gear backlash is plus or minus 1/4in. Who is right?

    a. A only   c. Both A and B
    b. B only   d. Neither A or B

18. An indication of excess chain slack in an overhead cam engine is
    a. more than 1/2in slack
    b. noise at idle or start up
    c. noise at high speeds
    d. more than .006in slack

19. Using a straight edge and feeler gauges, the average limit for cylinder head flatness is
    a. .002in
    b. .004in
    c. .006in
    d. .002in per foot of length

20. Machinist A says that overhead cam warpage causes cam binding. Machinist B says that binding causes bearing wear, cam drive failure, and cam breakage. Who is right?

    a. A only   c. Both A and B
    b. B only   d. Neither A or B

FOR ADDITIONAL STUDY

1. What temperatures do exhaust valves and cylinder head castings normally see?

2. How are valve guides checked using the no-go-go method?

3. List three points of inspection for valves.

4. List two points of inspection for valve springs.

5. How is cam lift measured?

6. What visual inspections are made for cam wear?

7. How is pushrod valve lifter wear inspected?

8. How are bucket followers and bores checked for wear?

9. How is timing chain slack checked in pushrod engines? How is it done in OHC engines?

10. How is camshaft alignment checked in overhead camshaft cylinder heads?

Chapter 10

INSPECTING ENGINE BLOCK COMPONENTS

Upon completion of this chapter, you will be able to:

*   Locate points of maximum cylinder wear.
*   Visually inspect pistons for damage.
*   Explain how and where to measure piston diameter.
*   Explain how to measure piston clearance in new and worn cylinders.
*   Compare methods of checking full floating and press-fit piston pin clearance.
*   Describe how to check block deck flatness.
*   Describe how to detect main bearing alignment problems in disassembly.
*   List the inspection steps required for crankshafts.
*   Explain how to detect connecting rod housing bore stretch.

INTRODUCTION

Once the engine block is completely disassembled and cleaned, make a detailed inspection of all parts to decide upon their condition and serviceability. Determine from this inspection if parts are serviceable, what particular machining operations are needed, or if they must be replaced.

Engine rebuilding generally implies that the engine is to be restored to new specifications in terms of fits and clearances. This usually requires reboring cylinders, new pistons and rings, new bearings, new valve lifters, and a new or reground camshaft. Also required, depending on condition, possible regrinding of the crankshaft, reconditioning of connecting rods, block resurfacing, and line boring or honing.

An engine overhaul is approached differently. The measurements of engine parts decide their serviceability and machining or parts replacement is not done unless necessary. Keep in mind that engine parts worn beyond new specifications may still be within service limits. An overhaul is less expensive than rebuilding and an overhauled engine will operate satisfactorily, although not with the longevity of new or rebuilt engines.

Disassemble each engine with care and inspect it thoroughly for wear. An engine with little wear may be overhauled with less labor, less machining, and fewer parts. Consider rebuilding when engine wear is beyond service limits or to obtain maximum longevity. Of course, it is the customer's decision as to which way the service will go.

MEASURING CYLINDER WEAR

Cylinder wear is perhaps the most important consideration in determining if an engine is within service limits for overhaul. Cylinders are especially important because the expense of reboring and installing new oversized pistons is a major part of the rebuilding cost. At the same time, shops must consider that replacing piston rings in excessively worn cylinders shortens ring life and leads to customer dissatisfaction.

As mentioned regarding ridge reaming, cylinder wear occurs primarily in the top inch of piston ring travel. Maximum wear is usually found across the top of the cylinder, 90 degrees across the piston pin. This is because the piston oscillates around the piston pin in the presence of hot combustion gases. While coolant circulation patterns and head bolts distort cylinders and cause variations in wear, check first 90 degrees to the piston pin and just below the ring ridge (see Fig.10-1).

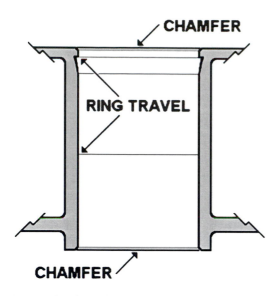

Fig.10-1　The location of maximum cylinder wear

The cylinder below the piston ring travel will not wear measurably. The difference between the measurements of cylinder diameter at the bottom and the most worn point in the top is cylinder "taper." Cylinder taper, and sometimes roundness, determines the serviceability of cylinders. A common method of measuring cylinder diameter is with inside and outside micrometers. First, set the inside micrometer to the cylinder size (see Fig.10-2). Second, find the diameter by measuring across the inside micrometer with an outside micrometer (see Fig.10-3). Read inside micrometers directly without an outside micrometer only if calibrated to the range of diameters being measured. This is usually not done, as it requires recalibrating inside micrometers with each change of extensions.

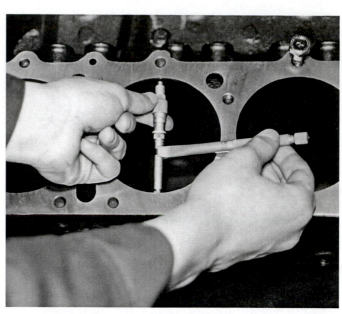

Fig.10-2 Measuring a cylinder with an inside mike

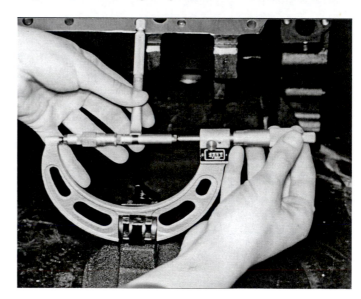

Fig.10-3　Transferring from an inside to an outside mike

A second method of measuring is with a dial bore gauge (see Fig.10-4). First, set the dial of the gauge to read the specified cylinder diameter using setting fixture or micrometer. The reading obtained in the cylinder is the difference between the cylinder diameter and the setting of the setting fixture or micrometer.

Fig.10-4 Measuring a cylinder with a dial bore gauge

When measuring cylinder wear, we are measuring diameters of out-of-round circles and therefore accuracy is poor. Actual wear typically exceeds measured wear. That is, it might be necessary to hone .005in (.13mm) from a worn cylinder to clean-up .003in (.08mm) measured wear. Specifically, smaller bores are less tolerant of wear than larger bores. Also, thin, low-tension rings are now commonplace and they are less tolerant of cylinder taper or out-of-roundness. In newer engines, cylinder out-of-roundness on the order of .0015in (.04mm), not taper, is a more suitable indicator of wear limits for small bore engines and perhaps slightly more with larger bores.

MEASURING PISTON CLEARANCE

Do not confuse piston clearance with cylinder taper. A cylinder with .005in (.13mm) wear does not necessarily have .005in (.13mm) piston-to-cylinder wall clearance. Piston clearance is the difference between the minimum cylinder diameter and the maximum piston diameter. As mentioned regarding cylinder wear, minimum cylinder diameter is below the ring travel. Generally, measurements of piston diameter are made midway between the piston pin

centerline and the tip of the skirt 90 degrees to the piston pin (see Fig.10-5). To be sure, check specifications for where to measure diameters.

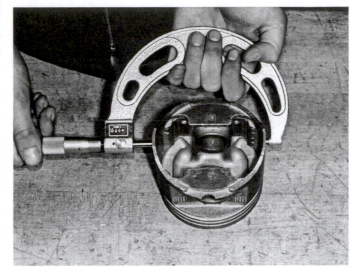

Fig.10-5 Measuring piston diameter between the pin centerline and tips of the skirts

While piston clearances are commonly between .001 to .003in (.02-05mm), engine and piston manufacturers make specific clearance recommendations for their products. It is essential that specifications be checked for service limits on piston-to-cylinder wall clearance. Excessive piston clearance causes noise, reduces piston ring life, and leads to piston failure.

CHECKING PISTONS

There are three basic checks for the serviceability of pistons. The first is for ring groove wear, especially on the top compression ring groove. The second is for worn or collapsed piston skirts. The third is for cracks in skirts or around piston pin bores.

Ring groove side clearance is critical for ring performance and longevity. No more than specified wear limits or double new clearances should be allowed. New clearances for passenger cars engines might be only .001 to .0015in (.03 to .04mm) allowing very little for wear. Check ring side clearance with a new or unworn ring and feeler gauges (see Fig.10-6). Wear generally occurs on the upper side of the top ring groove where exposure to combustion gases is greatest. Pistons with excessive ring side clearance are typically replaced.

Fig.10-6   Checking ring groove wear

Check pistons for wear or distortion at the skirts by measuring at two points on the piston skirts perpendicular to the piston pin (see Figs. 10-7 and 8). Many pistons have tapered skirts and diameters near the ends of worn skirts should be at least equal or possibly .001in (.03mm) greater than diameters near the ring lands. Do not measure across the ring lands because the piston is .020in or more (.50mm) undersize at that point.

Fig.10-7   Measuring piston diameter below the crown

Fig.10-8   Check for wear or distortion with a second measurement near tips of skirts

Should pistons exceed clearance limits by .002in (.05mm), consider expanding them by knurling. This keeps the engine quiet and permits normal ring life.

A new piston has a cam shape such that the diameter is greatest perpendicular to the piston pin (see Fig.10-9). When pistons overheat, or bind around pins because of too little pin clearance, they lose this cam shape. A piston that has lost this cam shape is "collapsed". Measurements taken 45 degrees to the piston pin are normally smaller than measurements taken 90 degrees to the piston pin. Collapsed pistons have a characteristic wear pattern and make noise called "piston slap" (see Fig.10-10).

Fig.10-9   Measuring piston cam diagonally across the pin

Fig.10-10 The wear pattern of a collapsed piston

CHECKING PISTON PIN CLEARANCES

For engine overhaul, it is necessary to decide whether clearances are within service limits, but checking some specifications presents problems. For example, in press fit assemblies, it is impractical to disassemble the piston from the rod. Efforts at disassembly and reassembly frequently result in distortion of the piston. Because of this, clearances are frequently not measured during overhaul. Clearances are checked by feel and obviously loose fits corrected by replacing worn piston pins or by fitting oversized piston pins.

Clearances for full floating piston pins are easier to check since pistons can be removed from connecting rods and diameters measured directly. Check clearances by comparing piston pin diameters to pin bore diameters in the piston and in the connecting rod. Specified limits for clearances are very small, approximately .0004in (.01mm) in connecting rods and .0002in (.005mm) in pistons. Because of the precision required, these measurements require special gauging equipment (see Fig.10-11).

Fig.10-11 Using a Sunnen AG300 gauge to measure pin bores

CHECKING CYLINDER BLOCK FLATNESS

Occasionally, the cylinder head gasket surfaces of an engine block warp. Check these surfaces for flatness to ensure good head gasket sealing, especially if the engine has a history of overheating or head gasket failure.

Check flatness with a straight edge and feeler gauges (see Fig.10-12). If cylinder heads are flat, a common limit for blocks is .004in (.10mm). That is, if a .004in (.10mm) thickness gauge fits between the straight edge and the block surface in any location, resurface the block. For this check, position the straight edge diagonally across corners of the block and then across the cylinders. Clean and deburr the block surface prior to gauging or the results will be inaccurate.

Fig.10-12   Checking block flatness

MEASURING MAIN BEARING BORES

It is not unusual to find main bearing housing bores in engines elongated or stretched beyond the limits of bore diameter specifications. The stretching occurs primarily in the main bearing caps and is caused by crankshaft loads during severe operation. The problem is more common in engine blocks used for heavy duty or performance engines.

The preferred method for measuring main bearing bore diameters is to use a dial bore gauge because of the close tolerance limits in the range of .0005 to .001in (.01 to .02mm) (see Fig.10-13). An inside micrometer or a telescoping gauge also may be used, but the accuracy of measurement by these methods is entirely dependent on the technician's skill. Clean and deburr the faces of the block and the bearing caps before torquing the bearing caps in place for measuring.

Failure to detect and correct discrepancies in bore diameters makes it impossible to maintain main bearing oil clearances within optimum or specified limits. Keep in mind that unequal clearances permit crankshafts to bend under load where clearance is excessive and breakage results.

Fig.10-13 Using a Sunnen dial bore gauge to check main bearing housing bores

Also check for misalignment along the main bearing bore centerline. If severe, misalignments are detectable with a straight edge and thickness gauge (see Fig.10-14). Another method of detecting this condition is to examine the wear patterns of the old main bearings. Look for signs of bearing bore or crankshaft misalignment in the bearing wear patterns (see Fig.10-15). A bent crankshaft will wipe the bearing surfaces all around the circumference. With a warped crankcase, bearings wipe along one side only.

Fig.10-14 Checking for warpage through main bearing housing bores

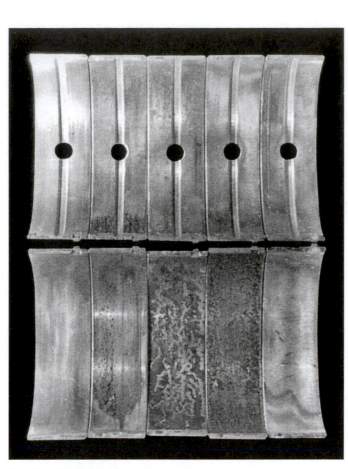

Fig.10-15 Main bearing wear patterns

There are other tests for fit and alignment of the crankshaft and crankcase used during engine assembly. These are covered in the chapter on assembly.

CHECKING THE CRANKSHAFT

Inspect the crankshaft for scoring, taper, or out-of-roundness. Scoring is a direct result of dirt in the oil. Journals and crankpins develop out-of-roundness because of high mileage and wear.

Evaluate scoring of crankpins or journals visually. Extreme cases present no particular problem in making a judgment. Cases of minor scoring require judgment and experience to decide whether to reuse or regrind the crankshaft. It is suggested here that if normal polishing does not bring the crankshaft into acceptable limits, regrind the crankshaft.

Check for taper with an outside micrometer by measuring diameters at each outside edge near

the fillet radius (see Fig.10-16). The difference in the diameters is the taper. While recommendations vary, taper exceeding .0005in (.01mm) allows oil to escape out the sides of bearings and bearing lubrication suffers.

Fig.10-16 Measuring journals near each fillet to check taper

Check for out-of-roundness with an outside micrometer by measuring the crankshaft diameter at three or more points around the circumference. The difference in these measurements is the amount of out-of-roundness. As with taper, use .0005in (.01mm) as the basic guideline for determining whether to regrind or not. Crankshaft diameters must be within specified limits but keep in mind that the crankshaft may have been ground before. If so, the diameters will be a precise .010, .020, or .030in (.25, .50, or .75mm) below specified size.

Check crankshaft straightness in V-blocks or in the engine block with only the end main bearings in place (see Figs. 10-17 and 18). The dial indicator should not read greater than half the main bearing oil clearance. Keep in mind that the total indicator reading (TIR) is a combination of roundness and straightness.

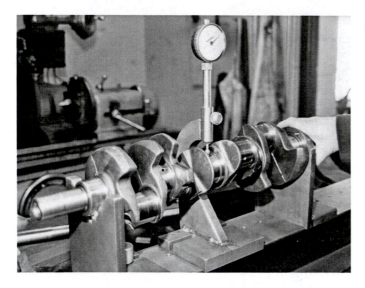

Fig.10-17 Checking a crankshaft for bend in V-blocks

Fig.10-18 Checking for crankshaft bend in the block with end bearings in place

Visually inspect for wear on the thrust face of the crankshaft as this has become increasingly common (see Fig.10-19). Some causes for wear have been around for some time including excessive thrust from overheated torque converters and inadequate oiling of bearing thrust faces. With today's engines, starting is instantaneous, sometimes before oil circulates to the bearing thrust faces. With manual transmissions, this problem is compounded by having to depress the clutch to crank the engine. Wear can be corrected by welding and grinding but it is essential that automatic transmissions cool properly and that main bearing designs assure adequate oiling to the thrust.

Fig.10-19 A worn crankshaft thrust face

A preliminary check for cracking can be made by holding the crankshaft suspended and "ringing" it by rapping the end of the crankshaft against the floor. If cracked, the shaft will not resonate or "ring" and magnaflux inspection is called for. All sprockets and keys must be removed for testing. If there were bearing, piston, or head gasket failures, test using magnaflux inspection. The next chapter covers magnaflux inspection procedures.

A crankshaft suitable for engine overhaul is straight, has a good surface finish, and taper or out-of-roundness does not exceed .0005in (.01mm). The diameter is within specified limits or an even amount undersize because of regrinding. Many engines, with reasonable care, have serviceable crankshafts after 100,000 miles or more.

MEASURING CONNECTING ROD BORES

The connecting rod housing bores of many large displacement, low speed, passenger car engines are lightweight relative to engine size. These rods work well so long as engine speeds remain low but stretch if run at high speeds. The cause of the stretching is the force acting on the connecting rod housing bores each time direction reverses at the top of intake strokes. Small displacement engines, because power is developed at higher engine speeds, are typically designed with relatively large housing bore ends and stretching is less frequent.

The amount of bore stretch varies, but it is not unusual to find bores as much as .002in (.05mm) out-of-round. Use care in measuring these bores. In service, they stretch perpendicular to the parting line, but when the rod cap is removed, they spring outward and measurements in the direction of load may be within specifications. The clue to stretching is the amount of out-roundness in the bore, not the diameter measured in the direction of load.

Special tools are available for checking connecting rod bores (see Fig.10-20). For reliable service, be sure that both out-of-roundness and diameter are within specified limits of tolerance. For example, if tolerance limits are 2.1000 to 2.1010in (53.34 to 53.37mm), diameters measured at any point should be within these. These bores can also be measured with an inside micrometer, but the accuracy of measurements depends upon the technician's skill.

Occasionally, one finds broken rods in the crankcase on tear down (see Fig.10-21). Failure often begins with excess bearing clearance caused by combinations of bore stretch or bearing and crankshaft wear. With excess clearance, the bearing shells pound against the shaft and deform. The deformation permits the bearing shells to wedge one beneath the other with the rotation of the shaft (see Fig.10-22). These "spun" bearings create a bursting force in the housing bore that breaks a rod bolt. The rod and cap then bend apart and are sometimes sheared off in the crankcase.

Fig.10-21 Broken connecting rods

Fig.10-22 Spun rod bearings in the rod housing

Fig.10-20 Gauging rod bores using a Sunnen
AG300 gauge

Considering how critical the rod bolts are, inspect visually for worn threads or other damage and replace them if in doubt. High strength alloy rod bolts are recommended for heavy duty and performance engines.

SUMMARY

In cases of normal wear, inspection is necessary to determine if short block components are within overhaul limits or if complete rebuilding is necessary. In cases of engine failure, inspection is necessary to determine the extent of damage.

Pistons and cylinders are first on the inspection list. Piston skirt wear, ring groove wear, and piston pin clearance must all be checked. Cylinder wear and piston clearance must also be measured. If any of these measurements are beyond service limits, rebuilding is likely necessary.

Connecting rod pin fits and housing bores also require measuring. There may be piston pin bushings that require replacement in full-floating assemblies or it may be necessary to check the amount of interference in press fit assemblies. Connecting rod housing bores are subject to stretching and may require resizing to restore diameters and roundness to specifications.

Crankshafts and crankcase alignments require visual inspection on teardown and measuring after cleaning. Especially when there has been poor maintenance or an oiling failure, crankshafts and bearings are subject to wear. Crankshafts are also subject to bending in cases of head gasket, piston, and connecting rod or bearing failures. Housing bores for main bearings also stretch or shift in alignment over time and can require align boring or honing.

Chapter 10

INSPECTING ENGINE BLOCK COMPONENTS

Review Questions

1. Machinist A says that head bolts and water circulation patterns effect the location of maximum cylinder wear. Machinist B says that in theory, maximum wear occurs 90 degrees to the piston pin at the top of the ring travel. Who is right?

   a. A only
   b. B only
   c. Both A and B
   d. Neither A or B

2. With high mileage, cylinder walls below the ring travel
   a. taper from wear
   b. wear oversize
   c. wear out of round
   d. remain unworn

3. Machinist A is checking cylinder wear with a dial bore gauge set to piston diameter and the gauge reads .004in at the top of the cylinder. He says that this is the clearance. Machinist B says that this is wear. Who is right?

   a. A only
   b. B only
   c. Both A and B
   d. Neither A or B

4. Piston clearance is the difference between
   a. maximum cylinder and minimum piston diameters
   b. minimum cylinder and maximum piston diameters
   c. piston crown and cylinder diameter
   d. piston skirt diameter and cylinder diameter below the ring ridge

5. Cylinder taper and roundness become more critical as
   a. diameters increase
   b. diameters decrease
   c. compression increases
   d. stroke lengths decrease

6. Machinist A says excessive ring groove wear causes ring wear and oil consumption. Machinist B says that ring groove wear should not exceed specified wear limits or double new clearance. Who is right?

   a. A only
   b. B only
   c. Both A and B
   d. Neither A or B

7. A piston skirt measuring largest in diameter 45 degrees across the piston pin centerline is
   a. collapsed
   b. expanded
   c. distorted
   d. slightly worn

8. Machinist A says that in overhauls, check pin clearance in press-fit piston and rod assemblies by disassembly and gauging. Machinist B says check for excess clearance in these assemblies by feel. Who is right?

   a. A only
   b. B only
   c. Both A and B
   d. Neither A or B

9. Machinist A says to check decks for burning from blown head gaskets and corrosion near cylinders or oil passages. Machinist B says that the limit for block deck flatness is .002in. Who is right?

   a. A only
   b. B only
   c. Both A and B
   d. Neither A or B

10. Machinist A says that stretched main caps cause unequal bearing clearance. Machinist B says that unequal main bearing clearance causes crankshaft breakage. Who is right?

    a. A only
    b. B only
    c. Both A and B
    d. Neither A or B

11. Measuring main bearing housing bore diameters detects
    a. stretching of main bearing caps
    b. block warpage
    c. wear from running a bent crankshaft
    d. damage from over-heated bearings

12. Machinist A says keep old main bearings in order during tear down to check for bent crankshafts. Machinist B says keep them in order to check for crankcase warpage. Who is right?

    a. A only   c. Both A and B
    b. B only   d. Neither A or B

13. To find taper in crankpins or journals, compare measurements taken at
    a. points near each fillet
    b. three points around the circumference
    c. journals at opposite ends of the shaft
    d. edges of worn bearings

14. Machinist A says that for overhauls, do not allow crankshaft taper and roundness to exceed .0005in. Machinist B says do not allow diameters to exceed specifications. Who is right?

    a. A only   c. Both A and B
    b. B only   d. Neither A or B

15. Machinist A says that out-of-round journals permit oil to escape the bearing area. Machinist B says tapered journals cause this. Who is right?

    a. A only   c. Both A and B
    b. B only   d. Neither A or B

16. Machinist A says check crankshaft straightness in the block by removing the center main bearing cap and placing an indicator on that journal. Machinist B says remove all caps and all but the end bearings for this check. Who is right?

    a. A only   c. Both A and B
    b. B only   d. Neither A or B

17. Maximum crankshaft TIR acceptable for assembly is
    a. .0005in
    b. .0010in
    c. .0015in
    d. half oil clearance

18. Machinist A says that in operation, rod bores stretch in the direction of load 90 degrees to the parting line of the rod and cap. Machinist B says that rods and caps spring apart when loosened and that housing bore out-of-roundness indicates stretching. Who is right?

    a. A only   c. Both A and B
    b. B only   d. Neither A or B

19. Rod bore stretch is caused by
    a. high speed operation
    b. defective forgings
    c. insufficient bearing clearances
    d. poor lubrication

20. Machinist A says that spun bearings break rod bolts. Machinist B says that the rod breaks first. Who is right?

    a. A only   c. Both A and B
    b. B only   d. Neither A or B

FOR ADDITIONAL STUDY

1. Compare the differences between an engine overhaul and an engine rebuild.

2. Where is cylinder wear likely to be greatest? Why do cylinders sometimes not wear uniformly or as expected?

3. List five points in piston inspection.

4. How is piston clearance measured in a worn cylinder?

5. How is crankcase distortion detected on teardown? What about a bent crankshaft?

6. Which crankshafts require magnaflux inspection?

7. Why do rod bores stretch?

Chapter 11

CRACK DETECTION AND REPAIR

Upon completion of this chapter, you will be able to:

- Describe procedures for dry-mag inspection.
- Describe procedures for wet-mag inspection.
- List difference in uses for wet and dry-may inspection.
- Describe the pressure testing procedure.
- List the advantages to pressure testing.
- Describe the dye penetrant test procedure.
- List advantages of dye penetrant testing.
- List the steps in repairing cracks using pins.
- Explain under what conditions to stop drill cracks.
- Explain the differences in welding aluminum and cast iron castings.
- Compare methods of sealing castings.

INTRODUCTION

Check for cracks before investing time in reconditioning cylinder heads, blocks, connecting rods, or crankshafts, especially where engine failures have occurred. For example, if an engine with a head gasket failure compressed coolant, inspect the cylinder, piston, rod, and crankshaft. The same is true for connecting rods and crankshafts when bearing failures occur. With cylinder heads, combustion chambers and the tops of pistons stripped of carbon deposits indicate that water entered the cylinder, possibly through cracks.

Cracked intake manifolds leak vacuum and potentially cause engine failures. Intake manifolds also crack across exhaust crossover passages and leak exhaust gas into the inlet flow or sometimes the crankcase. Exhaust manifolds crack but repair is seldom justified.

Flywheels often "heat check" or develop superficial cracks on friction surfaces and sometimes crack all the way through. Scrap flywheels showing cracks on the backside. Automatic transmission flywheel flexplates sometimes crack

around the crankshaft bolt circles and have to be replaced.

Cracked connecting rods or crankshafts are normally replaced. Only cracked cylinder heads and blocks are routinely repaired by proven procedures.

USING DRY MAGNETIC PARTICLE INSPECTION

Checking for cracks in iron castings and steel forgings requires magnetic particle inspection. To begin, clean the area to base metal, set up a magnetic field in the part and then dust the suspected area with magnetic powder. Interruptions in the magnetic field due to a crack cause magnetic lines of force to form in the crack. Look for magnetic powder collecting in these lines of force (see Figs. 11-1 and 2). Because the magnetic field is directional between the magnetic poles, it is necessary to test across the part in different directions

Fig.11-1   Cracks in a cylinder head made visible by magnetic particle testing

Fig.11-2 A crack in a cylinder block made visible by magnetic particle testing

Fig.11-3 Wetting a crankshaft with a solvent carrying the magnetic particles

Use dry magnetic particle testing for iron cylinder heads and engine blocks. Keep in mind however that to see cracks, they must be across visible parts of the casting. Common areas checked by this method include combustion chambers and ports, core holes, block surfaces, and main bearing webs. Internal cracks in castings cannot be detected by this method because they are not visible from the outside.

USING WET MAGNETIC PARTICLE INSPECTION

Use "wet magnaflux" methods to detect cracks in connecting rods and crankshafts. This process also uses magnetic particles but is more sensitive than dry magnaflux. The difference is that the magnetic particles are fluorescent and suspended in oil or water and applied wet by dipping or spraying parts (see Fig.11-3). As with dry mag inspection, clean parts to base metal before testing. To view possible cracks, parts are placed in a magnetic field under a black light (see Fig.11-4).

Fig.11-4 A cracked crankshaft viewed under a black light

As mentioned, this process is considerably more sensitive than dry testing as viewed in normal light and sometimes parts show "heat checking" from sudden heating and cooling cycles in machining. Most checking is superficial and does not necessarily lead to failure. It helps to clean, polish, and retest suspicious areas to eliminate false indications of cracks. Examples of unacceptable

cracks are those around rod bolt holes or in the fillet radii of crankshafts.

Because of the strength of the field set up in wet mag inspection, demagnetize parts after inspection. Consider the potential for problems should a magnetic crankshaft be installed in an engine and able to attract metallic particles into bearings. It is also true that magnetic parts are less resistant to stress.

USING DYE PENETRANTS

Testing by the dye penetrant method may be used for all materials, magnetic or nonmagnetic. Aluminum cylinder heads and blocks are nonmagnetic and require inspection by this process.

First, thoroughly degrease and remove carbon from parts and then apply penetrant by dipping, spraying, or brushing. Then wait a few minutes and allow the penetrant to enter any pores or cracks (see Fig.11-5). Wash off excess penetrant on the surface with a remover and wipe the casting dry (see Fig.11-6). Lastly, spray developer over the surface and wait for it to dry to a powdery film. The developer draws penetrant out of cracks and they appear as lines in the developer (see Fig.11-7). As in wet magnaflux methods, this process is more sensitive when used with fluorescent penetrant and a black light.

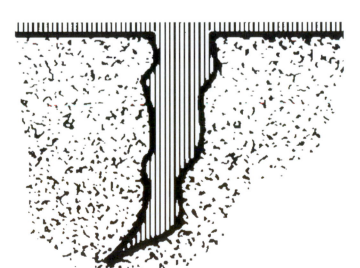

Fig.11-5   Dye penetrant applied to the surface

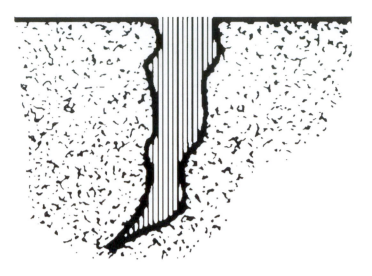

Fig.11-6   Dye penetrant removed from the surface

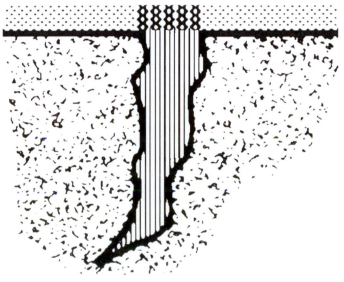

Fig.11-7   A crack showing through developer

PRESSURE TESTING CASTINGS

The most effective means of finding cracks in iron or aluminum cylinder heads and engine blocks is pressure testing. Testing requires blocking off water passages and pressurizing water jackets to three times cooling system pressure. To simplify testing, several manufacturers supply test plates to fit specific engine applications. These plates, with a rubber gasket, clamp or bolt to the face of the cylinder head using threaded rods, flat washers, and nuts (see Fig.11-8).

Fig.11-8 A cylinder head sealed for pressure testing

To inspect for leaks, spray surfaces with soap solution or immerse the head in water and watch for bubbles caused by air leaking through surface cracks (see Fig.11-9). This is the best way to locate cracks in oil passages, oil return holes, lifter bores or other areas not visible from the outside.

Fig.11-9 Checking for an air leak with the head submerged

USING CRACK REPAIR PINS

Threaded pins are used to repair cracks in cylinder heads and blocks. The first requirement is that the full length of the crack must be accessible with drills and taps used to make repairs. Iron castings are repaired with iron or steel pins and aluminum castings with aluminum pins. Repair requires drill-

ing, tapping, and installing pins along the length of the crack.

Before beginning any repair, it is necessary to locate exactly where cracks begin and end using magnetic particle testing or dye penetrants. Also, repairs must extend past the visible ends of the crack to protect against the possibility that the crack extends farther on the interior of the casting than on the exterior.

The prevalent system of crack repair is one made available by Lock-N-Stitch. This system of repair uses threaded steel pins for repairing iron castings and aluminum pins for aluminum castings. They are available in various sizes and lengths to match casting thicknesses (see Fig.11-10). These pins have straight threads with a unique design that seals using a combination of mechanisms. The threads have an interference fit and a thread design that clamps onto the casting walls as they pass through. There is also taper at the top end of the thread that gives them a third sealing mechanism (see Fig.11-11). As with other pins, they are over-lapped on installation.

Fig.11-10 A selection of Lock-N-Stitch L6 series pins. Lengths are selected to extend through the casting wall (all photos courtesy Lock-N-Stitch).

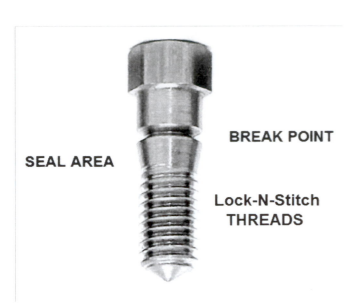

Fig.11-11 Details of a Lock-N-Stitch repair pin

Repairing Cracks Across a Valve Seat

The first example of crack repair with Lock-N-Stitch pins is a crack across a valve seat (see Fig.11-12). As with any crack repair, it begins with crack detection to be certain that the full length is visible.

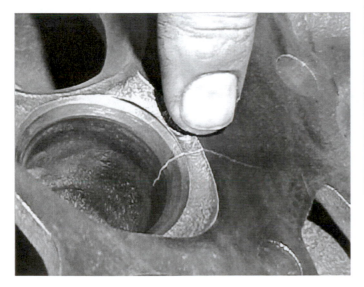

Fig.11-12 The crack extends across the valve seat and down into the port

Begin by drilling just past the lower end of the crack. Because it is necessary to drill deep into the port, an extended length drill and drill guide are used for the first few holes (see Fig. 11-13, 14 and 15).

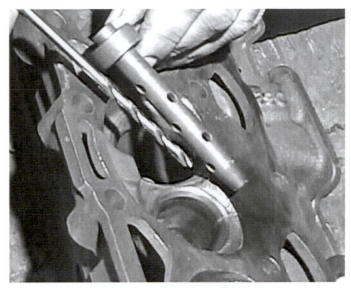

Fig.11-13 The extended length drill and drill guide

Fig.11-14 Drilling the first hole

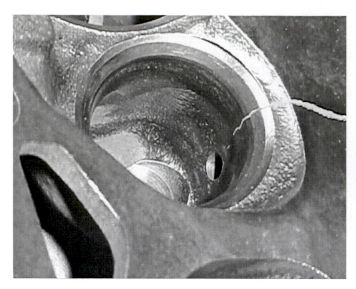

Fig.11-15 Note that first hole is past the visible end of the crack

After drilling, the hole is slightly counterbored to add sealing at the taper above the thread on the pin. This is advised here to protect against combustion pressure and extreme heat. The depth of the counterbore must be at least 1/3 of the tapered shoulder length. A stop is adjusted on the counterboring tool to set the depth of counterbore (see Fig.11-16).

Fig.11-16 The stop is set on the counterboring tool to the correct counterbore depth

After counterboring, the hole is tapped for a repair pin. A tapping attachment on the drive motor limits torque and reverses by pulling back on the drive. Tapping attachments and tapping fluids greatly reduce tap breakage (see Fig.11-17).

Fig.11-17 Tapping first hole

Before installing each repair pin, the hole is blown clear of chips and a sealer is applied to the threads. The first pin is then threaded into place and tightened until the head of the pin snaps off (see Fig. 11-18).

Fig.11-18 The first pin is in place and the head snapped off. The excess length is cleaned up later.

The following pins are installed is the same way until the pins extend up to, but not onto, the valve seat. The excess length on the lower pins was left in place to help position the drill for the next hole. The upper pins are partially cleaned up (see Fig.11-19).

Fig.11-19 The upper pins are trimmed with care to not touch the valve seat

As mentioned, repair at the valve seat was skipped until pins were started across the deck. The repair of the valve seat begins with hole centered on the valve seat and overlapping the pins on either side (see Fig.11-20).

Fig.11-20 The hole though the valve seat is drilled and counterbored.

After installing the pin at the valve seat, it is roughed and then carefully peened to blend the seat and repair pin into each other. This assures that the finished valve seat will fully clean up (see Fig.11-21). Should the seat be badly worn, a valve seat insert can be installed through the installed repair pins.

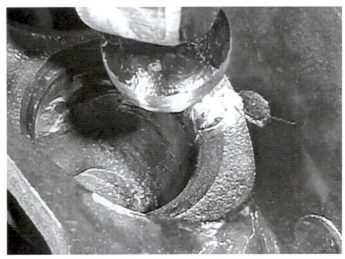

Fig.11-21 Peening the pin at the valve seat

The remaining pins continue across the deck. Each pin is ground down before installing the next pin. Much care is taken not touch the deck as contact would require extra surfacing in the final clean-up (see Figs.11-22 and 23).

Fig.11-22 Each pin is ground down before installing the next pin

Fig.11-23 All pins are peened after rough grinding and before surfacing

In the port. clean up begins by roughing with a carbide burr and then semi-finishing with a mounted stone (see Figs, 11-24 and 25).

Fig.11-24 Rough finishing in the port with a carbide burr

Fig.11-25 Semi finishing in the port with a mounted stone

A cosmetic touch up is done in the port with a peening attachment. Appearance displays craftsmanship and is important to the customer. Peening leaves a finish matching original cast surface (see Fig.11-26).

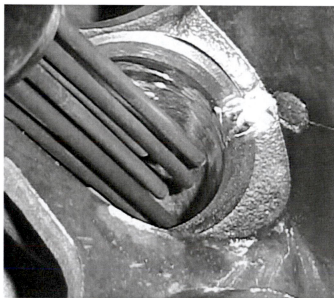

Fig.11-26 The peening attachment

Last is machining the valve seat. This could be done by three angle cutting or by seat grinding. Shown in Figures 27 and 28 are seat grinding and the finished seat.

Fig.11-27 Valve seat grinding over the repair pin

Fig.11-28 The finished seat shows only a color change at the repair pin

Repairing a Cracked Water Jacket

The next example uses Lock-N-Stich pins to repair a cracked water jacket. This is a common repair resulting from freezing coolant. As in the prior example, crack detection comes first to make visible the full length of the crack (see Fig.11-29).

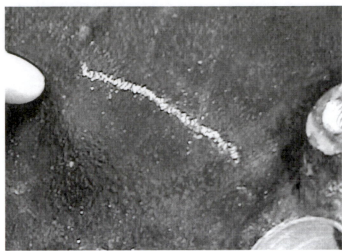

Fig.11-29 A cracked water jacket

To be sure that the repair gets the full length of the crack, repair begins by drilling the first hole just before the beginning of the crack (see Fig.11-30). These holes are not counterbored since sealing is against cooling system pressure, not combustion.

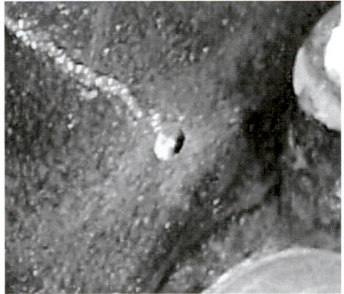

Fig. 11-30 The first is drilled ahead of the crack

Next, a drill jig is used to quickly drill alternate holes with the precise spacing. With the dowel pin inserted into the predrilled hole, the drill is run through the first position in the drill jig (see

Fig.11-31). This process is continued until alternate holes are drilled full length of the crack (see Fig.11-32).

Fig.11-31 Drilling the next hole through first position in the drill jig

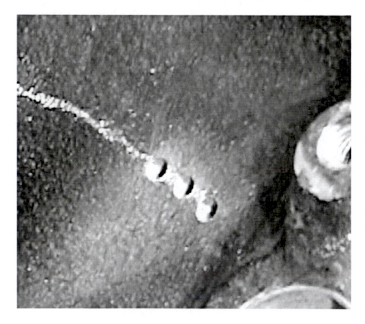

Fig.11-32 The drill jig is moved along until all alternate holes are drilled

After the alternate holes are drilled, they are tapped, blown clear, sealer applied and pins installed (see Fig.11-33).

Fig.11-33 Installing the plugs

The tops of pins are snapped off and the stubs ground down close to the casting surface taking care not to grind into the casting (see Fig.11-34).

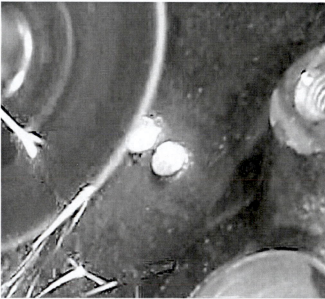

Fig.11-34 Using a small disc grinder to cut down the pins

Next is the second set of alternate pins. Begin by drilling between the first pair of pinned holes and insert the dowel pin on the drill jig into the hole. Continue drilling but this time use the second position in the drill jig (see Fig. 11-35).

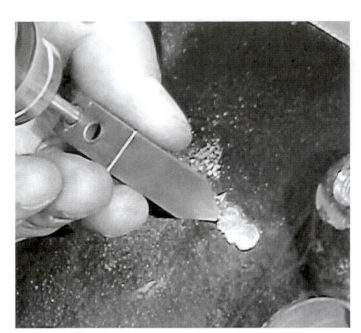

Fig.11-35 The second set of alternate holes uses the second position in the drill jig

Last, clean up the appearance of the repair. First by grinding pins down just flush with the casting and then peening to restore the as-cast finish (see Fig.11-36 and 37).

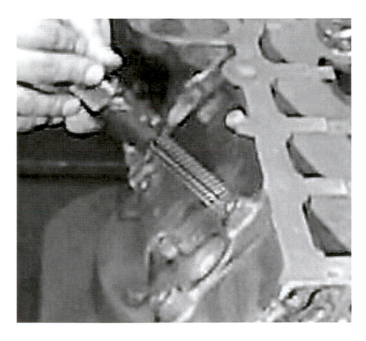

Fig.11-36 Peening over the repair area

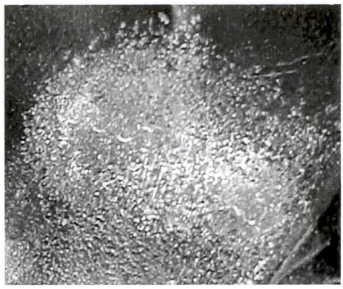

Fig.11-37 The finished repair

Of course, no crack repair is complete without pressure testing. Some shops also seal castings when porosity is found or following crack repair.

STOP DRILLING

Because some cracks across cylinder head and block decks do not cause problems, some are not repaired. These are cracks that do not cross oil passages, bolt holes, or seal surfaces. Still, it is best to take steps to stop the growth of such cracks. Stop the growth by simply drilling a small hole about 1/8in (3mm) in diameter just past each end of the crack. Cracks starting outward from water circulation holes are examples of cracks repaired by stop drilling (see Fig.11-38).

Fig.11-38 Stop drilling to prevent the extension of a crack

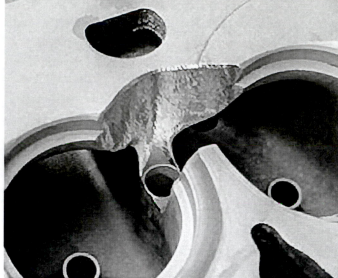

Fig.11-39 Aluminum gouged away in preparation for TIG welding

WELDING HEAD AND BLOCK CASTINGS

Cylinder head and block castings are sometimes repaired by welding but be sure to evaluate whether the investment in time is worthwhile. Welding aluminum cylinder heads is common. Welding iron castings is less common and less likely to be cost effective except for expensive or hard to replace parts.

The process used for welding aluminum is Tungsten Inert Gas, or "TIG" welding. Preparation requires degreasing and gouging out contaminated base metal with a rotary file (see Fig.11-39). The technique is much like oxyacetylene welding except that the heat source is an electric arc struck between the work and a non-consumable Tungsten electrode (see Fig.11-40). The welder feeds the aluminum filler metal into the puddle by hand and an inert gas such as argon or helium shields the molten metal.

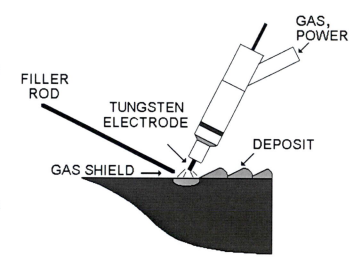

Fig.11-40 ATIG torch, tungsten electrode, and filler rod

It is helpful to preheat castings before welding. Concerns include distortion of the casting and porosity in the weld. If welding near a valve seat insert, remove the insert and fit another in its place after welding to protect against loosening from distortion. Porosity from gas and contaminants in the base metal of the casting show up occasionally and it is sometimes necessary to gouge out the first weld and make another.

Unlike aluminum, iron castings are subject to cracking during welding and during cooling after welding. One proven method of welding iron castings is oxyacetylene welding using an iron filler rod. The difficulty is that this process requires heating the casting to approximately 1300 degrees F, welding, and returning the casting to an oven or somehow controlling the cooling rate (see Fig.11-41). This is a painstaking process that not every shop can undertake and this work goes to rebuilders or welding shops experienced in these techniques.

Fig.11-41 Gas welding an iron casting while red hot (Courtesy Diesel Cast Welding)

Another method of repair for iron castings is flame spraying. Maintenance Welding Alloys provides a specially designed oxyacetylene torch and alloy metal powders applicable to automotive casting repair. With acetylene pressure set at 15 PSI and oxygen at 20 PSI, a mixture of nickel powder and flux is siphoned into the gas flow. In preparation, the damaged area is gouged out and the edges radiused (see Fig.11-42). Cylinder head and block castings are preheated to approximately 650 degrees F and manifolds to 350 degrees F. The immediate area is then heated to a dull red color before introducing the powder mix into the

gas flow (see Fig.11-43). The heat "wets" the surface and the flux cleans it so that the nickel fuses readily with the cast iron. Once cool, the buildup is machinable or can be cleaned up with a carbide burr (see Fig.11-44).

Fig.11-42 Iron gouged away for metal spray (Courtesy Maintenance Welding Alloys)

Fig.11-43 Powdered metal feeds from the bottle into the torch's gas flow

Fig.11-44 A flame spray repair on an iron head
(Courtesy Maintenance Welding Alloys)

Some simple repairs, if well removed from water jackets, may be arc welded using a flux coated nickel electrode. If water jackets or thin sections are too close, cracking on cooling will follow. Precautions include preheating the area of the weld and peening the weld on cooling to relieve stress. Damaged areas around pan rails or edges of cylinder heads under valve cover gaskets are areas sufficiently removed from water jackets to allow repair by arc welding (see Fig.11-45). While the weld itself is machinable, the fusion zone between the base metal and the filler metal is usually hard and the repair requires finishing by grinding.

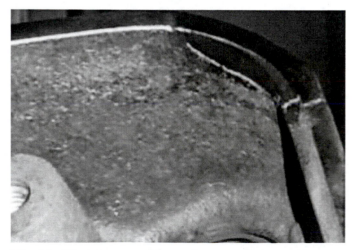

Fig.11-45 An arc weld repair on an iron casting

SEALING CASTINGS

There are several systems in use for sealing castings. Some are adaptable to occasional use in small shops and others are intended for the sealing of castings in production.

One common aftermarket system uses a ceramic sealer known as sodium silicate or "water glass." Ceramic sealer cures in air although the actual stimulus to curing is carbon dioxide. It has good high temperature performance but effectiveness is influenced by factors in the process. First, for best results, castings require thorough cleaning including stripping away rust and scale. Second, the sealer deteriorates if left sitting unused in circulating equipment, an important consideration because many shops only use this process only intermittently. With proper preparation of the casting and fresh solution, sealing is effective. As mentioned earlier, no crack repair or sealing job is complete without a final test.

While ceramic sealer can be added to an engine cooling system, many shops prefer sealing cylinder heads and blocks before assembly and in equipment where there is greater process control. Ceramic sealing in the shop often uses the pressure test equipment to seal the water jackets and a circulation system for the ceramic sealer solution. The circulator both heats the solution and pumps it through the casting under pressure for up to an hour (see Fig.11-46). Finally, the circulator inlet and outlet hoses are then closed, one hose disconnected and replaced by the pressure tester line, and the casting pressurized to approximately 55 PSI. The high pressure forces the ceramic sealer into pores or holes in the casting. Curing requires draining the casting allowing the sealer to cure in air for 24 hours. Ceramic sealing often follows other methods of crack repair but, in some situations, it is the only repair. For example, ceramic sealing works well for casting porosity in areas not subjected to combustion pressures.

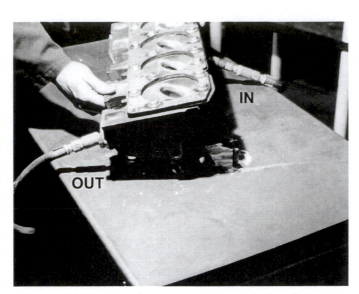

Fig.11-46 Connecting a ceramic seal circulator to a cylinder head

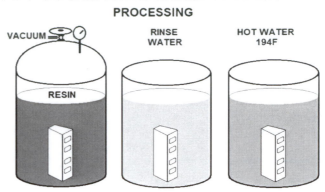

Fig.11-47 Vacuum sealing castings

Loctite has a hot water curing resin sealer of the type used in original equipment manufacturing. In "Wet Vacuum Impregnation," the casting is first placed into a resin filled tank and then vacuum is applied. Air evacuates from pores or holes in the castings under vacuum, and when positive pressure is applied to the tank, resin fills the pores and holes. Following impregnation, the casting is drained and washed in plain water. In LocTite's equipment, the casting is run in a centrifuge inside the impregnation tank to remove and recycle excess resin. The casting is then immersed in water at 194 degrees F for 10 minutes to cure the resin. Per Loctite, the sealer is non-hazardous and presents minimum problems in waste handling.

The place of Loctite or competitive processes and products in the automotive aftermarket naturally follows their use by original equipment manufacturers. Equipment similar in function to that illustrated in Figure 11-47 is available to the aftermarket from a various manufacturers. On the left side of this equipment, castings are immersed in resin and pressure reduced one atmosphere to evacuate pores. Following this, positive atmospheric pressure is applied forcing resin into the pores. The castings are then rinsed and placed in the hot water to the right to cure the resin.

SUMMARY

The need for crack detection occurs in cylinder heads primarily because of overheating. Cylinders sometimes crack as a result of head gasket or piston failures. Water jackets crack typically because of freezing.

Crankshafts crack for a variety of reasons. In heavy-duty applications, lugging could be the cause. In any engines, detonation or attempting to compress liquid in cylinders stresses the crankshaft beyond limits. Of course, liquid can only find its way into cylinders as a result of cracks in water jackets or head gasket failures.

Cracks in connecting rods are more rare but because of the potential engine damage should they fail they are also inspected. This is especially so for heavy-duty or performance engines.

Familiarity with a number of test methods is required. In the case of iron cylinder head or block castings, magnaflux or dye penetrant testing works only if cracks are in visible areas. For aluminum castings, magnaflux is not an option and dye penetrants must be used. To locate internal cracks or porosity in any casting, iron or aluminum, pressure testing is necessary.

Wet magnaflux testing is necessary for the inspection of connecting rods and crankshafts. Crankshafts and connecting rods that have been subjected to strains caused by head gasket, piston, or bearing failures should be inspected. If these components are removed from engines with normal wear only; the need for inspection becomes a matter of judgment.

Repair methods range from cylinder sleeving to pinning cracks to spray or torch welding to the sealing of castings. Automotive machinists must be familiar with all of these methods of repair.

Chapter 11

CRACK DETECTION AND REPAIR

Review Questions

1. Machinist A says that when failures occur, inspect affected components for cracking. Machinist B says that one indication of cracked heads is carbon removal from combustion chambers. Who is right?

   a. A only
   b. B only
   c. Both A and
   d. Neither A or B

2. Machinist A says that dry magnetic particle testing is required for connecting rods and crankshafts. Machinist B says that wet magnetic particle testing is required for iron heads and blocks. Who is right?

   a. A only
   b. B only
   c. Both A and B
   d. Neither A or B

3. Machinist A says that dye penetrants are limited to aluminum. Machinist B says that pressure testing is the only means of detecting cracks inside castings. Who is right?

   a. A only
   b. B only
   c. Both A and B
   d. Neither A or B

4. Machinist A says clean parts to base metal before magnetic particle testing. Machinist B says demagnetize crankshafts after testing. Who is right?

   a. A only
   b. B only
   c. Both A and B
   d. Neither A or B

5. A black light is required for viewing cracks when using
   a. dry mag inspection
   b. wet mag inspection
   c. wet mag or fluorescent dye penetrants
   d. pressure testing

6. Machinist A says that to use dye penetrants, clean and then apply penetrant. Machinist B says apply developer or cracks will not show. Who is right?

   a. A only
   b. B only
   c. Both A and B
   d. Neither A or B

7. Pressure testing has advantages over other methods of testing castings because it is good for
   a. all castings
   b. non-magnetic castings
   c. connecting rods
   d. crankshafts

8. The first requirement for crack repair by pinning is that the
   a. full length of crack be accessible with tools
   b. crack not exceed one inch in length
   c. crack not be in ports
   d. repair cost not exceed 50 percent of a replacement

9. Crack repair using Lock-N-Stitch repair pins begins by tap drilling
   a. at one end of the crack
   b. just past one end of the crack
   c. at both ends of the crack
   d. just past both ends of the crack

10. Machinist A says cut off the heads of Lock-N-Stitch repair pins after installing them. Machinist B says that the head of each repair pin snaps off when the torque limit is reached on installation. Who is right?

   a. A only
   b. B only
   c. Both A and B
   d. Neither A or B

11. Machinist A says that if a valve seat is in acceptable condition, a Lock-N-Stitch crack repair pin can be installed through the seat and the seat remachined. Machinist B says if a seat is badly worn, a seat insert can be installed through the repair pins. Who is right?

    a. A only
    b. B only
    c. Both A and B
    d. Neither A or B

12. A crack extends across a deck surface from a water passage but does not cross oil holes, head bolt holes, or the combustion chamber. Machinist A says that crack repair is not required. Machinist B says drill a small hole at the end of the crack to keep it from growing. Who is right?

    a. A only
    b. B only
    c. Both A and B
    d. Neither A or B

13. Machinist A says that in welding on iron, problems include shrinkage cracks and porosity. Machinist B says that aluminum is successfully TIG welded. Who is right?

    a. A only
    b. B only
    c. Both A and B
    d. Neither A or B

14. Machinist A says heat iron castings before and after welding. Machinist B says that iron castings are more successfully arc welded. Who is right?

    a. A only
    b. B only
    c. Both A and B
    d. Neither A or B

15. Machinist A says that ceramic sealing requires a circulator. Machinist B says run the sealer in the cooling system. Who is right?

    a. A only
    b. B only
    c. Both A and B
    d. Neither A or B

FOR ADDITIONAL STUDY

1. Where are cracks likely to occur in iron heads?

2. List the steps in dry magnaflux inspection.

3. List the steps in wet magnaflux inspection.

4. List the steps in dye penetrant inspection.

5. What requirements must be met for crack repair by pinning?

6. How can the progress of cracks be stopped? What requirements must be met for repair by this method?

7. How is a casting prepared for welding?

8. How are iron castings welded? How are aluminum castings welded?

Chapter 12

RECONDITIONING VALVE TRAIN COMPONENTS

Upon completion of this chapter, you will be able to:

- Compare methods for removing and replacing valve guides.
- Explain precautions in removing and replacing valve guides in aluminum heads.
- List the steps in replacing integral guides with valve guide bushings.
- List the steps in knurling valve guides.
- List the steps in refacing valves and valve stems.
- Describe under what conditions a valve seat must be replaced.
- Explain two possible sequences of steps for regrinding a valve seat.
- Explain the procedure for three-angle seat cutting.
- List the steps in removing aList the steps in replacing a valve seat.
- Describe the procedure for replacing integral valve seats.
- Describe the procedure for fitting positive valve seals.
- Explain how valve spring height is corrected.
- Explain how valve stem-height is corrected.
- Describe the process for straightening aluminum cylinder heads.
- Explain under what conditions a cylinder head must be line bored.

INTRODUCTION

Valve service operations include multiple variations in tools, equipment, and procedures. Covered here are some prevalent practices that prepare cylinders for longevity of service comparable to original equipment. At the same time, it must be accepted that customers' budgets sometimes dictate compromises.

REMOVING AND REPLACING VALVE GUIDES

Aluminum heads, and some iron heads, come with replaceable valve guides that are driven in and out without machining the cylinder head. Still, take the necessary precautions so that these repairs go as planned. For example, some guides are driven out in only one direction. Although this is usually from the valve port side, check manufacturers' service references first. Also measure the height of the guides above the spring seats and install replacement guides at the same height.

Occasionally, removing the valve guide "broaches" the hole oversize and this requires reaming for oversized valve guides. The problem is that these are not available for all applications. To avoid complications, heat these heads to 250 degrees F and squirt oil around the valve guides before driving them out or back in. Especially with aluminum heads, heat expands the castings, reduces interference, and allows oil to penetrate guide bores. Be especially careful to follow these procedures for iron guides in aluminum heads as they commonly gall and tear the bore on removal.

Another method of making removal easier is to tap threads half way through the valve guide and install a capscrew. It is then possible to remove the guide by driving against the capscrew with a punch and avoid driving against the shoulder of the guide (see Fig.12-1). Bronze guides can deform when drivers impact directly against the shoulder of the guide and they broach the holes oversize on removal. Using a capscrew greatly reduces problems with these guides.

Fig.12-1 Drive out guides against a capscrew threaded into the guide

Another method is to "core" drill the guide three-quarters of the way through leaving a shoulder to drive against (see Fig.12-2). Core drills with pilots, such as those used in guide and seat machines, are recommended to keep the drill centered in the guide.

Fig.12-2 A "core" or a step-drilled guide and driver

Observe precautions when installing new guides such as measuring guide outside diameters and bore inside diameters to be sure the fit is correct. Heat aluminum heads as recommended and lubricate both guides and bores to prevent galling on installation. Also install guides to the height specified or recorded on removal (see Fig.12-3).

Fig.12-3 Checking valve guide height above the spring seat

For heads with guides square to the deck, it is possible to use a press for installation and avoid driving them with a hammer. Other head configurations require press fixtures to adjust for the angle of the guides through the head.

An air hammer and driver work well but be sure to allow the guide to align itself with the bore before using full force (see Fig.12-4). Always ream guides after installation in case drivers deform the guides and to make sure that inside diameters are correct.

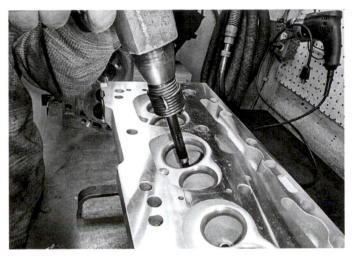

Fig.12-4 Using an air hammer and driver to remove guides

KNURLING VALVE GUIDES

Although many rebuilders have adopted other repairs in the search for maximum longevity, knurling guides is still common especially when repairs are limited to a valve grind or minor engine overhaul. Knurling is best suited to "integral" valve guides; those that are part of the cylinder head casting and non-removable. There are two basic methods of knurling, fixed arbor and roller knurling. Keep in mind however, that knurling by any method will not correct extreme wear. By using a wear limit of approximately .004in. (.10mm), original equipment longevity is possible.

Fig.12-5 Setting a telescoping gauge to guide "no-go" limits

Fixed arbor knurling calls for running a cold forming tap with a pilot, called a knurling arbor, through each valve guide (see Fig.12-6). The knurling arbor produces a thread in the guide by deforming the metal and forcing the crest of the thread inward, making the inside diameter of the guide smaller. It is important to note that the thread is "formed" and not cut as with a conventional tap. Turn the knurling arbor through valve guides at low speeds in a guide and seat machine or with a drill motor and speed reducer (see Fig.12-7). Always use a good extreme pressure lubricant for knurling.

Fig.12-6 A solid knurling arbor has a pilot but no cutting edges

Fig.12-7 A knurled valve guide

The knurling process reduces the inside diameter and reaming straightens guide bores tapered from wear. Reaming also provides the necessary valve stem-to-guide clearance. A valve guide reamer is special in that it has a pilot to keep the reamer in alignment with the guide (see Fig.12-8). The reamer is typically .001 to .002in (.03 to .05mm) greater than the stem diameter and produces a valve guide very nearly the same diameter as the original. A shallow thread or spiral groove remains in the guide and helps longevity by retaining oil and reducing the chance of sticking a valve.

Turn reamers clockwise only as reversing direction dulls the cutting edges. Cast iron requires no cutting oil or lubricant although a continuous flow extends tool life by flushing away chips. Bronze guides require cutting fluid for all operations.

The basic knurling arbor diameters for domestic passenger cars are 5/16, 11/32, and 3/8in. Some metric sizes are available, but because so many of these valve guides are removable, most shops replace them. Take care to select the correct size tool for the guide diameter. A 3/8in knurling arbor will start in an 11/32in valve guide, but it will break before completing the operation. Be sure to discard dull knurling arbors as they also readily break off in valve guides. Reamers are available in standard valve guide diameters for each application and in oversizes to provide specified clearance for oversized valve stems.

Because tools wear, measure the diameters of reamed valve guides before assembly. Failure to check clearance on assembly could result in insufficient clearance and a stuck valve.

Knurled valve guides are especially difficult to measure because of the internal grooves. For improved accuracy in measuring, use a valve guide bore gauge for checking the finished job. First set the gauge probe to the diameter of the valve stem and the indicator then reads the clearance in the valve guide (see Fig.12-9 and 10). The tool shown works especially well for knurled guides because the probe spans the internal grooves in the valve guide (see Fig.12-11).

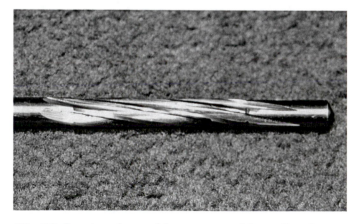

Fig.12-8 A valve guide reamer with a pilot for alignment

Fig.12-9 Setting the Sunnen P310 gauge to the stem diameter

Fig.12-10 Reading valve guide clearance on the bore gauge

Fig.12-11 Note the length of the Sunnen P310 gauge probe

The Sunnen Products Company manufactures a roller knurler and diamond hone system (see Fig.12-12). The knurler runs through the guide with a continuous flow of oil, and because it is a roller, it forms a thread with a minimum of metal fracturing (see Fig.12-13). After knurling, the guide is finished to size with a diamond hone (see Fig.12-14). Diamond hones have minimum wear and adjust to compensate and therefore obtain consistent sizes.

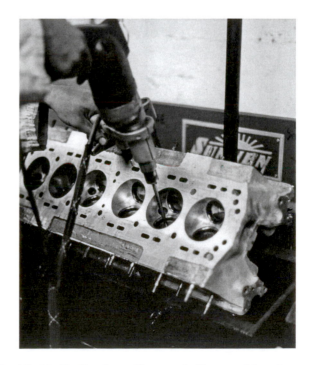

Fig.12-12 Roller knurling and diamond honing

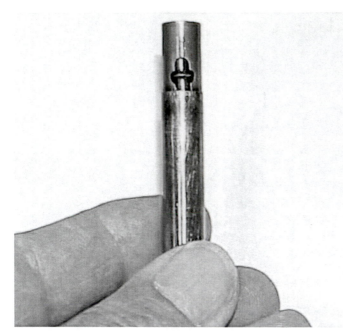

Fig.12-13 The Sunnen roller knurler

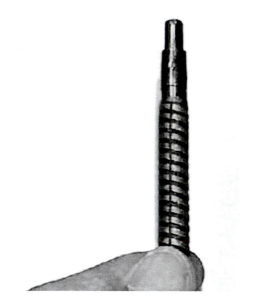

Fig.12-14 A close-up of the diamond hone

Be sure to scrub the valve guides clean after knurling and reaming or honing. Chips left in the grooves from machining are removed only by thoroughly scrubbing with a bore brush. Failure to clean valve guides causes stuck valves and rapid guide and stem wear.

FITTING OVERSIZED VALVE STEMS

For correcting wear in integral valve guides, most engine manufacturers call for reaming oversize and installing valves with oversize stems. Available sizes include .003, .005, .010, and .015in (.075 to .380mm) over standard. This repair works especially well when valves have stem-wear or heat damage and must be replaced anyway. Reamers are available for all required sizes and clearances.

Keep in mind that a valve guide reamer has a pilot to keep the reamer in alignment. Because of the pilot, each reamer is intended to remove only .003 to .005in (.07 to .13mm). When fitting stems beyond the first oversize, it is necessary to ream in incremental steps or use a combination drill and reamer called a "dreamer". Dreamers can finish guides .015in oversize in one pass.

Valve seals may have to be changed when using oversized valve stems. Most umbrella and positive seals are acceptable for oversizes up to .015in (.38mm). However, use solid Teflon seals for standard stem sizes only because they sometimes swell with heat and grip the valve stem too tightly.

REPLACING INTEGRAL VALVE GUIDES

For cylinder heads with integral valve guides, especially if wear is extreme, many shops prefer to install valve guide bushings called "false guides." One advantage is that installing false guides enable upgrading the original valve guide material from cast iron to bronze for added longevity.

While there is a variety of machinery in the field, in the guide and seat machine shown, correct guide alignment in the head requires precise alignment of the original guide to the centerline of the machine spindle. This is done by placing a precision level in the guide, taking readings in two directions, and tilting and swiveling the head to level it in both directions (see Fig.12-15). The head fixture then floats on air to center the pilot in the guide (see Fig.12-16).

Fig.12-15 Squaring the guide and spindle in guide and seat machine with a level

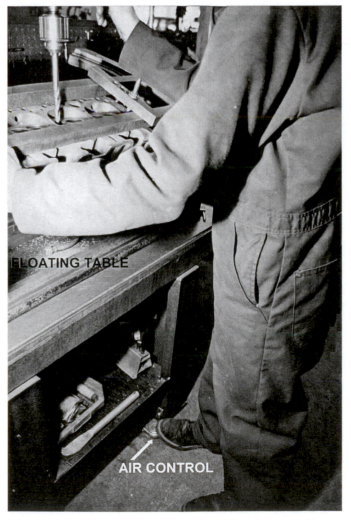

Fig.12-16 Centering the guide under the spindle

False guides are bushings similar to replaceable original equipment guides used in other cylinder heads. They are available in basic inside diameters; the most common for domestic passenger cars being 5/16, 11/32, and 3/8in inside diameters. Valve guides for 5/16in valve stems commonly have 7/16in outside diameters and the 11/32 and 3/8in guides commonly have 1/2in outside diameters. Larger outside diameters are available but not recommended given the thin walls in some castings. As mentioned, they are available in cast iron or bronze should the shop wish to upgrade materials.

The first step in machining requires core drilling the original guide approximately .025in (.64mm) under the outside diameter of the replacement guide (see Fig.12-17). Next, ream the drilled hole to final size, typically an exact size such as 7/16 or 1/2in (see Fig.12-18). The outside diameter of the false guide is approximately .0015in (.04mm) larger to provide an interference fit. Both core drills and reamers have pilots that center in the original valve guides (see Fig.12-19).

Fig.12-17 Core drilling through an integral guide

Fig.12-18 Reaming for a valve guide bushing or a false guide

Fig.12-19 Core drill (L) and reamer (R) with pilots

False valve guide bushings are generally driven into place with an air hammer and driver (see Fig.12-20). The outside diameter of the driver is slightly smaller than that of the false guide and the driver pilot centers in the false guide. Oil the guide and the bore to prevent galling on installation. Select guides that are the exact length needed or longer so that they can be cut to length after installation.

Fig.12-20 Driving in a false guide with an air hammer

Because false guides are often stocked in universal lengths to reduce inventory, an extra length of guide extends from one side of the head or the other after installation. Remove this extra length from the port side with a cutoff tool or, from the topside of the head, a seal cutter (see Fig.12-21 and 22). Pilots for the cutoff tool or seal cutter are selected to fit the inside diameter of the guides. Clean up guide bores by reaming after installation and cutoff steps.

Fig.12-21 Cutting a guide to length with a cutoff tool

Installs with auto-installer

Spiral groove for lubrication

One-piece design

Bullet nose for installtion

Fig.12-23 A K-line thin wall bronze liner

Fig.12-22 Using a seal cutter to cut a guide to length

BUSHING

REAMER

CENTERING CONE

PILOT

Fig.12-24 Guide Liner boring tool, bushing and centering cone

Another method of repair is the installation of K-line thin wall bronze liners (see Fig.12-23). These liners are made of a durable alloy that reduces wear rates significantly and are available from a number of suppliers. There are both production and non-production installation procedures. For non-production, as discussed here, the guide liner boring tool is one of the tools included in a portable kit Fig. 12-24).

The first step is to bore the guide .030in oversize using a high speed, coated reamer that runs through a bushing centered on the valve seat with a pilot centered in the guide (see Fig.12-25). Iron guides are machined dry while bronze lube is used on alloy guides. The guide is brushed clean after boring (Fig. 12-26) and the liner is placed on an installation tool and driven into place (see

Figs. 12-27). Any excess guide liner material that extrudes from the guide is trimmed using a guide top cutter (see Fig.12-28).

Fig.12-27 Installing the liner (Goodson)

Fig.12-25 Reaming the guide .030in over. Note the centering cone on the valve seat (Goodson)

Fig.12-28 Trimming the excess length with a guide top cutter (Goodson)

The liner is then simultaneously expanded and sized in the guide using one of two burnishing tools. One has burnishing tools fixed on an arbor (Fig. 12-29) and the other a selection of balls within in a size range that are driven through the guide (see Fig.12-30). Because bronze "springs back" when burnished, use burnishing tools approximately .001in larger than the desired finished size. One advantage to burnishing is that tool wear is minimal when compared to reamers.

Fig.12-26 Brushing the guide clean before installing the liner (Goodson)

Fig.12-29 The burnishing arbor (K-Line)

Fig.12-30 Burnishing with a ball and driver
(Goodson)

Following valve seat machining, the guide is given a crosshatch finish by running a brush type hone through the guide (see Fig.12-31). This step is left for last to preserve the finish. All required tools are air motor driven and use "Bronze-Lube" for all machining steps except through cast iron.

Fig.12-31 A flex-hone for valve guides (K-Line)

Precise valve guide size control is often required to meet engine specifications, and, in today's machine tools, for the proper fit of "live-pilot" seat cutting systems. Should it be necessary, the required precision can be easily obtained by honing with a variety of tools including the Sunnen single stroke diamond hone, portable Sunnen hone or Serdi hone (Figs.12-32, 33, 34).

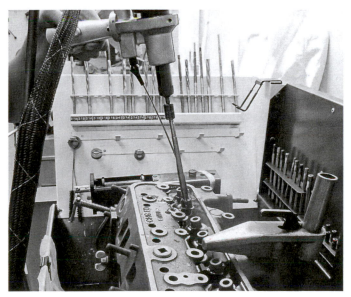

Fig.12-32 A Sunnen single stroke diamond hone

Fig.12-33 A Sunnen P190 portable hone (Goodson)

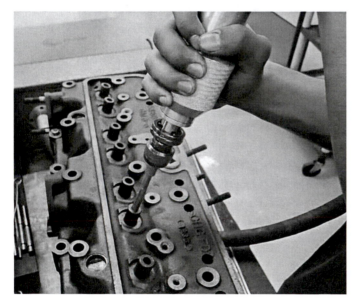

Fig.12-34 A Serdi guide hone

where there are pits. The surface finish quality must be as fine as possible to ensure full area contact on the valve seat. Dress grinding wheels frequently and direct clean oil between the grinding wheel and valve face to obtain good finishes.

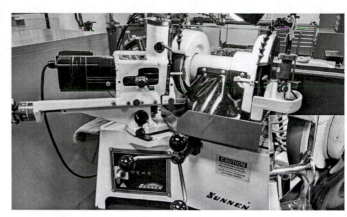

Fig.12-35 Refacing a valve in a Sunnen grinder

Although some valve seats are ground at 30 degrees, 45-degree seats are more common. Valves are typically refaced to provide a slight interference angle between the valves and seats (see Fig.12-36). The interference angle provides for a narrow line of contact when the valve first contacts the seat. Manufacturers typically specify 45 degrees for the valve face and 46 or 47 degrees for the valve seat providing a one or two-degree interference angle.

REFACING VALVES AND VALVE STEMS

Once it is decided that valves are serviceable, they are refaced in valve grinders (see Fig.12-35). Remove all pits in grinding because any pits that remain cause hot spots on the valve face and burning results. The burning occurs because heat cannot transfer from the valve through the seat at

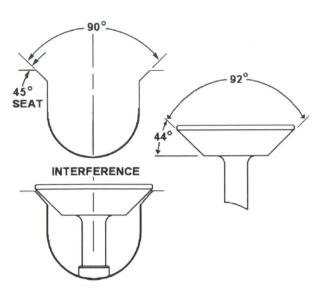

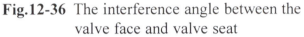

Fig.12-36 The interference angle between the valve face and valve seat

Many machinists use a 44-degree face angle and a 45-degree seat angle, except for 30-degree seats for which they use 29 and 30-degree angles. This practice provides for one-degree interference and the smaller face angle allows grinding valve faces with less thinning of the margins (see Fig.12-37). Grinding with a smaller face angle saves more valves and the slight variation from specifications does not reduce longevity or affect sealing.

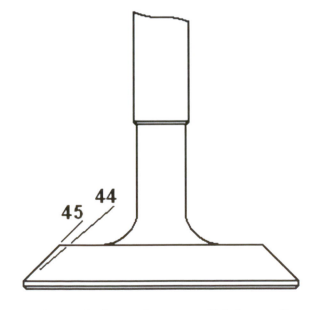

Fig.12-37 Refacing valves at a slightly smaller angle thins the margin less

While the practices above are common in regard to interference angles, there are cases where no interference angle is used. This is in the case of very hard valve seats and valve faces such as found in CNG and some other particular applications. In these cases, the valve and seat materials will not deform on break-in resulting in narrow seat contact areas and overheated valves. Check specifications for the particular application.

Warped or bent valves frequently show up during refacing. A warped valve cleans up on one side only (see Fig.12-38). Replace these valves and do not attempt grinding them to a cleanup.

Fig.12-38 A warped valve shows clean up on only one side of the valve face

Also clean up the tips of valve stems. First, chamfer the tips to a width of 1/32in (.7mm) or enough to keep sharp edges from forming after refacing stems (see Fig.12-39). Then reface stems at the back end of the valve grinder (see Fig.12-40). Remove .003 to .005in. (.1mm) from the tip of the valve stem to provide a smooth, flat surface for the rocker arm. Chamfering prevents sharp edges from shaving the rocker arm faces and helps prevent damage to seals on assembly.

12-13

Fig.12-39 Setting up to chamfer a valve stem in a Sunnen grinder

Fig.12-40 Setting up to reface a valve stem in a Sunnen grinder

The instructions for each valve grinder specify a sequence for refacing and chamfering steps. Be sure to check because the sequence affects results. For example, because some grinders use the chamfer on the stem for centering the valve in the chuck, chamfer stems before refacing the valve. Other grinders do not use this centering system and chamfering can be done before or after refacing.

GRINDING VALVE SEATS

It is essential that valve guides be in good condition before seat grinding because a pilot in the guide centers the seat grinding stones (see Fig.12-41). The pilot cannot properly locate in a badly worn valve guide. Incorrect location of the pilot causes the valve seat to grind off center or at angles to the valve guide. Valves cannot seal under these conditions.

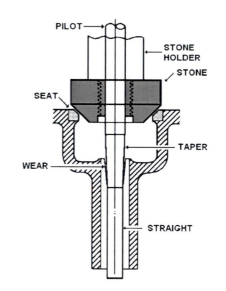

Fig.12-41 Positioning of a valve seat stone over a pilot in a valve guide (Sioux)

Check valve seat inserts at this time for visible cracks or looseness. If in doubt, check for looseness by placing a finger on the seat insert and rapping the insert lightly on the opposite side with a hammer (see Fig.12-42). A loose insert moves when hit and the motion can be felt through your finger on the opposite side. Replace loose or cracked seats (see the section on Installing Valve Seat Inserts).

Fig.12-42 Checking for a loose seat insert

Keep in mind that the object of valve seat grinding is to obtain a valve seat of the specified width, with uniform width all around, and is positioned correctly on the valve face. Valve seat width is critical. Too wide a valve seat traps carbon and causes burning and too narrow a valve seat does not transfer sufficient heat and causes burning. In two-valve per cylinder heads, seat widths range from .060 to .090in (1.5 to 2.0mm). In small engines or in three and four valve per cylinder heads, seat widths are closer to .040in (1.0mm). Seat widths must be equal all around as variations cause unequal cooling, distortion and warpage, and subsequent burning. A good valve seat contacts the valve slightly above center on the face.

Remember that while seats are ground at room temperature, valves operate at 1500 degrees F. The expansion of the valve at operating temperature causes seat contact to move closer to the center of the valve face. Valve seating too high on the valve face causes burning because the heat transfers through the thinnest part of the valve.

While not recommended, valve seats are sometimes lapped by placing abrasive compound between the valve face and the seat and turning the valve back and forth (see Fig.12-43). Because heat causes growth in valve diameter in running

engines, lapping is not generally recommended. While lapping provides a seat at room temperature, as the valve expands, it moves away from the lapped location and does not seal well. However, if lapping is necessary, the author recommends refacing an old valve and using it to lap the seat. In this way, the seat finish is improved without undercutting the face of the valve to be used in assembly. Of course, if lapping improves the seat finish, there is a machining problem that ought to be addressed directly.

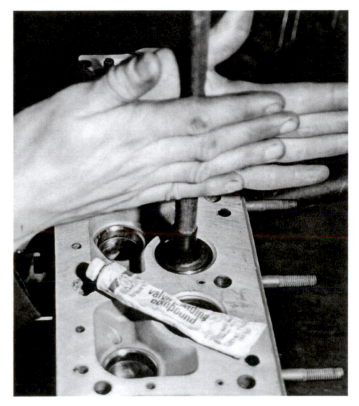

Fig.12-43 Lapping a valve seat

Stones are available in different diameters, angles, and compositions for roughing, finishing, and different seat materials. To prepare for seat grinding, select three stones with 30, 45, and 60 degrees angles. Be sure that these stones are approximately 1/8in (3mm) larger in diameter than the valve. If smaller, the 30-degree stone forms a groove near the corner of the valve that traps carbon and interferes with seating. A groove at this point also is not good for airflow across the seat.

Fixed diameter seat grinding pilots are available in common guide diameters and oversizes. Adjustable pilots are also available in common sizes and are particularly useful for slightly worn guides or odd sizes (see Fig.12-44). Select a pilot that fits snugly in the guide and press it firmly into place using the pilot wrench (see Fig.12-45). Be sure that the pilot does not move in the guide during grinding or the seat will not be round.

Fig.12-45 A solid pilot pressed firmly into the guide with a wrench

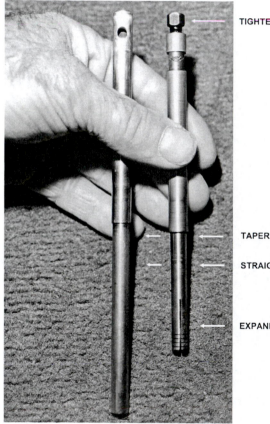

TIGHTEN TO EXPAND

TAPER

STRAIGHT

EXPAND HERE

Fig.12-44 Solid and adjustable seat grinding pilots

There are two basic approaches to valve seat grinding. In the first method, the primary seat angle, usually 45 degrees, first gets ground to clean base metal. Redress this grinding stone frequently to obtain good finishes and grind the minimum required to remove all pitting (see Fig.12-46). The high-speed drive motors also help obtain high quality valve seat finishes, especially important on the primary seat angle (see Fig.12- 47). As with valve facing, remove pits or burning results. Remember that grinding the 45-degree angle moves the valve deeper into the head causing changes in installed valve spring and stem heights. Second, grind a top angle of 30 degrees, or the angle specified if different. This smaller angle moves the outer edge of the valve seat toward the center of the valve face. Check the position of seat contact on the valve face and make corrections as needed with 30 or 45-degree stones before going on to the next step. Third, grind with a 60-degree stone, or the angle specified if different. This larger angle narrows the seat to specifications and makes the seat uniform in width.

Fig.12-46 Dressing a seat grinding stone

Fig.12-47 Seat grinding with a high speed drive motor

ified seat width. This method quickly restores the seat to its original width and position.

Of these two methods, the author recommends the first in cases where seat wear is significant. With wear, it often necessary to replace seats and, by grinding the 45-degree angle first, it is possible to measure installed stem height and decide whether to continue machining the seats or replace them.

Both methods of seat grinding call for checking the position of seat contact on the valve face after grinding 30 and 45-degree angles. Do this by applying Prussian blue paste to the valve face and tapping the valve lightly on the seat. Remove the valve and read the seat position on the valve face by looking at the point of contact for the upper edge of the seat (see Fig.12-48). Check the position of the upper edge of the seat on the valve face after grinding with the 30 and 45-degree stones and before going on to the sixty-degree angle.

Fig.12-48 Using Prussian blue paste to read the valve seat position on the valve face

In the second method of seat grinding, the top angle of 30 degrees gets ground first. Grind at 30 degrees until the old seat begins to narrow. Second, clean up the 45-degree primary angle. As before, check the position of seat contact on the valve face and make corrections as needed before going on to the next step. Third, grind the 60-degree angle until the seat narrows to the spec-

Do not attempt to read the seat width or the location of the lower edge of the seat on the valve face as the interference angle makes these readings inaccurate. Also, do not rotate the valve on the seat or the relationship of the seat and valve guide cannot be accurately checked; the contact pattern appears all around the valve face even with poor seat-to-guide concentricity.

A check for concentricity can also be made with a dial indicator. Read seat-to-guide concentricity by positioning the dial indicator over a valve seat pilot and rotating the indicator on the seat (see Fig.12-49). However, conditions other than concentricity enter into the total indicator reading. Should the reading exceed .002in (.05mm) TIR, concentricity is poor or the seat is tilted or out of round.

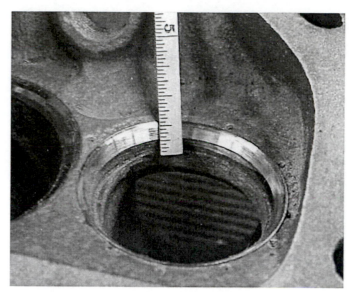

Fig.12-50 Using a scale to read valve seat width

Fig.12-49 Checking valve seat concentricity

For best accuracy, read seat width with a scale at the seat, not the valve face (see Fig.12-50). Remember, because of the interference angle, the pattern made on the valve face with Prussian blue paste does not accurately show seat width.

By either of the two approaches to grinding, three grinding angles assure that desired width, uniformity, and position are obtained (see Fig.12-51). These requirements are critical to longevity of valve service. Somehow the notion has become popular that "three angle valve jobs" are "racing valve jobs." In truth, it is the passenger car, not the racecar, which runs 100,000 miles or more and all manufacturers therefore specify three angle valve grinds.

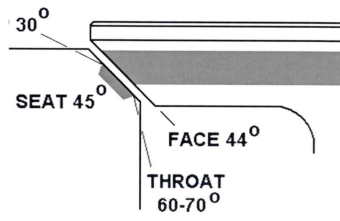

Fig.12-51 A three-angle valve seat necessary for proper seat to valve contact

CUTTING VALVE SEATS

Many shops use seat cutters, not grinding stones, to restore valve seats to new condition. There are two basic systems; one uses carbide cutters mounted in holders at fixed angles similar to grinding stones, and the other using carbide cutters to cut all three seat angles at one time.

In the first system, the carbide inserts adjust for diameter by sliding them up or down in their slots (see Fig.12-52). As in seat grinding, cut the 30-degree angle first, the 45-degree angle second, and the 60-degree angle only after checking seat positioning with Prussian blue paste. Cutting the 45-degree angle after the 30-degree angle allows for finishing the primary angle after it is narrowed. This reduces "chatter" and makes it easier to obtain good seat finishes. The cutter centers over a pilot, as with a grinding stone. Although seats may be cut by hand, the best cutting action is obtained with a low speed, high torque drive motor (see Fig.12-53).

Fig.12-53 Cutting a valve seat using a drive motor (Neway)

For improved precision and productivity, many shops have adopted three angle seat-cutting systems. The first step in set up requires placing the valve in a setting fixture and positioning a pointer to the proper position for seating on the valve face (see Fig.12-54). Second, remove the valve and place the cutter assembly in the setting fixture. Then adjust the carbide insert to the same position on the valve face (see Fig.12-55). Seat width and positioning are very precise when cut in this way.

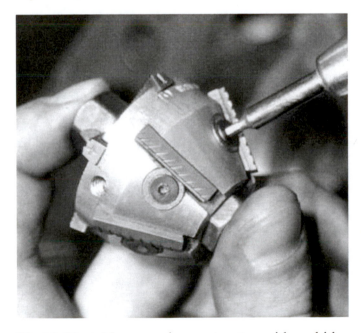
Fig.12-52 A Neway valve seat cutter with carbide inserts

Fig.12-54 Setting the pointer to the proper seat to valve contact position (Serdi)

12-19

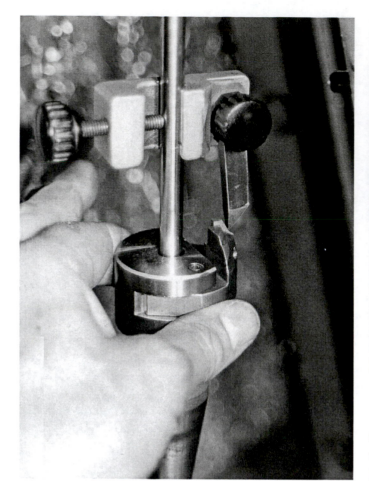

Fig.12-55 Adjusting the cutter to match the seat position (Serdi)

Fig.12-56 Form cutting three seat angles at the same time (Serdi)

While seat all the cutting tools are similar, there are important variations in equipment. In some, the pilot is fixed in the guide and the cutter body runs on the pilot. In others, the pilot is fixed in the cutter body and runs live in the valve guide. With live pilots, a selection of sizes is necessary to keep clearance in the guide to a minimum (approximately .0004in or .01mm).

Because of the torque required to form cut three angles at once, these tools require machines with rigid spindles and work holding systems (see Fig.12-56). If there are problems with surface finishes, check for tool wear or any looseness of the cylinder head in the set up. As a last resort, touch up seats with a grinding stone.

The industry is now adopting computer numerically controlled machine tools for production and performance machining of valve seats (see Fig.12-57). Machining in these machine tools is done with a single pointed tool following a programmed path beginning in the combustion chamber and continuing across the valve seat and down into port bowl (see Fig.12-58). The results are precise and absolutely consistent, a big advantage in performance engine building.

Fig.12-57 The Newen single pointed tool system for their CNC equipment

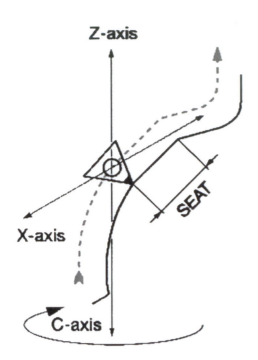

Fig.12-58 A programmed tool path in the Newen system

INSTALLING VALVE SEATS

Valve seat inserts are used to replace worn original inserts or to repair or upgrade integral seats. Replacement is also sometimes required when welding aluminum heads or pinning cracks in iron heads.

Aside from crack repair, the need to replace valve seats becomes apparent during seat grinding when it is necessary to grind deeply into a cylinder head to obtain a good seat. There are however, often no specifications to help decide at exactly what point a seat requires replacement. In the author's experience, if valve spring installed height cannot be corrected with a .060in (1.5mm) shim, install a valve seat insert. Use a new valve, or one with a thick margin, to measure installed height to obtain an accurate estimate of seat depth (see Fig.12-59).

Fig.12-59 Measuring valve spring installed height to determine valve seat recession (BHJ)

While shims correct the installed height of springs, keep in mind that valve stems also extend further above the spring seat. As mentioned in the engine disassembly section, excessive stem height interferes with rocker arm and cam follower operation and can hold valves open. Although refacing stems reduces the stem height, stock removal is limited to approximately .020in (.50mm). Seat inserts are sometimes required because this is not enough to restore rocker arm geometry or allow for lash adjustments in overhead cam engines.

Valve seat inserts are available as replacements for original inserts or for repairing cylinder heads with integral seats. Parts catalogs list inserts by application and by dimensions and materials. The dimensions used are outside diameter, inside diameter, and depth. The available materials include cast iron, high nickel-chrome alloys, and hard alloys or iron. Because the seat cutter diameters and the actual outside diameter of seat inserts are in catalogs, it is possible to calculate the interference fit by subtracting the cutter diameter from the actual seat diameter.

For integral seats, use cutters and seats as specified for the particular engine. To avoid cutting into water jackets, do not select seats according to cutters or seats on hand (see Fig.12-60). If there are no specifications available, use an insert the next size larger than the valve and limit the depth of the insert to a shallow depth of 7/32 or 1/4in. (6mm).

Fig.12-60 An oversize cutter can break into water jacket in thin castings

In addition to numerical specifications, catalogs also recommend seat materials that are compatible with valves. Valves and seats, even though heavy duty, do not hold up in service unless compatible. Use hard or alloy seats for gasoline engines and only hard seats for propane or CNG (compressed natural gas) engines. In the opinion of the author, alloy seats are more machinable and hold up well in gasoline engines and hardened seats should be limited to applications that absolutely require them.

The procedure for repairing a cylinder head with original seat inserts begins with removing the worn or damaged seats. If possible, remove the original seat without damaging the bore. Do this by welding the seat to an old valve of a slightly smaller diameter. After welding, use the valve as a driver for removing the seat (see Fig.12-61). Another method is to weld a bead inside the seat, wait for it to cool and shrink, and then pry it out

(see Fig.12-62). Note however that welding may not work on powdered metal seat inserts.

Fig.12-61 A small valve welded to a seat insert for a driver

Fig.12-62 Welding around the inside of a seat insert causes it to shrink

Other methods for removing inserts include pullers but they often do not fit small passenger car engines and they sometimes dig into castings below the inserts (see Fig.12-63). As an option, bore them to a thin shell using a cutter one size smaller than the insert and then pry the shell out of the bore. Remember that cutting hard seats dulls cutters.

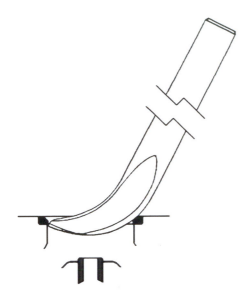

Fig.12-63 A pry bar used to remove seat inserts

Fig.12-64 The lower corner of a seat insert has a radius or chamfer

With the valve seat removed, inspect the bore for damage and measure the inside diameter. Compare the inside diameter of the bore to the outside diameter of the replacement seat. If the bore is clean, and the difference in measurements is within specifications, install an insert in the existing bore. It is sometimes better to use the original bore because boring oversize thins the water jacket. If the interference fit with a standard replacement seat is less than specified, oversizes of .005, .010, .015in are available. Check catalogs for availability.

Use care when installing seat inserts. Clean the bore and use lubricant to prevent galling on installation. The radius on the lower outside corner of the seat insert must face down on installation to prevent broaching the bore on the way in (see Fig.12-64). As with valve guides, it is helpful on aluminum heads to heat the cylinder head to reduce interference on installation. A driver, adapter, and hammer are used to install the seat. The adapter fits the driver and is selected to be slightly smaller than the insert. Leave the seat cutter pilot in the guide to center the driver and adapter and then drive the insert into place with a hammer (see Fig.12-65).

Fig.12-65 Installing a seat insert

Except for removal, the procedure for replacing integral seats with seat inserts is similar to replacing inserts. First, look up the specified seat insert and the required cutter. Second, select and install a cutter pilot that fits the valve guide (see Fig.12-66). Third, place the cutter over the pilot and engage the machine spindle to the cutter (see Fig.12-67). Fourth, set the depth stop to the thickness of the seat insert (see Fig.12-68). Lastly, set the spindle speed to the correct RPM and bore. Be sure to lubricate the cutter when machining aluminum to prevent galling.

Fig.12-66 Installing the pilot used to guide the seat cutter and driver

Fig.12-67 The pilot, cutter, and machine spindle

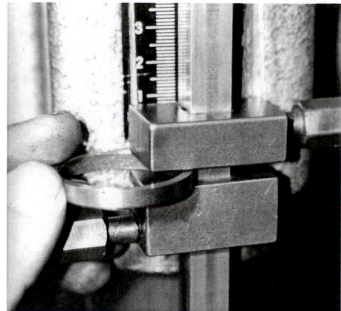

Fig.12-68 Setting the depth to match the seat insert

When both intake and exhaust seats are required, the inserts sometimes intersect. If this is the case, install a machinable insert in the intake seat first. This permits boring slightly into the intake seat for the exhaust seat insert. The reverse may not work since exhaust inserts are often harder and not as machinable.

Because the new seat insert is likely high in the cylinder head, begin valve seat grinding or cutting by machining the primary seat angle to the specified installed valve spring or stem height. For example, for a 45-degree seat, machine the primary angle to installed height, position the seat with a 30-degree angle, and then narrow the seat with a 60-degree angle.

FITTING VALVE SEALS

First, consider that poor valve guide sealing may be a result of excess valve stem-to-guide clearance and that excess clearance must be corrected before any valve seal can be expected to function properly. Otherwise, valve seals fail in service because of engine overheating or long-term exposure to high heat. This becomes apparent when umbrella type seals breakup or slide freely up and down valve stems instead of remaining fixed to the valve stem (see Fig.12-69).

Fig.12-69 A heat damaged valve stem seal

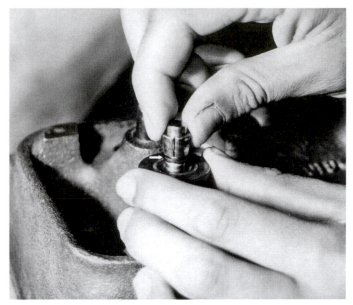

Fig.12-70 The correct assembly of an O-ring valve stem seal

Valve seals are also crushed during cylinder head assembly. Do not be surprised if defective seals are found in new or overhauled cylinder heads. The damage occurs when valve springs compress too far during assembly and the spring retainer crushes the seals against the top of the valve guide. To avoid this, adjust the stroke of the valve spring compressor to allow for the seal plus additional clearance to allow installation of valve keepers.

Another common problem is the incorrect assembly of the O-ring type of seal. These seals, when used, are installed in the lower of the two grooves on the valve stem (see Fig.12-70). First compress the valve spring and retainer until lower groove becomes accessible; install the seal, and then the valve keepers. Some of these same engines also use a steel oil shield over the spring and under the retainer to deflect oil from the guide (see Fig.12-71).

Fig.12-71 This steel shield sheds oil from around the valve guide

Umbrella type valve seals are sometimes upgraded by installing "positive seals." The seal is retained on the valve guide by a ring and the valve travels through the bushing (see Fig.12-72). Positive seals are sometimes the only seals that clear the inside diameters of valve springs.

Fig.12-72 A positive valve stem seal

For valve guides not machined for positive seals, a multipurpose cutter reduces the guide outside diameter, lowers the height of the guide above the spring seat, and chamfers the outside corners in one machining operation (see Fig.12-73). It is sometimes necessary to cut down the height of the guide to allow space for the seal between the guide and the underside of the retainer. Allow 3/16in (5mm) between the underside of the retainer and the top of the guide in the valve open position to prevent crushing the seal when compressing the valve spring for assembly (see Fig.12-74).

Fig.12-73 Cutting for positive valve seals

Fig.12-74 Allow clearance between the retainer and guide in the valve open position

Seal cutters and pilots are available in 1/32in (.77mm) increments. Different valve stem and valve guide outside diameters call for different combinations of cutters and pilots. Run the cutters with a 1/2in drill motor or in a valve guide and seat machine. Always clean and lubricate the guide and cutter pilot.

For assembly, install a plastic sleeve over the valve stem to prevent damaging the seal as it passes over the valve keeper grooves in the valve stem (see Fig.12-75). One sleeve only comes with each set of seals and it must be cut to length just below the keeper grooves. Unless cut to length, the sleeve sticks inside the first seal installed and pulling it free ruins it. Take care also to press seals closely against the valve stem as the seal slips over to the valve guide. Tools are available to press the seal over the valve guide (see Fig.12-76).

Fig.12-75 A plastic sleeve over the valve stem protects the seal from keeper grooves

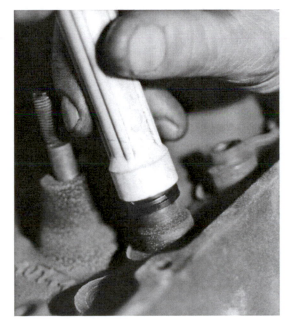

Fig.12-76 A tool used to press the seal over the guide

Some engines call for sealing both intake and exhaust valves and others provide only for sealing intake valves. Gasket manufacturers sometimes package only the number and type of seals used in original equipment although they often upgrade seal materials. Original equipment seals made of Nitrile, for example, do not survive past 250 degree F while Polyacrylate seals survive to 350 degrees F and Viton seals up to 440 degrees F. Many

machinists go ahead and machine both guides for positive seals and upgrade materials as well.

REPLACING ROCKER ARM STUDS

To prevent failure, replace rocker arm studs with cuts on the sides to prevent failure. Also replace rocker arm studs that are pulling out of the cylinder head because of cam action against rocker arms. There are stud removal tool kits that operate by placing a spacer over the stud and tightening a nut against the spacer but in the author's experience, these tools are satisfactory only for occasional repairs (see Fig.12-77). Much more efficient and durable are hydraulic pullers (see Fig.12-78).

Fig.12-77 Removing a rocker stud

Fig.12-78 Removing a rocker stud with a hydraulic puller

Replacement press-fit studs come in standard sizes and .003in or larger oversizes. The stud bore must be reamed for oversized studs (see Fig.12-79). After reaming, there is an interference fit of .001 to .002in (.04mm). Check sizes of old and new parts carefully as attempts at installing an oversized stud in a standard hole will crack the head casting.

Fig.12-80 A driver used to install studs to correct depth.

Threaded studs are available but these require tapping the hole to a thread size matching the lower end of the stud, commonly 7/16-14 (see Fig.12-81). Be careful to tap the threads square with the existing hole and install the studs using sealer because they sometimes go into the water jacket.

Fig.12-79 Reaming for the oversize rocker stud

Coat the replacement stud with anti-seize lubricant or sealer before driving it into the bore. Remember to install the stud at the correct height (see Fig.12-80).

Fig.12-81 Threaded rocker arm studs

CORRECTING INSTALLED SPRING HEIGHT

After refacing valves and grinding valve seats, valves sink into the cylinder head and the installed height of the valve spring extends further on the topside of the head. It is important to restore the valve spring installed height to specifications to maintain specified spring pressure.

Correct installed height by placing shims under the valve spring (see Fig.12-82). The installed height must first be measured to Figure out the thickness of shims required. A recommended procedure for selecting shims is to use a .060 or .030in shim (1.5 or .75mm), or no shim at all, whichever brings the installed height nearest to specifications. Following this practice typically places spring heights within plus or minus .015in (.04mm) of specifications (see Fig.12-84). Keep in mind that aluminum cylinder heads already have a steel shim under the spring to prevent the spring from working into the soft casting. Keep this original shim in place when measuring installed height.

Fig.12-83 Measuring installed spring height

Fig.12-82 The location of valve spring shims

Understand that the purpose of spring shims is to restore spring installed height to specifications and not to increase spring pressure. Do not shim springs below installed height because, besides raising spring pressure, it introduces the possibility of "coil bind." Coil bind occurs when coils bottom out against each other in the valve open position. This also overloads the valve train and bends pushrods or flattens cam lobes. With cam changes, make sure that springs compress .100in (2.5mm) past the valve open position to guard against binding and to allow for keeper installation.

CORRECTING INSTALLED STEM HEIGHT

Do not confuse valve stem installed height with spring installed height. Valve stem installed height extends from the spring seat to the tip of the valve stem and is critical on engines with nonadjustable rocker arms or cam followers. Height increases because of valve and seat grinding and can cause hydraulic valve lifters or lash compensators to hold the valves open.

As a recommended procedure, measure the stem height of the valves at each end of the cylinder head (see Fig.12-84). Correct the height on these two valves by facing the stems in the valve grinder. Then place a straightedge across the two end valves, and face the other valve stems until they align with the straightedge. Of course, stem height can be measured and corrected individually for each valve but this is much slower.

Fig.12-85 Comparing installed stem heights with a straightedge

Fig.12-84 Measuring installed stem height with a gauge

Keep in mind that specifications for this dimension are not always readily available. Recall the emphasis on measuring stem height during disassembly of the cylinder head. Keep a notebook of these dimensions to save time and trouble on each future job.

If the head is disassembled and the installed stem height unknown, at least make all installed stem heights equal. The fastest way of doing this to reface the valve stem at each end of the head as required to equalize heights. Then use the straightedge to check the heights of the remaining valves and make corrections just as before (see Fig.12-85).

It is most serious when no attention at all is given to installed stem height on those engines with non-adjustable rocker arms. This oversight often leads to those valves with tall stems remaining open. In such cases, identify the open valves and decide what can be done. It may be necessary to remove the cylinder and reface the valve stems. In some engines, it is possible to shim rocker arm assemblies, use shorter pushrods, or change over to adjustable rocker arms depending on engine design and parts availability. Ultimately, it is faster to follow correct procedures, take the time to measure, and correct stem heights before assembly.

Installed stem height is critical on overhead cam heads as well because lash adjustments may be impossible unless stem height is within specifications. For example, some overhead cam engines use shims of varying thickness between the camshaft and follower (see Fig.12-86). If the installed height is wrong, shims of the required thickness will be unavailable.

Fig.12-87 Refacing rocker arms in a Sioux grinder

Fig.12-86 Installed stem height is critical for OHC heads using shims for adjustment

REFACING ROCKER ARMS

Rocker arm wear on surfaces in contact with valve stems can sometimes be refaced easily and quickly in valve grinders. However, stamped steel rocker arms cannot be refaced and are discarded if worn excessively.

Check the clearance between the rocker arm and rocker shaft before going ahead with the refacing operation. Sometimes clearance is excessive (over .003in or .08mm) because of wear in both the rocker arms and the shaft. Reface rocker arms only if clearance is acceptable or can be restored to specifications by replacing the shaft. Since excessive clearance affects oil pressure, replace the entire assembly if the shaft only will not restore clearance.

If reconditioning rocker arm and shaft assemblies, first set up of the valve grinder and reface the rocker arms. In a Kwik-Way grinder, the grinding wheel moves back and forth with the rocker arm face held against it. A light pressure at the opposite end of the rocker arm keeps the rocker arm face in contact with the wheel. In a Sioux grinder, the wheel is stationary and the rocker arm moves on an arc against the wheel (see Fig.12-87). In either machine, grind only enough to clean up visible wear. If reusing rocker shafts, remove oil plugs, scrub passages clean, and polish them for reassembly.

STRAIGHTENING ALUMINUM HEADS

Anyone who services engines with aluminum cylinder heads experiences problems with cylinder heads warped as much as .040in (1mm). The cause of warpage is traceable to a combination of conditions, the first being the assembly of aluminum cylinder heads to iron engine blocks. When an engine overheats, the cylinder head expands in all directions, including total length, but it is restrained by cylinder head capscrews. Because it cannot expand along its length, the cylinder head buckles upward in the center.

A thorough inspection of a badly warped cylinder head shows a variety of conditions consistent with the preceding sequence of events. Most obvious is the warpage of the cylinder head upward in the center. The warpage upward occurs simply because the stresses in the casting have no other place to go. To verify this, check the surfaces on the topside of the cylinder head under the heads of the center capscrews. In cases of severe warpage, there are depressions at these points showing the upward force concentrated at the center of the cylinder head (see Fig.12-88).

Fig.12-88 Depressions under the center capscrews

Warpage on the topside of the cylinder head is especially troublesome for overhead cam engines. Warpage along the camshaft centerline, if not corrected, causes camshaft binding, breakage, scuffed bearing bores, and timing chain or belt failures. On tear down, check for binding in cam rotation (see Fig.12-89). It is also possible to check alignment with a straightedge or by rocking the camshaft in the bearing saddles. Most importantly, check alignment before beginning any other work on the cylinder head.

Fig.12-89 Check OHC heads for cam binding

Another subtle change is the alignment of valve seats to valve guides. A quick check of seat concentricity reveals misalignments of .010in TIR or more caused not by wear, but by shifts and tilting of the deck surfaces. Seat grinding, if done in this state, requires "burying" the valves beyond preferred limits.

Obviously, it is desirable to straighten these cylinder heads as a first step in reconditioning. Straightening reduces surfacing, improves seat-to-guide alignment, and corrects or improves camshaft alignment. In short, a cylinder head that at first glance appears hopeless becomes a routine overhaul. Various techniques have been and continue to be used to straighten cylinder heads. They vary from pressing head castings cold, to heating parts of the head castings with a torch, and combinations of these.

In the author's experience in process development, the most effective technique draws from conventional heat-treating; a process called stress relieving. Stress relieving does just what the name implies. Consider that new aluminum cylinder heads are finish machined and assembled to an engine block in perfect alignment. However, to emphasize a point made earlier, they are often bolted to iron cylinder blocks with half the expansion rate. If overheated, the aluminum heads cannot expand along their length because of the head bolts and instead bow upward in the center. This stresses the head internally and, if it is to be brought back into alignment with any confidence that it will stay that way, the stress must be relieved.

Ordinarily, such aluminum castings are stress relieved by "soaking" them at varying temperatures according to alloy and cross-sectional thickness. By experimentation, the author has found that no more than 500 degrees F and four hours are required. A cleaning oven, even a commercial baking oven, is adequate and specialized heat-treatment furnaces are not required. Also important is that while the process requires several hours, it is not labor intensive and two or more jobs may be batched together for cleaning and straightening simultaneously.

There are concerns about heating alloy aluminum heads to such temperatures because hardness can be lost. This occurs with high silicon alloys when silicon coalesces into nodules resulting in some loss of hardness. From experience, the degree of loss is however only a factor in severe duty applications, not our typical OHC passenger car cylinder head with greatly reduced valve spring pressures. Regarding concerns that valve seat inserts might fall out, in practice it was found that few did. Those that did fall out were the brass-bronze seats that many shops already have problems with at hot tank temperatures under 200 degrees F. It is the judgment of the author that should a valve seat insert fall out at 500 degrees F, it already had too little interference. In either case, such seats are best replaced as a part of cylinder head overhaul.

To clarify further the effects of heating cylinder heads to these ranges, calculations of thermal expansion are helpful. Calculating thermal expansion requires multiplying the diameter, or thickness, by the temperature change by the coefficient of expansion.

Seat Interference at 70 Degrees F:
    Seat insert diameter   1.506in
    Seat counterbore diameter   1.500in
    Interference   .006in

Seat Expansion at 500 Degrees F:
    1.506in x 430F x .000006 = .0039in

Seat Counterbore Expansion at 500 Degrees F:
    1.500in x 430F x .000012 = .0077in

Seat Interference at 500 Degrees F:
    1.506in + .0039in = 1.5099in
    1.500in + .0077in = 1.5077in
    Interference   .0022in

NOTES
    500 - 70 Degrees F = 430 Degrees F
    Coefficient for iron and steel = .000006
    Coefficient for aluminum   = .000012

Shown in the example, at least .002in (.05mm) interference is retained at 500 degrees F. The example uses average dimensions but exceptions will be few. At .002in (.05mm) interference, there is considerable allowance for variables, and it can be generally assured that properly fitted seats will remain in place through stress relieving.

Straightening cylinder heads requires minimum tooling. A flat base used to hold the cylinder head in alignment in the oven. Shown is an iron casting approximately 3x8x20in with 3/8-16 threaded rods inserted through the center head bolt positions (see Fig.12-90). In the author's experience, this fixture is long enough even for six cylinder heads. If the oven is large enough, and if the engine block is on hand, remove dowel pins from the deck and use the block for a fixture.

Fig.12-90 A base plate holds the cylinder head in alignment for straightening

To prepare a pushrod cylinder head for straightening, check the amount of warpage by holding a straightedge against the deck and measuring in the center with feeler gauges. Next, place shims of this thickness under each end of the cylinder head and tighten the center studs (see Fig.12-91). Do not torque to more than 25 ft/Pd to bend the head casting into alignment. While this may seem low, remember that the aluminum head expands more than the steel studs and clamping force increases as the temperature rises.

Fig.12-91 A pushrod cylinder head shimmed, tightened in place and ready for stress relieving

Fig.12-92 Measuring the amount of warpage on the topside

Preparation for overhead cam cylinder heads is different. For these heads, correct alignment in reference to the camshaft bores and not the deck surface. First, find the amount of warpage by holding a straightedge across the topside of the head and checking under one end with feeler gauges (see Fig.12-92). The valve cover surface is parallel with the cam bores and can also be used to measure warpage. Second, place shims one half this thickness under each end of the cylinder head and tighten the center studs (see Fig.12-93). Check alignment after tightening the head to the fixture and then readjust shims as needed. On twin cam 16-valve heads, watch for additional warpage on the exhaust side. It may be necessary to use extra shims on the exhaust side. Over straightening up to .002in in set up is acceptable. As mentioned earlier, tightening beyond 25 ft/Pd is unnecessary.

Fig.12-93 Shimming an OHC head for straightening

Then place the cylinder head in the oven and allow it to soak at temperature for four hours (see Fig.12-94). After this cycle, shut the oven off and allow the cylinder head to cool slowly in the oven. Slow cooling prevents inducing stresses from non-uniform cooling within the casting. With oven controls, place cylinder heads in the oven at closing time, set the timer to shut down in four hours, and remove cylinder heads the next morning. A second cycle through the oven may be required to straighten severely warped heads.

Fig.12-94 Placing the cylinder head and fixture in the oven

Most pushrod cylinder heads clean up with minimum surfacing after straightening, approximately .010in (.25mm). Because overhead cam heads are straightened in reference to the camshaft centerline and not the deck surface, slightly more surfacing is required but still less than .020in (.5mm). At .025in (.069mm), surfacing remains within most limits for overhead cam cylinder heads.

CORRECTING OVERHEAD CAMSHAFT CENTERLINES

For overhead cam heads with removable bearing supports, it is best to surface the topside of the head after straightening. By stripping the topside of bearing supports, dowels, and valve guides, and then surfacing under the bearing supports, the cam centerline is made perfectly straight. For some engines, 015in (.4mm) shims are placed under cam bearing supports on reassembly to help compensate for surfacing both sides of the cylinder head (see Fig.12-95). Thicker replacement head gaskets also help. Also remember that keeping the cam centerline elevated tightens the timing chain tight and keeps valve timing from retarding.

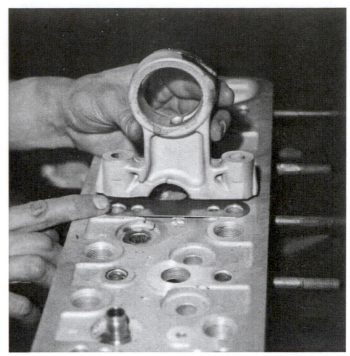

Fig.12-95 Shimming up cam bearing housings after surfacing the topside of the head

Some overhead cam heads have removable bearing caps with bearing saddles integral in the head casting and lend themselves to line boring. The bearing caps are cut down, the boring tool set to size, and bearing housings line bored to standard (see Figs. 12-96, 97, and 98). In extreme cases, bore saddles are welded and bored to standard diameters or oversize for retrofit bearings. Prepare the head by straightening the topside to within .004in (.10mm) and then cutting the caps .006in (.15mm). The scuffing common in the center saddles cleans up with minimal drop in the cam centerline. Also, because of straightening, the required surfacing is still typically within limits.

Fig.12-96 Milling bearing caps for line boring

Fig.12-97 Setting the boring tool to size

Fig.12-98 Line boring cam bores

Should there be major scuffing of integral cam bores, line-boring oversize is necessary (see Fig.12-99). Some manufacturers provide cams with oversize journals and others provide retrofit bearings for these heads. It is possible to bore heads for new cams with oversize journals or for bearings to fit standard or reground cam journals. Should bearings be unavailable, journals can be hard chromed and ground to oversizes. Straighten the head and line bore oversize, insert bearings, or build up cam journals according to parts availability.

Fig.12-99 A cylinder head with integral cam bores

If wear on journals and in cam bores is minor, clearance can be restored in heads with removable bearing caps without major work. Prepare by tightening the caps in place and polishing the bores. It also helps to polish the cam journals. Then place the cam in the head and check clearances with Plastigage. If clearance exceeds specifications, lightly lap each cap down until Plastigage readings indicate approximately .002in (.05mm) clearance (see Figs. 12-100 and 101). While engines with minor wear in this area may run fine, keeping clearance within limits assures good oil pressure, quiet operation, and eliminates another source of customer complaints.

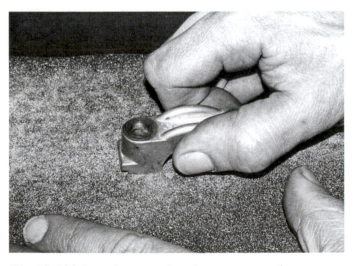

Fig.12-100 Lapping cam bearing caps to reduce clearance

Fig.12-101 Checking cam bearing clearance with Plastigage

Of course, all of these heads require additional checks before considering them finished. Pressure testing and new core plugs are recommended in every case.

REGRINDING CAMSHAFTS, LIFTERS, AND FOLLOWERS

Regrind or replace camshafts worn beyond usable limits. Before regrinding, inspect cam lobes, fuel pump eccentrics, and distributor drive gears and determine the extent of wear. If the value justifies the extra work, badly worn cam lobes can be built up by welding (see Fig.12-102). Fuel pump eccentrics can also be welded up and reground, but if the distributor drive gear shows wear, scrap the camshaft (see Fig.12-103).

Fig.12-102 A welded cam lobe

Fig.12-103 A worn distributor drive gear

Each cam lobe on a stock camshaft is reground to specified lift and duration (see Fig.12-104). To maintain specified lift, the difference in measurements across the nose of the cam and across the base circle diameter, must be maintained (see Fig.12-105). Grinding the camshaft reduces base circle diameters different amounts depending

on the extent of wear. Increasing lift or duration for performance camshafts requires reducing the base circle even further. For pushrod engines with non-adjustable rockers, or overhead cam engines with hydraulic lash compensators, changes in base circles are limited. If base circles are ground too small, there will be excess valve lash and noise. On pushrod engines, there is an approximate .002in (.05mm) taper on the lobes that also requires regrinding.

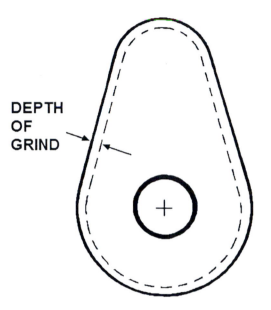

Fig.12-104 While lobes become smaller, lift and duration are unchanged

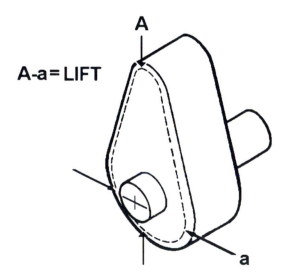

Fig.12-105 To maintain lift, dimension "A" minus "a" must remain unchanged

Preparation for grinding begins by checking and correcting camshaft straightness. With the camshaft in V-blocks or between centers, read the run out on a dial indicator (see Fig.12-106). Then position the camshaft with the low side down and hit it with a hammer and a chisel next to the journal (see Fig.12-108). This causes the camshaft to rise or exactly the reverse of what seems logical. Limit the TIR of the camshaft run out to less than .001in (.02mm) before grinding to prevent base circle run out and timing errors in assembly.

Camshaft grinding masters are then selected and set up in the grinder (see Fig.12-108). If the camshaft is "dual pattern," this means making one set-up for intake lobes and another grinding master and set-up for exhaust lobes.

Fig.12-108 Setting up cam grinding masters in a Berco grinder

For lobe taper, the wheel or table is swiveled to the required angle or the face of the grinding wheel dressed to the angle. The machinist first grinds those lobes with the same taper and then changes the angle to grind the remaining lobes (see Fig.11-109).

Fig.12-106 Checking cam alignment in V-blocks

Fig.12-109 Grinding cam lobes (Berco)

Fig.12-107 Straightening a cam with a punch and hammer

After grinding each lobe, the machinist indexes the camshaft to the next lobe position. To hold specifications, the machinist must know the angle between lobes, the amount and direction of taper, and tolerance limits for base circle diameters.

In the simplest possible terms, the rotation of the camshaft is geared to the rotation of the master cam. The master cam rocks the table and camshaft in and out against the face of the grinding wheel to produce the desired cam lobe profile.

Most grinders use wheel diameters in the 16 to 18in range (400-450mm) for grinding flat tappet camshafts. With some roller camshafts now in production, grinders must use smaller grinding wheel diameters as larger grinding wheels will not fit the contour of the lobes (see Fig.12-110).

Fig.12-111 Parkerizing or applying the scuff resistant coating

Fig.12-112 Polishing cam bearing journals

Fig.12-110 The inside radius of some roller cam lobes require small grinding wheels

Cam lobes are treated with a scuff-resistant coating following grinding (see Fig.12-111). The scuff-resistant properties of the surface treatment add to the service life of the camshaft and are especially important during break-in when wear patterns first develop. Because cam journals run in soft Babbitt or aluminum bearings, a highly polished surface finish is preferred (see Fig.12-112).

The crown on the base of some solid valve lifters is sometimes restored in a tappet grinder. The lifter rotates against the contoured face of a grinding wheel that reproduces the specified crown (see Fig.12-113). Regrinding bucket followers for overhead cam engines requires special set-ups such the one shown in a tool and cutter grinder (see Fig.12-114). Crown specifications vary according to lobe taper; lifter bore offset from the cam lobe, and lifter diameter. The arc of the crown is typically comparable to circles with diameters ranging between 40 and 100in with a 60in diameter being most common (1.0 to 1.5M). Hydraulic lifters are generally replaced and not reground.

Fig.12-113 Regrinding a valve lifter base

Fig.12-114 Regrinding a bucket follower in a tool and cutter grinder

SUMMARY

Before rebuilding a cylinder head, it should be inspected for cracks, corrosion, and broken fasteners or stripped threads. Aluminum heads should also be straightened if necessary. Unless these things are done, labor may be wasted on a head that is not serviceable.

Valve guides are repaired first. This means resizing by knurling, reaming for oversize valve stems, removing and replacing guides, drilling and reaming for "false" guides or installing thin-wall bronze liners.

Valves are cleaned, inspected, and then refaced. The valve stems must also be refaced and chamfered. At the assembly stage, it may also be necessary to return the valve grinder and again grind valve stems to restore them to the specified installed height.

Valve seats sometimes require replacement but are always remachined. Processes for machining seats include both grinding and cutting. With multi-valve combustion chambers, seat cutting with three angle cutters is clearly the way of the future.

Assembly methods are an important area of training. Specified stem-to-guide clearance and installed spring and stem heights must be maintained. Those assembling heads must be sure to lubricate guides, cam bearings, and cam lobes.

They must also protect and properly install valve seals.

Overhead cam cylinder heads require special skills in straightening castings and line boring camshaft bearing housings. Since not all shops are equipped to perform these operations, machinists must be alert to problems and set aside those heads that require such repairs.

Chapter 12

RECONDITIONING VALVE TRAIN COMPONENTS

Review Questions

1. Machinist A says remove guides by driving against a capscrew in the guide. Machinist B says to remove them by core drilling part way through and driving against the shoulder. Who is right?

   a. A only   c. Both A and B
   b. B only   d. Neither A or B

2. Machinist A says heat aluminum heads 250 degrees F for removing guides. Machinist B says to chill these heads to install guides. Who is right?

   a. A only   c. Both A and B
   b. B only   d. Neither A or B

3. Before removing valve guides, check the
   a. height of valve guides above the spring seat
   b. valve stem diameter
   c. valve guide diameter
   d. valve guide to stem clearance

4. A knurling arbor or roller knurler
   a. cuts threads
   b. forms threads
   c. finishes valve guides to size
   d. finishes valve guides to an oversize

5. A telescoping gauge set .006in over the stem diameter enters a valve guide. The guide is worn _____ in.
   a. less than .004
   b. more than .004
   c. more than .006
   d. less than .006

6. Machinist A says that knurling and reaming valve guides reduces their diameter from original specifications. Machinist B says that knurling and reaming straightens guides tapered from wear and restores diameters to specifications. Who is right?

   a. A only   c. Both A and B
   b. B only   d. Neither A or B

7. Machinist A says that engine manufacturers recommend correcting wear in integral guides by knurling or installing false guides. Machinist B says that they recommended oversize stems. Who is right?

   a. A only   c. Both A and B
   b. B only   d. Neither A or B

8. Machinist A says that integral guides are repaired with false valve guide bushings. Machinist B says that guide materials can be upgraded by installing false valve guide bushings. Who is right?

   a. A only   c. Both A and B
   b. B only   d. Neither A or B

9. Bronze liner installation requires boring guides _____ in. oversize.
   a. .003 or .005
   b. .010
   c. .015
   d. .030

10. Machinist A says that valve seat angles are ground one degree less than valve face angles to improve seating on break-in. Machinist B says that valve faces are ground one degree less than original so that refacing thins margins less. Who is right?

a. A only
b. B only
c. Both A and B
d. Neither A or B

11. The minimum margin acceptable after refacing valves is
a. 1/64in or one half of new
b. 1/32in or one half of new
c. 1/16in
d. 3/32in

12. Machinist A says that grinding a 45-degree valve seat with a 30-degree stone narrows the seat. Machinist B says that this moves the seat toward the valve margin. Who is right?

a. A only
b. B only
c. Both A and B
d. Neither A or B

13. Machinist A says that grinding a 60-degree angle on a 45-degree valve seat narrows the seat. Machinist B says that grinding this angle last makes the seat uniform in width. Who is right?

a. A only
b. B only
c. Both A and B
d. Neither A or B

14. Machinist A says that when seat grinding stones contact one side of seats, the seat is out-of-round. Machinist B says that concentricity is poor. Who is right?

a. A only
b. B only
c. Both A and B
d. Neither A or B

15. Machinist A says that seat indicators read concentricity. Machinist B says that these indicators read tilting or out-of-round conditions in the seat. Who is right?

a. A only
b. B only
c. Both A and B
d. Neither A or B

16. Machinist A says that seat cutters are used in the same sequence of 45, 30, and then 60 degrees. Machinist B says that the primary seat angle is cut after narrowing to reduce chatter and improve surface finishes. Who is right?

a. A only
b. B only
c. Both A and B
d. Neither A or B

17. Per recommendations, install a seat insert when
a. seats are burned
b. guides are worn
c. valve stems are oversize
d. installed height cannot be corrected with a .060in spring shim

18. Machinist A says remove hard seats by boring. Machinist B says to remove them by welding them to a smaller valve. Who is right?

a. A only
b. B only
c. Both A and B
d. Neither A or B

19. Welding around the inside of a seat insert causes it to
a. expand
b. shrink
c. break
d. distort and break

20. Machinist A says that CNG engines require hard seats. Machinist B says that alloy seats are satisfactory for gasoline engine exhaust valves. Who is right?

a. A only
b. B only
c. Both A and B
d. Neither A or B

21. Machinist A says that after inserting seats, first machine seat angles to specified installed spring or stem height. Machinist B says first grind seats to specified width and position on the valve faces. Who is right?

    a. A only   c. Both A and B
    b. B only   d. Neither A or B

22. Select inserts from catalogs by application so that
    a. outside diameters are not undersize
    b. inside diameters are not oversize
    c. depths are not shallow
    d. boring does not break into the water jacket

23. Machinist A says check that positive valve seals remain fixed to the valve stem. Machinist B says check that umbrella or oil shedder valve seals remain fixed to the valve guide. Who is right?

    a. A only   c. Both A and B
    b. B only   d. Neither A or B

24. Machinist A says adjust the stroke of valve spring compressors to protect valve seals. Machinist B says protect positive seals from keeper grooves on assembly. Who is right?

    a. A only   c. Both A and B
    b. B only   d. Neither A or B

25. Protect positive valve seals from damage during assembly by
    a. oiling valve guides
    b. oiling valve stems
    c. deburring stems
    d. using plastic sleeves over valve stems

26. Repair press fit rocker arm studs by removing old studs then
    a. press fitting standard size studs
    b. press fitting .001in oversize studs
    c. reaming and press fitting .001in oversize studs
    d. reaming and press fitting .003in oversize studs

27. Machinist A says measure installed spring height from the spring seat to the tip of the valve stem. Machinist B says measure installed stem height from the spring seat to the underside of the spring retainer. Who is right?

    a. A only   c. Both A and B
    b. B only   d. Neither A or B

28. Machinist A says that installed spring heights measuring .020in over specifications require .030in shims. Machinist B says that because the valve springs are being reused, a .060in shim is required. Who is right?

    a. A only   c. Both A and B
    b. B only   d. Neither A or B

29. Machinist A says that 3/16in clearance is required between the guide and underside of retainers in the valve open position. Machinist B says springs must compress .100in more than the valve lift. Who is right?

    a. A only   c. Both A and B
    b. B only   d. Neither A or B

30. Machinist A says that springs must compress beyond the valve open position to guard against coil bind. Machinist B says that extra compression is required to allow for keeper installation. Who is right?

    a. A only   c. Both A and B
    b. B only   d. Neither A or B

31. In non-adjustable valve trains, failure to reduce the installed stem height to specifications causes valves to
    a. remain open and burn
    b. remain closed
    c. wear valve guides
    d. make noise

32. Machinist A says that if valves remain open and lash is non-adjustable, grinding valve stems corrects the problem. Machinist B says that in pushrod engines, short pushrods solve the problem. Who is right?

   a. A only
   b. B only
   c. Both A and B
   d. Neither A or B

33. With shim-adjustable OHC followers, failure to reduce installed height of valve stems may
   a. not allow for thickness of available shims
   b. cause guide wear
   c. cause noise
   d. cause excess valve lash

34. For overhauls, worn stamped steel rocker arms are
   a. refaced
   b. reconditioned
   c. replaced as a set
   d. replaced as required

35. For straightening, clamp OHC heads in the center and shim at each end
   a. until the cam centerline is straight
   b. until the deck is straight
   c. an amount equal double the warpage
   d. an amount equal to the warpage

36. Machinist A says that straightening heads helps restore guide to seat alignment. Machinist B says that straightening keeps combustion chamber volumes equal. Who is right?

   a. A only
   b. B only
   c. Both A and B
   d. Neither A or B

37. Straightening at no more than 500 degrees F is recommended because beyond this temperature
   a. press fit seats and guides loosen
   b. strength is lost
   c. castings anneal
   d. aluminum melts

38. Machinist A says that, prior to straightening, remove worn guides, core plugs, and broken studs or screws. Machinist B says to complete any welding beforehand. Who is right?

   a. A only
   b. B only
   c. Both A and B
   d. Neither A or B

39. Raising the cam by shimming under removable cam bearing supports
   a. retards valve timing
   b. advances valve timing
   c. tightens chain slack
   d. advances valve timing and tightens chain slack

40. Machinist A says that OHC heads with scuffed cam bores require line boring. Machinist B says that worn integral cam bores require boring for bearings or oversize cam journals. Who is right?

   a. A only
   b. B only
   c. Both A and B
   d. Neither A or B

41. No economic repair is available when excessive wear is found on camshaft
   a. lobes
   b. fuel pump eccentric
   c. bearing journals
   d. distributor drive gears

42. Machinist A says that regrinding a bent camshaft causes base circle run-out. Machinists B says that straightening the camshaft after grinding corrects base circle run-out. Who is right?

   a. A only
   b. B only
   c. Both A and B
   d. Neither A or B

43. Machinist A says that flat tappet camshafts for pushrod engines are ground without lobe taper. Machinist B says that overhead camshafts with rocker followers require lobe taper. Who is right?

    a. A only   c. Both A and B
    b. B only   d. Neither A or B

44. Valves lifters are ground
    a. flat
    b. with a 60in radius crown
    c. with a 150in radius dish
    d. with a crown .0002in high

45. Machinist A says that for nonadjustable valve trains, the allowable base circle reduction is limited. Machinist B says that the allowable variation in these base circles is limited. Who is right?

    a. A only   c. Both A and B
    b. B only   d. Neither A or B

FOR ADDITIONAL STUDY

1. Compare methods for removing valve guides from iron and aluminum heads.

2. What limits the knurling of valve guides?

3. Describe three methods of repair for integral valve guides.

4. List in sequence the steps in grinding a valve seat.

5. Describe two methods of removing valve seats.

6. How are valve seals protected during assembly?

7. What are the consequences of not correcting installed valve stem height?

8. What are your options in correcting installed valve spring height?

9. Which rocker arms are refaced? Which ones are not?

10. List in sequence the steps in straightening an aluminum cylinder head.

11. How is cam bearing clearance in an overhead cam cylinder head corrected?

12. How is cam bearing alignment corrected in an overhead cam cylinder head?

13. When regrinding a camshaft to original specifications, how is lift restored?

14. Which cams require lobe taper?

Chapter 13

RECONDITIONING ENGINE BLOCK COMPONENTS

Upon completion of this chapter, you will be able to:

- Describe methods of glaze breaking for purposes of installing new piston rings.
- List the precautions and requirements in reboring cylinders.
- Describe the cylinder honing process for new or oversize cylinders.
- Explain cylinder surface finish requirements for different piston ring materials.
- Explain the cylinder sleeving procedure.
- Compare line-boring and line-honing procedures.
- List the steps in preparing press-fit pistons and connecting rods for assembly.
- Describe the process for replacing and fitting full floating piston pin bushings in connecting rods.
- Explain the process for resizing connecting rod housing bores.
- Calculate bearing clearances, housing bore diameters, and crankshaft diameters.
- Describe the assembly process for press-fit piston and connecting rod assemblies.
- Describe the assembly process for full floating piston and connecting rod assemblies.
- Describe procedures for aligning pistons and connecting rod assemblies.
- List the standards and requirements to be met in regrinding a crankshaft.
- Describe the camshaft grinding process.
- List the requirements to be met in regrinding a flywheel.
- List the requirements to be met in regrinding a stepped flywheel.
- Describe the procedure for replacing a starter ring gear.
- List the steps in overhauling an oil pump.

INTRODUCTION

The selecting of engine repair procedures follows parts inspection and depends upon the extent of wear, damage, parts availability, and equipment or abilities of shop personnel. Selected for discussion here, are some of the more prevalent procedures and practices used in the engine building industry.

HONING CYLINDERS FOR OVERHAUL

After an engine runs for several hours, a glaze forms on the cylinder walls. This glaze is very smooth and not suitable for the seating of new piston rings. The finish lacks sufficient abrasion for rings to seat and has poor oil retention properties. An engine selected for overhaul typically also has a certain amount of cylinder wear. Because of bore distortion from head bolts, small wear pockets generally remain under the ring ridge after glaze breaking (see Fig.13-1). The honing does however produce the quality of the surface finish required for ring seating over 90 to 95 percent of the cylinder wall area.

Fig.13-1 A small wear pocket typically remains after glaze breaking

Many piston ring sets are sensitive to the quality of surface finish obtained in honing. For example, too fine a finish slows the seating of chrome rings and too coarse a finish can cause moly ring sets to wear. Some machinists contend that cast iron rings are less likely to have seating problems due to finish or wear and prefer them for overhauls. Rings seat, when properly matched with cylinder wall finishes, within the first few hours of operation. If ring seating does not occur quickly, or if rings wear prematurely, it is likely that the cylinder wall finish and piston rings are poorly matched.

General repair shops not specialized in automotive machining sometimes use portable hones when overhauling engines. One type of these hones is a "glaze breaker" intended only to restore a suitable surface finish for new rings. Because the hones follow the shape of the worn cylinders, hone only enough to break the glaze and restore the finish (see Fig.13-2). No amount of honing with glaze breakers will improve cylinder roundness or reduce taper.

Fig.13-2 Honing with a glaze breaker

PISTON INSPECTION AND KNURLING

Other operations used for overhauls compensate for piston-to-cylinder wall clearance and sometimes ring grooves clearances exceeding specifications. For example, with worn pistons and rehoned cylinders, piston-to-cylinder wall clearance is sometimes excessive. Unless corrected, excess clearance increases the possibility of piston noise and reduced ring life because of the rocking of the piston in the cylinder (see Fig.13-3).

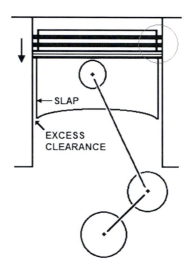

Fig.13-3 Excess clearance leading to piston "slap" and reduced ring life

Depending upon the diameter and rigidity of the particular pistons, expanding the piston diameter by knurling corrects piston-to-cylinder wall clearances exceeding specifications by as much as .003in (.07mm) (see Fig.13-4). The piston knurler adjusts for the length and width of the pattern and the amount of expansion (see Fig.13-5).

Fig.13-4 A knurled piston

Fig.13-5 Knurling a piston

Practices for fitting knurled pistons vary. One method specifies knurling pistons and polishing them to fit cylinders. Another method is to knurl the piston, lubricate the piston skirts, and stroke the piston through the cylinder before honing. Stroking the piston through the cylinder matches the piston size to the cylinder, and honing then provides clearance. Knurled pistons fit with as little as half specified clearance because of the reduced surface area in contact with the cylinder wall and the enhanced oil retention quality of the knurl.

Piston top compression ring grooves wear in high mileage engines. Check wear using a new or unworn ring and a feeler gauge (see Fig.13-6). The "no-go" limit is equal to the ring thickness plus the specified wear limit, typically not more than double new clearances. Excessive ring groove wear causes rapid ring wear and increased oil consumption because of oil "pumping." That is, oil accumulates under compression rings on the down stroke and squeezes around the top of the rings on the upstroke (see Fig.13-7).

Fig.13-6 Checking for excessive clearance in top ring grooves

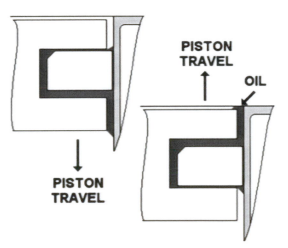

Fig.13-7 "Oil pumping" causes by excessive ring groove clearance

Prior practice included cutting worn top ring grooves oversize and fitting them with spacers but current practice is to replace pistons. If ring wear is a common problem in particular applications, replacement pistons might be selected with more wear resistant alloys.

REBORING AND HONING CYLINDERS

As recommended earlier, bore cylinders exceeding recommended limits for taper or roundness for oversized pistons. Common oversizes for domestic engines are .020, .030, .040, and .060in. Common oversizes for import engines are .50, .75, and 1.0mm. With newer lightweight castings, use the minimum oversize that removes all cylinder wear. This practice leaves the maximum cylinder wall thickness and allows for future rebuilding.

To follow piston manufacturers' recommendations, finish cylinders to the exact oversize. Typically, for a .030in oversize 4-inch bore, bore and hone to exactly 4.030in and the pistons will fit with minimum specified clearance up to maximum clearance. The measured piston diameter is therefore undersize from the finished bore diameter by the amount of specified piston clearance. Before finishing cylinders, check by measuring each piston in the set to be sure that this is true. If diameters vary beyond acceptable limits, hone cylinders selectively to fit pistons within specifications.

Production shops rough bore cylinders and then hone to finish size. Maximum cuts are possible because of the rigidity of the boring bar and work holding system (see Fig.13-8). Air-float and air-clamping devices also speed up setups. One distinct advantage to these machines is that the boring bar positions from a table parallel to the main bore centerline and above the block deck. This ensures that cylinders bore 90 degrees not to the deck, but the main bore centerline. Be sure to bore 1/2in past the end of the cylinder to allow the cylinder hone to pass through the bottom without interference.

Fig.13-8 Boring cylinders in production equipment

Because of cost, portable boring bars are popular for occasional jobs and they perform satisfactorily if set up and run properly (see Fig.13-9). Because the boring bar clamps directly to the block deck, first carefully clean and deburr the deck. Note that the bar bores cylinders 90 degrees to the deck and not the main bores, and unless the deck is parallel to the main centerline, cylinder alignment with the crankshaft will be incorrect. Check deck alignment before boring by comparing measurements taken from the main bearing bores to the deck at each end of the block (see Fig.13-10). As an alternative, check by making a small cut of .003in (.08mm) in the first cylinder bore and check alignment with the original visually. If properly aligned, the bore will clean up on all sides and not pass though at an angle. Again, bore 1/2in past the cylinder to allow for honing.

Fig.13-9 Boring cylinders with a Kwik-Way boring bar clamped to the block deck

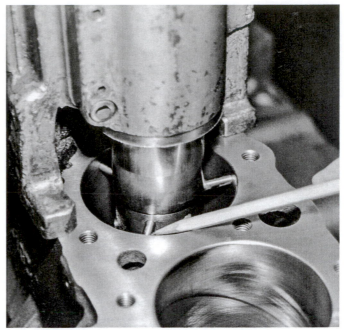

Fig.13-11 Expanding pins center a Kwik-Way bar in the cylinder

Fig.13-10 Measuring block parallelism

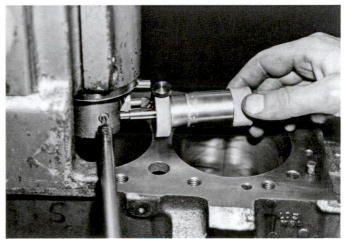

Fig.13-12 Setting the boring tool to size (Kwik-Way)

Boring bars center in the bore by uniformly expanding fingers positioned in an unworn portion of the cylinder (see Fig.13-11). Boring tools must also be set to size for boring (see Fig.13-12). Bore cylinders .003 to .005in (.08 to .13mm) under finish size to allow for honing. If surface finishes and size control are good, it is possible to leave the minimum for honing, if in doubt, leave more.

Cylinder honing by hand produces a high quality cylinder finish if done properly. The honing unit is a rigid assembly that, if used properly, produces round and straight bores (see Fig.13-13). Obtaining the desired surface finish requires careful selection of honing stones. While possible to rough hone dry, finish honing requires honing oil. The honing oil cools the cylinder and cleans the honing stones so that they cut more efficiently. Follow the specific instructions of the tool and abrasive manufacturer.

Fig.13-13 Honing cylinders with portable hone after boring

Honing machines have clear advantages over honing by hand (see Fig.13-14). First, since no physical effort is required, cylinders can be bored further undersize allowing more for finish honing. Second, in non-production shops, boring can be skipped altogether and cylinders honed from standard to oversize. Cylinder finishes are obtained by abrasive selection, honing speed, and stroking rate. Honing RPM and strokes per minute are part of the machine setup and adjusted according to cylinder diameter and length.

To save time, machinists often hone .0005in (.01mm) oversize when finishing to allow for shrinkage on cooling. They check the cylinders 20 minutes later and find them close to finish size. Experience is required to better estimate the amount of shrinkage in different blocks.

An experienced machinist can hone eight cylinders .030in (.75mm) oversize in 40 minutes or even faster if roughing with diamond abrasives. Diamond "stones" are not vitrified bonded abrasives like those discussed so far but instead are made up of diamond particles embedded in bronze. They wear very little, operate at high honing pressures and have high rates of metal removal. Although finish honing is possible, common practice is to rough hone undersize with diamonds and finish with the conventional abrasives.

With the lightweight castings now in production, cylinder distortion is a problem. Both main and head bolts can distort cylinders. To help minimize this effect, torque all main bolts for honing. If uncertain about the effect of head bolts on distortion, torque the cylinder head to the bare block and measure cylinders from the bottom side of the block. If cylinders distort, use a torque plate during honing to simulate the distortion caused by the cylinder heads (see Fig.13-15). The use of torque plates was once limited to performance engines but increasing numbers of passenger car blocks now require the same attention.

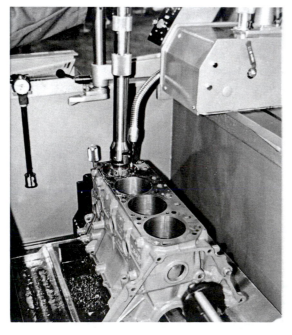

Fig.13-14 Honing cylinders in a Rottler hone

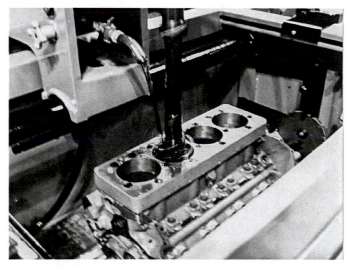

Fig.13-15 Honing with a torque plate to simulate distortion caused by head bolts

Another finishing process may follow honing. Here, this process is referred here as "brush honing" and consists of running bristled brushes through the cylinder to remove debris that adheres loosely to cylinder walls after honing (see Figs. 13-16 and 17). This process further improves the plateau finish and prevents debris coming off cylinder walls from contaminating the engine oil during break-in. The brushes are run in the cylinder for a limited number of strokes honing speed using light pressure. Care in this process must be taken to not to polish the cylinder to too fine a finish as this will interfere with ring seating.

Chamfer the top edge of each cylinder to a width of at least 1/32in (.8mm). This allows piston rings to slip easily into the cylinders on assembly. Chamfer cylinders after boring by hand feeding a 60-degree tool into the top edge of the cylinder bore (see Fig.13-18). Because the exact angle and width of the chamfer are not critical, chamfering with an abrasive cone in a drill motor produces satisfactory results (see Fig.13-19). Also break sharp edges on the lower edge of the bore to prevent shaving metal from the piston skirts as they pass by.

Fig.13-16 Note brush bristles

Fig.13-18 Chamfering cylinders with a boring bar

Fig.13-17 Honing head with brushes

Fig.13-19 Chamfering with a sanding cone

Special emphasis must be placed upon the final surface finishes. Surface finishes are measured according to scales such as Arithmetical Average and Roughness Average. The numerical value of these two scales is not greatly different. To put these scales in perspective, the numerical value of units is roughly equivalent to one-third of the depth of surface variations (see Fig.13-20). Because Rz scales compare extreme highs and lows, the numerical value can be 5-10 times greater.

SURFACE CROSS-SECTION

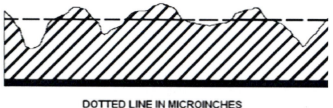

DOTTED LINE IN MICROINCHES
(1/3 PEAK-TO-VALLEY DEPTH)

Fig.13-20 Surface finish scales

Chrome and cast-iron piston rings require an average surface finish of 25 RA and moly faced piston rings require a surface finish of approximately 15 RA microinches. Specified surface finishes are generally becoming finer and most problems result from finishes that are too coarse. When too rough, moly rings wear quickly and thin, low-tension rings sometimes bounce over the surface and fail to seat. Although cylinder wall material, RPM, honing pressure, and honing oil are all factors, some general correlations between abrasive grit size and surface finish are as follows:

Ring Type	Grit Size	Hone Operation	RA Finish
-------	70	Roughing	--------
Iron, Chrome	220	Semi-finish	25-35
Moly	400	Fine finish	10-20

To obtain the specified surface finishes requires that the machinist follow procedures rec-

ommended by the equipment manufacturers. These manufacturers check piston ring requirements and run tests to decide what operating parameters produce the desired surface finish. As mentioned earlier, this is not a substitute for measuring finishes but it does give the machinist a place to start. The machinist can adjust the process as necessary to ensure that the finish conforms to specifications. Following are key elements of the recommended honing procedure for a particular engine:

Engine Specifications:
　　Bore diameter 3.875in
　　Cylinder length...6in

Basic Machine Set-Up:
　　Stone length.....................................2 3/4in
　　Stone overstroke at each end 7/16in
　　Rotational speed (RPM) 155
　　Strokes per minute (SPM) 61

Rough Honing:
　　Stock removal below finish size ..003-.005in
　　Grit size (Sunnen EHU 123) 70
　　Feed rate; load meter reading 70-80%

Finishing for Chrome or Iron Rings:
　　Stock removal; minimum .... .003in
　　Grit size (Sunnen EHU 525) ......　220
　　Feed rate - load meter reading.........40-45%
　　Approximate surface finish 25-30RA

Finishing for Moly Rings:
　　Minimum stock removal .... .0005in
　　Grit size (Sunnen JHU 820) 400
　　Feed rate; load meter reading 25%
　　Approximate surface finish 15-20RA

The included angle of the crosshatch pattern is approximately 60 degrees and is an important aspect of surface finish (see Fig.13-21). The RPM requires adjusting for changes in cylinder diameter to keep the surface speed of the stones constant. The number of strokes per minute requires adjusting for RPM and cylinder length. The correct combination of adjustments to RPM and stroking rate produces the correct crosshatch. If the angle is too small when honing by hand, increase the stroking rate, if too large, decrease the stroking rate.

Fig.13-21 The cross-hatch pattern in properly honed cylinders

To summarize what is expected in rebuilding newer engines, consider that most engines now use moly faced or thin, low tension rings that reduce friction but require greater precision. In general, expect that cylinders must be held to within .0005in roundness and taper and that required surface finishes be in the 15 to 20 RA range.

SLEEVING CYLINDERS

Common causes of cylinder damage are piston ring breakage and piston seizure caused by engine overheating or lack of lubrication. Cylinders also crack or break when pistons fail or compress coolant that has leaked into cylinders.

Scored cylinders are repaired in several ways depending on the extent of cylinder damage and general engine condition. For example, if not scored too deeply, boring to an oversize is possible. While it seems strange, engines operate perfectly with only one cylinder bored oversize providing the weights of the original and replacement pistons are close or at least within 10 grams. If cylinders are scored and the engine is worn, it makes sense to bore all cylinders oversize and replace the pistons. If scoring is too severe on

only one or two cylinders, those cylinders can be sleeved. In cases of cylinder damage on low mileage engines, it is sometimes desirable to sleeve one damaged cylinder to match the diameters of other serviceable cylinders.

This limits the cost of piston replacement, sleeving, and parts replacement to perhaps only one cylinder. Only sleeving repairs deeply scored or cracked cylinders. The strength of sleeved cylinders compares to that of the original cylinder block. Sleeves are available in 3/32 and 1/8in wall thickness (2.4 and 3.2mm) and in various lengths. Some metric sizes are also available. The sleeves come approximately .015in (.38mm) undersize on the inside diameter so the specified wall thickness is obtained only after installation, boring, and honing to size.

Boring for the sleeve requires going approximately 3/16in oversize for a 3/32in wall sleeve or 1/4in oversize for a 1/8in wall sleeve (4.8 and 6.4mm). The bore diameter should allow for a .003in (.08mm) press-fit of the sleeve. A shoulder is sometimes left at the bottom of the cylinder to retain the sleeve. The shoulder is approximately 3/16in (4.8mm) wide (see Fig.13-22). The shoulder forms when the machinist disengages the power feed on the boring bar short of passing through the cylinder and hand feeds the boring tool to the desired position. Deburr the top of the cylinder after boring and the bottom edge of the sleeve in preparation for installation.

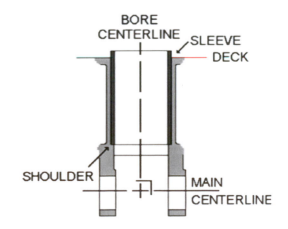

Fig.13-22 A shoulder at the bottom of the cylinder stops the cylinder sleeve

Methods of installing the sleeve vary. One method calls for using a sleeve at least 1/8in (3mm) longer than the cylinder and pressing or driving it into place using a hammer and driver (see Fig.13-23). The sleeve extends above the deck of the block after installation and the extra length is removed using a facing tool in the boring bar. The machinist hand feeds the facing tool down until the sleeve almost blends with the deck of the block and then draw-files the deck to a cleanup. Machinists sometimes cut the extra sleeve length after installation and true the block by resurfacing.

Fig.13-23 Driving a cylinder sleeve into place

Another method of installation is to heat the block in a cleaning oven. Using a 4-inch bore as an example, a temperature difference of 175 degrees F between the sleeve and the block changes a .003in (.08mm) interference fit to a slip-fit. This requires heating to not more than 300 degrees F. Just don't hesitate on installation; the sleeve must be in place before temperatures equalize.

If "shrink fitting" the sleeve, clean up the deck after honing. This is because the sleeve grows and protrudes from the deck .002-.003in after being heated and "massaged" by honing.

LINE BORING AND HONING

Main bearing housing bore diameters and alignments change because of block warpage, stretching of caps, or cap replacement. Line boring or honing corrects both the alignment of the bore centerline and the diameters of the bores.

Line honing is done with a long and rigid honing unit running through all housing bores at the same time (see Fig.13-24). Line boring has typically meant boring with a single pointed tool on a boring bar running on center through all housing bores in one set-up (see Fig.13-25). Computer numerically controlled (CNC) machining centers are now used for a number of block machining operations including line boring. The difference in the CNC operation is the ability to move from the centerline of one housing bore and precisely relocate on center in each of the remaining bores (see Fig.13-26).

Fig.13-24 Line honing main housing bores (Sunnen)

Fig.13-25 Line boring main bearing housing bores (Berco)

Fig.13-27 Grinding the sides of main caps to establish a reference for positioning

Fig.13-26 Line bore in a CNC machining center

Fig.13-28 Grinding main cap parting faces

The first step in preparing an engine block for line boring or honing is the grinding of the main caps. Grind cap parting lines the amount required to reduce bore diameters below specifications by approximately .002in (.05mm). Grinding the cap for the thrust bearing requires particular care because tilting of this cap decreases crankshaft end-play. In a Sunnen cap and rod grinder, first grind one side of the cap square to the parting face (see Fig.13-27). Then position the ground side of the cap against the fixed jaw of the grinder vise and grind the parting line (see Fig.13-28).

When overloaded in operation, main caps sometimes stretch. This pulls the ends together causing fit them to fit loosely and not locate properly in the crankcase registers (see Fig.13-29). Cutting the cap parting lines corrects for stretch but does nothing for the location problem. Peening the housing bore of the cap with an air hammer relieves stress and restores the fit but this must be followed by line boring. If the error is small enough, placing the cap in its register, peening it with a brass hammer and staking the block against the caps with a blunt chisel may solve the problem (see Fig.13-30). In extreme cases, build up the ends by welding and machining to fit or simply replace the cap. These measures also require line boring.

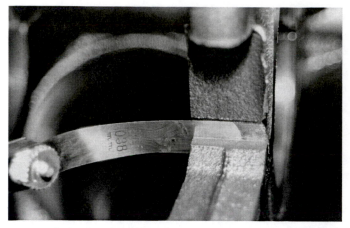

Fig.13-29 Watch for caps that fit loosely in their registers

Fig.13-30 Staking the web against the cap

In line honing, all main bearing bores are machined simultaneously thus the centerline becomes an average of all bore positions. The centerline does move down into the block approximately half the amount that bores are reduced below specifications in preparation for honing. This is not equal to the amount ground from the caps because they are sometimes stretched. Take care to reduce all bore diameters .002in (.05mm) undersize and the centerline will move only .001in (.025mm). This amount of change in centerline position causes no problems.

Align boring, as the name implies, involves setting up a boring bar with a single pointed tool and running it through the main bearing bores.

Unless the particular bores are damaged, the boring bar positions from the two end housing bore saddles and thus determines the centerline. As with line honing, the centerline moves into the block slightly but cap grinding has no influence on the amount. With careful set up, the depth of cut into the saddle is no more than .001in (.25mm).

Use caution when line boring or line-honing aluminum engine blocks. The tightening of fasteners in assembly causes the crankcases of these engines to distort. In the engine shown in Figure 13-31, the crankcase is not in alignment until cylinder heads and intake manifolds are assembled to the block. When assembling these engines, the crankshafts bind unless main bolts are tightened after the cylinder head bolts. Such distortion is common with all aluminum engine assemblies and both technicians and machinists must be aware of such possibilities. If necessary, attach cylinder heads and tighten main bolts for line boring.

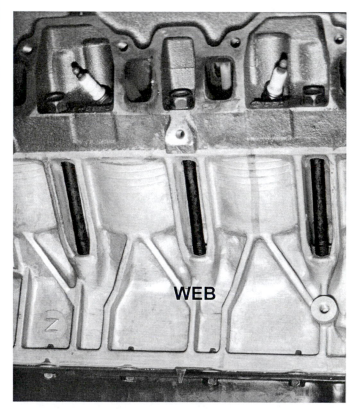

Fig.13-31 Note that head bolts in this engine pull directly on main bearing webs

There are advantages to each of these machining processes. Line honing is fast and precise and ideally suited to correcting problems of warpage and stretching of the caps. Line boring better handles severe misalignments such as when changing caps or building up damaged saddles. Multiple diameter or "stepped" main bearing bores can only be remachined by boring. In addition, main bearing thrust faces can only be machined in line boring machines. To summarize, line boring is clearly more versatile but line honing is very fast for those jobs within its capacity.

Occasionally, housing bore specifications are not available or the engine builder wants to check oil clearance before machining. In such cases, the housing bore diameter with a specific oil clearance is calculated as follows:

Shaft diameter	2.3750in
Bearing thickness	+.0615
Bearing thickness	+.0615
Oil clearance (average)	+.0020
Housing bore diameter	2.5000in

Keep in mind that in addition to any original variations in tolerance, cutting the rear main caps distorts the seal bore, groove, or flange used for locating the seal in assembly (see Fig.13-32). Remember too, that the crankcase centerline moves up in the block and adds to errors in concentricity at both rear main and timing cover seals. It may be necessary to remachine the rear seal location to restore roundness and concentricity. In the front, it is usually possible to remove locating dowels, or drill locating holes oversize, to allow moving the timing cover and seal onto center. Failure to recognize these seal location problems leads to some very persistent oil leaks.

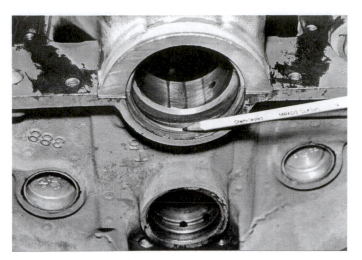

Fig.13-32 Watch for distortion in seal bore location

FITTING PISTON PINS

Most current engines use press fit piston pins. With this design, the piston pin is press-fit through the connecting rod with approximately .001in (.025mm) interference and fits with approximately .0005in (.01mm) clearance through the pin bores in the piston.

With press-fit assemblies, it is difficult to determine piston pin clearance in the piston while assembled and disassembly for measuring followed by reassembly is difficult to do without damaging pistons. Excessive clearance is therefore detected only by feel; not a very may reliable method. Be careful because loose piston pins may have been quiet in the running engine only because the rings were worn and the pistons were sliding freely in the cylinder. These same piston pins however may make noise when new rings add drag to the pistons. Although availability is limited, oversized pins might be fitted or new standard pins might restore clearances. Considering ring groove and piston skirt wear, and the labor required to fit pins, replacing piston assemblies is likely the best option.

Should the fitting of piston pins be required, the connecting rod pin bores must be honed to the specified press-fit (see Fig.13-33) and the piston pin bores honed to the specified clearance. Honing the pin bores in pistons requires a honing unit with sufficient length to hone both bores simultaneously to maintain a true centerline (see Fig.13-34).

Fig.13-33 Honing a rod to the correct press-fit

Fig.13-34 Honing a piston to the correct clearance

Checking clearances as close as these requires highly accurate measuring methods. Sunnen manufactures a precision gauge that uses two of the new piston pins to set the gauge to size (see Fig.13-35). Once set, the gauge measures the amount of interference in the connecting rod or clearance in the piston pin bores (see Figs. 13-36 and 37).

Fig.13-35 Setting a Sunnen AG300 gauge to the piston pin diameter

Fig.13-36 Gauging the rod for correct pin interference

Fig.13-37 Gauging the piston for correct piston pin clearance

Fig.13-38 Press fitting a bushing for a full floating piston pin

Checking new pin fits by feel is highly unreliable with such small specified clearances. Full-floating piston pins allow the free rotation of the piston pin through bores in the piston and the rod. There is less wear of the piston pin because wear distributes around the circumference. Although oversized pins are sometimes available, fitting full-floating piston pins is often limited to replacing the bushings in the connecting rods. Replacing the bushings reduces pin noise in overhauled engines and is part of complete engine rebuilding.

The process begins by pressing out the old bushings and pressing new bushings into place (see Fig.13-38). Press fit bushing with predrilled oil holes so that they align with the oil holes in the connecting rod. If not predrilled, drill them after bushing installation and before finish machining.

After pressing, seat the bushings in their bores by expanding them using an expanding unit in the honing machine (see Fig.13-39). Thin wall bushings require expanding because the bores in the connecting rods are sometimes rough and not well suited to holding the bushings in place (see Fig.13-40). The bushing expander unit includes a tool for facing off any bushing material extending from the bore (see Fig.13-41). This step is omitted for thick wall bronze bushings or bushings with steel backs.

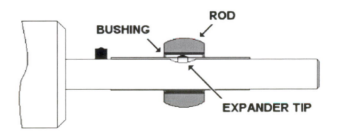

Fig.13-39 Expanding a rod bushing into place with a bushing expander (Sunnen)

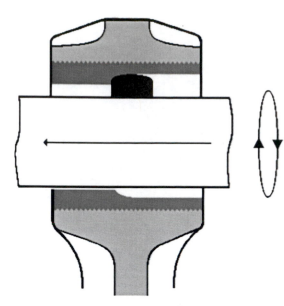

Fig.13-40 Expanding seats the bushing into rough finished bores (Sunnen)

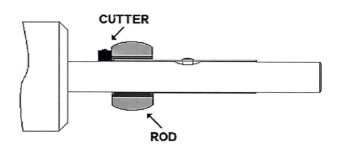

Fig.13-41 A facing tool for bushings (Sunnen)

Fig.13-42 Boring piston pin bushings to size (Berco)

Next, rough hone the bushing to within .002in (.05mm) of size using a coarse abrasive stone and then change to a fine finishing stone and hone to size. Again, obtaining specified clearances of approximately .0004in (.01mm) requires precision gauging.

Piston pin bushings are also resized in boring machines (see Fig.13-42). The advantage to boring machines is that the centerline of the bushing is bored parallel to the centerline of the rod housing bore. This alignment corrects for bend or twist in the connecting rod and eliminates the need to align rods. With sufficient wall thickness in the bushings, boring machines can also equalize or correct center-to-center spacing of the rod bearing and piston pin bores.

RESIZING CONNECTING ROD HOUSING BORES

Rod housing bores frequently stretch .001in (.025mm) or more oversize. To guarantee normal bearing service requires restoring housing bores to original specifications for roundness and diameter.

Reconditioning begins with removing the connecting rod bolts by clamping the rod in a vise and driving out the bolts with a brass hammer (see Fig.13-43). If replacing rod bolts, discard the old hardware and reassemble with new parts. Always replace bolts in heavy duty and performance engines or rods that have had bearing failures.

To seat new rod bolts, torque, loosen, and retorque three times. Also, if inspecting for cracks, mag the rods with the bolts removed.

Fig.13-43 Driving out rod bolts using a brass hammer and a rod vise

Fig.13-44 Grinding the parting faces of rods and caps in a Sunnen cap and rod grinder

The author has specific recommendations in regard to TTY cap screws in connecting rods. Since these fasteners are normally replaced after using them once, how can rods be resized, checked for clearance with Plastigage, and then assembled with the same fasteners? Consider using aftermarket high performance fasteners that can be tightened repeatedly. Be aware however that tightening to the higher torque values typically recommended for these fasteners can cause distortion in housing bores so be sure to install them when resizing.

Next, grind .002in (.05mm) from the parting faces of both the rod and the cap (see Fig.13-44). If the parting faces do not clean up at .002in, grind until clean and flat. Try to keep the total amount ground from each set of rods and caps equal so that changes in rod lengths are uniform. While not necessarily recommended, be aware that some shops grind the cap only and save the time required for removing rod bolts and grinding the rod. Clean the grit out of the rods and caps and oil the rod bolts prior to reassembly.

A variation in procedure is needed for "cracked" rods. The housing bore ends of these rods are machined in a full-round piece, scored at the parting line, and the cap removed by cracking at the parting line. The rough surfaces at the parting line permit the cap to fit the rod in only one position (see Fig.13-45). The parting lines cannot be cut and housing bores are therefore honed to fit oversized bearings. Be especially careful to clean and blow grit from these surfaces during assembly.

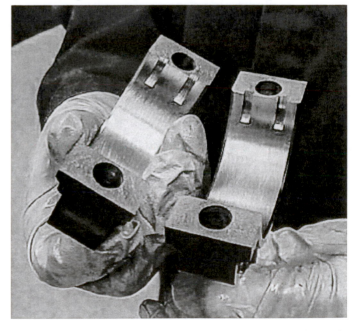

Fig.13-45 The rough parting line of a "cracked" rod

Once reassembled, resize housing bores by honing. In power stroking hones, place the rod over the honing unit and inside the power-stroking attachment (see Fig.13-46). The RPM for the honing unit requires adjustment according to the housing bore diameter and the stroking rate requires adjustment according to the RPM. As with most grinding and honing operations, the rods travel approximately 1/3 of their width past the stones at each end of the stroke. The correct combinations of RPM, stroking rate, and overstroke keeps the stones sharp and true.

Fig.13-46 Resizing housing bores in a Sunnen power stroking hone

Hone two rods at a time if honing without the power-stroking attachment (see Fig.13-47). Placing two rods side by side, and periodically reversing their position, keeps the bores round and straight. Only RPM and honing pressure require adjustment but the machinist must still overstroke the stone slightly at each end of the stroke.

Fig.13-47 Honing two rods at a time without the power stroking attachment

A relatively new variation in honing rod housing is the use of diamond stones. Preparation for resizing is the same as for traditional honing. In the Petersen-Kansas Instruments system shown, the honing unit has two carbide shoes and five diamond abrasives. These keep the unit well centered in the housing bore to achieve roundness quickly and easily (see Fig.13-48). With diamonds, the near absence of wear and the predictability of stock removal simplify size control. The machines are operated in a similar manner to other hones with adjustment for RPM and honing pressure (see Fig.13-49). Power stroking and automatic size control attachments are available for use in production.

Fig.13-48 A diamond abrasive honing unit in a Petersen-Kansas Instruments machine

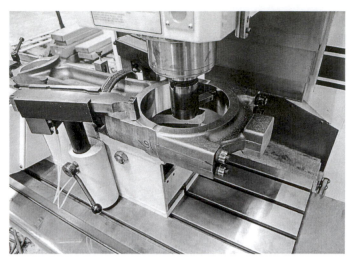

Fig.13-50 Boring a rod in a Rottler machine

New housing bore specifications call for roundness within .0003in (.008mm). Keep housing bores within tolerance limits for diameter and for roundness. Because these specifications are very close, measuring requires precision gauging (see Fig.13-51). As with main housing bores, rod housing bore diameters may be checked by calculation before machining.

Fig.13-49 Honing in a Petersen-Kansas Instruments machine

Rod housing bores are also bored to size (see Fig.13-50). The rod positions from its pin end and housing bore end is bored on a centerline parallel to the piston pin. Done properly, this eliminates the need for aligning the rod later.

Fig.13-51 Gauging housing bores in a Sunnen AG300 gauge

The center-to-center distance between the housing bore and the pin bore shortens after resizing. This situation is comparable to line boring or honing. The amount of change is approximately half the amount of housing bore reduction after cap and rod grinding before resizing. For example, with the bore diameter reduced .004in after cutting the rods and caps, the rods shorten .002in after machining.

Rod center-to-center distances can vary over a .010in (.025mm) range in most gasoline engines without noticeable changes in performance. Of course the usual practice for diesel and performance engines is keep such variations to a minimum. It is easy to keep rod center distances equal within .003in (.08mm). First, make sure bores are within specifications and then measure edge to edge between the pin bore and housing bore with a vernier caliper (see Fig.13-52). Using the shortest rod as a standard, recut the caps only and resize any rods that are more than .003in (.08mm) longer. Keep in mind that, after cutting the cap another .004in (.10mm); resizing again shortens the rod .002in (.05mm). Do not cut parting lines excessively in this process as bearing lock grooves become smaller and can interfere with bearing tangs and distort the bearings on assembly. Also, as mentioned earlier, rods with piston pin bushings may be bored to specified or equal center distances.

Rod reconditioning sometimes requires other minor cleanup operations. One of these is deburring the sides of the rods by draw filing or lightly sanding the surfaces (see Fig.13-53). Do not narrow the rods excessively or rod side clearance and oil throw-off is increased.

Fig.13-53 Deburring the sides of connecting rods

Another operation is the rechamfering of the rod housing bores. In a hone, replace the honing stone with a chamfering tool (see Fig.13-54). The deburring operation prevents scuffing between rods on the crankshaft and rechamfering restores clearance over the crankshaft fillets.

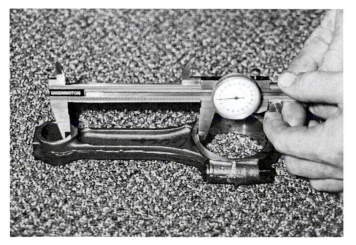

Fig.13-52 Comparing rod lengths with a dial caliper

Fig.13-54 Chamfering rod housing bores with a Sunnen chamfering tool

ASSEMBLING AND ALIGNING PISTONS AND CONNECTING RODS

To prepare for assembly, first Figure out the position of each connecting rod and piston. Most pistons have a notch, arrow, the letter F, or FRONT on the piston. These marks show that the piston is to be installed with the mark pointing to the front of the engine. Connecting rods are positioned in respect to an oil spurt hole, a chamfer on the rod housing bore, rod numbers, or bearing tangs (see Fig.13-55). For example, a typical assembly calls for the notch in the piston to face the front and the connecting rod oil spurt hole to point toward the camshaft.

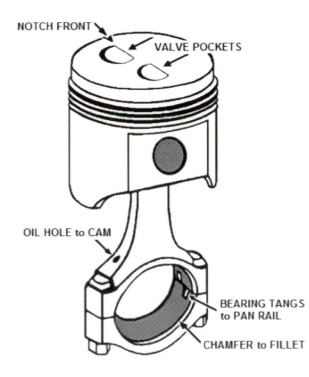

Fig.13-55 Instructions for piston and rod assembly

Check specifications because correct assembly for one engine may be incorrect for another. Also, assemblies for most V-block engines assemble one way for one bank and another way for the opposite bank.

Reversing piston directions places piston pin offset on the wrong side of the cylinder and causes piston noise and possible engine failure. In proper assembly, the piston pin is offset to the side of the cylinder opposite the crankpin on the power stroke (see Fig.13-56). Offset keeps the piston from rocking severely at the top of stroke when the piston reverses direction, reducing noise and extending piston and ring life. Also, if positioned wrong, valve reliefs in the pistons misalign with the valves resulting in piston-to-valve interference.

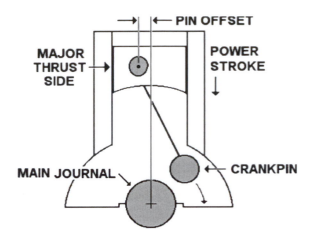

Fig.13-56 Piston offset to the major thrust side

With oil spurt holes, reversing connecting rod in assembly reduces cylinder lubrication. In V8's, connecting rods usually have a large chamfer on one side only to clear the fillet on the outside edge of the crankpin. The rod bearing also offsets to the center of the crankpin to clear fillet radii. If reversed in assembly, the bearings run on the fillet and fail.

Once oriented for assembly, putting full-floating pistons and connecting rods together is simple. Because the piston pin is full floating, it slips through both the piston pin bores and connecting rod bushing. Simply clean and oil the parts before slipping them together. It is best practice to replace used lock rings or snap rings as a precaution against failure. With lock rings, needle nose pliers work while snap rings require snap ring pliers (Fig. 13-57 and 58).

Fig.13-57 Installing lock rings with needle nose pliers

Fig.13-58 Installing snap rings with snap ring pliers

Performance pistons often use more positive "spiral locks" instead of snap rings. For installation, spread them apart approximately 3/8in (10mm) and spiral them into the groove (see Fig. 13-59 and 60).

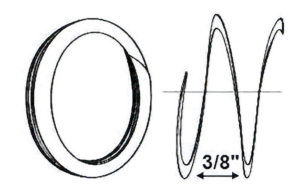

Fig.13-59 Spreading the spiral lock

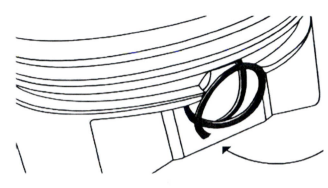

Fig.13-60 Spiraling the lock into place

The assembly of press fit piston pins requires great care. Unlike full-floating assemblies, they cannot be taken apart put back together again without risking damage. Begin as with full-floating assemblies; clean, lubricate, and lay out parts out in position for assembly. The necessary press tools are available from the engine or tool manufacturers. Cautions include lubricating the piston pin to prevent galling as it moves through the rod and using the correct tooling to avoid damaging the piston (see Fig.13-61). Look for a chamfer on one side of the pin bore and press pins through that side. Use caution to get the job right the first time because attempts at centering piston pins in the rod by pressing from both sides distorts the pistons and binds the piston pins. If assembled incorrectly, press the pin all the way through from the same direction and reassemble correctly.

Fig.13-61 Press fitting piston and rod assemblies

An alternative to pressing piston pins into assembly is to heat the rod to expand it and then slip the assembly together. Heating the small ends of the rods to 425 degrees F changes the press-fit of the piston pins to a slip fit (see Fig.13-62). With the piston pin at room temperature, quickly insert it through the heated rod. If errors in centering should occur, oil the parts and use the press to position the pin. Beware of overheating rods, especially small, lightweight rods as the small ends may not return to size on cooling allowing the pins to loosen in operation. It is also recommended that powdered metal (PM) rods not be heated.

Fig.13-62 Assembling press fit pins in a Sunnen rod heater

Check rod alignment after machining or when bearing wear patterns show the need for alignment. There are two conditions frequently found called "bend" and "twist" (see Figs. 13-63). Both conditions cause abnormal wear and stress on connecting rod bearings, pistons, and piston pins (see Figs. 13-64 and 65).

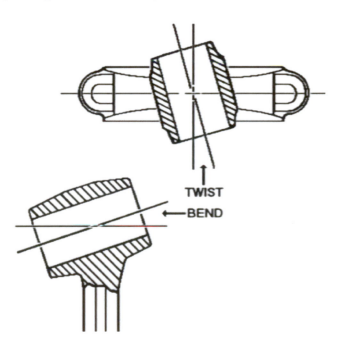

Fig.13-63 Bend and twist conditions in rods

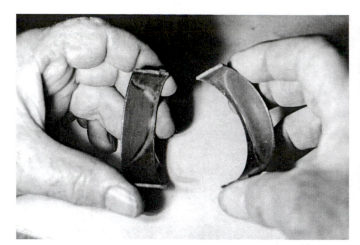

Fig.13-64 Bearing wear caused by a bent rod

Fig.13-66 Gauging and correcting bend in a K.O.Lee rod aligner

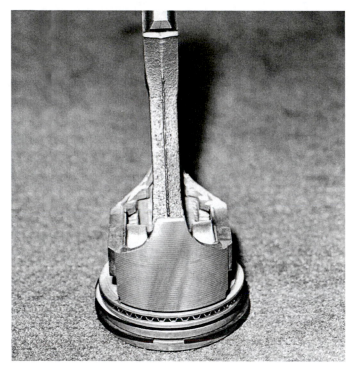

Fig.13-65 Piston skirt wear pattern caused by a bent rod

With press fit pins; align the rods after assembly with the pistons. First check and correct bend using a notched pry bar (see Fig.13-66). Correct bend first because a bent rod always checks out as if twisted. Note that checking twist requires turning the piston to one side and correcting alignment with the same notched pry bar (see Fig.13-67). Alignment should be within .001in per six inches. Both checks and corrections can be made in reference to the piston pin for full-floating designs (see Figs. 13-68 and 69).

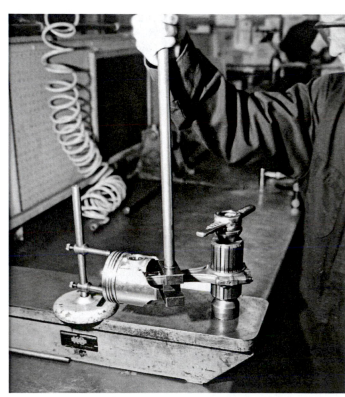

Fig.13-67 Gauging and correcting twist in a K.O.Lee rod aligner

Fig.13-68 Checking bend against a piston pin

Fig.13-69 Checking twist against a piston pin

Occasionally, there is a third condition called "offset." Offset can be part of the design but can also be the result of incorrect straightening. By design, offset positions the rod and bearing on the cylinder centerline and, with two rods per crankpin, offset keeps rod bearings away from fillets (see Fig.13-70). However, correcting bend at the wrong point also creates an offset. It is important to install rods designed with offset in the correct position to prevent wear caused by interference with the piston pin bosses and fillets of the crankpin (see Fig.13-71). An incorrectly offset rod causes the same problems.

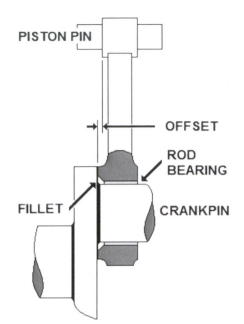

Fig.13-70 A connecting rod designed with offset

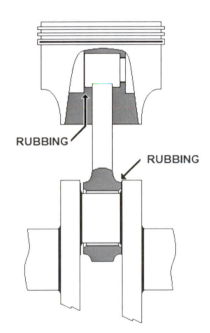

Fig.13-71 Problems caused by incorrect offset

Check rod offset by measuring the distance between the rod and the face of the rod aligner before and after reversing the rod in the set up (see Fig.13-72). The difference in the two measurements is the amount of offset.

Fig.13-72 Checking for offset in a Sunnen rod aligner

REGRINDING AND POLISHING CRANKSHAFTS

Worn, scored, or otherwise damaged crankshafts require regrinding. Wear shows as out-of-roundness caused by bearing loads on one side of the shaft. Scoring is caused by dirt or metallic particles carried through the engine in the oil. Other defects or conditions such as cracking or bending require closer inspection.

Crankshafts require cleaning before mag inspection or regrinding. While crankshafts are cleaned in the same way as other engine parts, they require particular care in handling to prevent further surface damage or bending. For mag inspection, first spray the shaft with the fluid containing the magnetic particles, magnetize the shaft, and then view it under a black light (see Fig.13-73). If present, cracks show up as sharp, clear lines typically around fillet areas.

Fig.13-73 Wet "mag" inspection of a crankshaft

Before regrinding, check crankshafts for straightness in V-blocks with a dial indicator (see Fig.13-74). Automotive crankshafts can usually be straightened in the V-blocks using a punch with the end rounded to match the fillet curvature and radius. The V-blocks are positioned to each side of the bend and the shaft straightened by rapping the fillets of the journal or crankpin with the punch (see Fig.13-75). Using a punch also stabilizes or relieves stresses in the shaft causing the shaft to return to its original centerline. The trick is to isolate the location of the stresses and this often requires shifting the V-blocks and striking the fillets of adjacent journals or crankpins. For purposes of grinding, straighten the shaft to within .002in TIR (.05mm). Heavy duty crankshafts may require a press for straightening.

Fig.13-74 Checking a crankshaft for straightness in V-blocks

Fig.13-75 Straightening a crankshaft

Crankshafts reground while bent have at least three potential problems. First, they spring out of alignment later because of internal stresses. Second, they are out of balance because the counterweights are out of position. Third, crankpins also shift from location causing stroke length variations.

Building up badly damaged journals by welding reduces the need to grind shafts further undersize. The welding is done by either "submerged arc" or "gas shielded" processes in specialized crankshaft welding machines. The submerged arc process involves covering the welding arc with flux powder to protect the metal deposit from the atmosphere (see Fig.13-76).

Fig.13-76 Submerged arc welding

Carbon dioxide and Argon gases perform this function in the gas-shielded process (see Fig.13-77). Both processes aid in producing welds that are clean and free of porosity (see Fig.13-78). Because of the limited availability of oversized thrust bearings, badly damaged thrust faces are often built up by welding (see Fig.13-79). Of the two processes, gas shielded welding lends itself to welding thrusts because powdered flux falls away from the vertical sidewall of thrust faces.

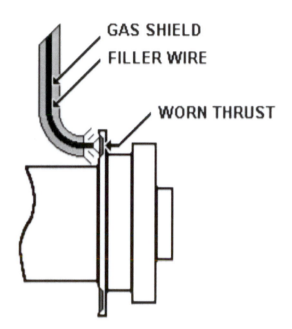

Fig.13-77 Gas shields weld deposited on the vertical thrust face

Fig.13-78 A damaged crankpin built up by welding

Fig.13-79 A welded thrust face; gas shielded

When welding on crankshafts, avoid welding fillets if possible to minimize distortion. Weld deposits in the fillet areas pull the shaft out of

alignment as they shrink on cooling. For heavy duty shafts, fillets are sometimes welded with soft alloy material to minimize this effect. When fillets are welded, restoring alignment requires three intermediate steps in grinding. First, straighten the shaft immediately after welding while it is still hot. Second, rough grind the weld to within .030in (.75mm) of size and straighten again. Remember that it is the deposited metal that pulls the shaft out of alignment. Third, grind the mains last so that the main bearing centerline is straight when finished.

If welded crankpins are roughed and finished without the intermediate straightening steps, the shaft bends along the main centerline. Straightening at this late stage causes unequal stroke lengths and nonparallel centerlines between rods and mains. Do not grind the mains until straight as this requires grinding to greater undersizes and stresses remain in the shaft. Check for unequal stroke lengths when mains are ground smaller than crankpins.

When welding more than one crankpin, stress relieving is recommended. This requires heating the shaft to between 900 and 1,100 degrees F and allowing it to soak for two to three hours, or one hour per inch of thickness. Media blasting may be necessary as scaling develops at this temperature. The heat treater can "stop off" the scaling in critical areas such as threads by painting the areas with a high temperature metallic paint, or even better, stress relieve in a controlled atmosphere type of furnace.

Stress relieving is especially important for shafts scheduled for later nitriding. Unless stress relieved first, the shaft distorts during the nitriding operation.

Hardened crankshafts overheated by bearing failures require testing to see if they have softened. Testing is commonly done with a portable hardness tester. First, a spring loaded punch makes an impression in the shaft using a "penetrator" of a specified diameter; for Rockwell "C" scale, a "C" diameter penetrator is used. The diameter of the impression is then read using a scope: the wider the impression, the softer the shaft (see Figs. 13-80 and 81).

Fig.13-81 The diameter of impression made by the penetrator to evaluate hardness

Fig.13-80 Using a penetrator when testing crank shaft hardness

Crankshafts hardened by nitriding have ferric nitride in the surface and may be tested chemically. Testing for the presence of ferric nitride requires a 10 percent aqueous solution of copper ammonia chloride. A drop of the solution "plates" and turns copper colored on non-nitrided shafts within approximately 10 seconds (see Fig.13-82).

Fig.13-82 Copper solution on a nitrided shaft

Hardened shafts often require rehardening after grinding. The most prevalent method of hardening is nitriding by one of two processes, salt bath or gas atmosphere. Although manufacturers use the salt bath process, it requires immersion in cyanide salts at approximately 1,100 degrees F and many heat treaters prefer using gas atmosphere furnaces. In the gas atmosphere process, a clean shaft is placed into a furnace at 900 to 1,000 degrees F with ammonia gas for several hours, depending upon materials and the depth of hardening desired.

Because journal diameters grow in nitriding, and sometimes with a lobe shape, some shops prefer pregrinding to mid or high limits, nitriding to maximum depth, and then regrinding to finish size. It is important to follow this practice when nitriding nodular iron shafts because they tend to form pits and blister as excess graphite burns off. Others treat the shaft for fewer hours to a shallow depth and simply polish the shaft after nitriding. As with stress relieving, the heat treater can stop off or protect critical areas from the atmosphere.

As an alternative to nitriding, rebuilders sometimes grind the shafts .006in or more undersize and plate with industrial hard chrome. The shafts are "dried" in an oven at approximately 350 degrees F and then ground to size slowly. Rushing this process heats the shaft and results in the chrome separating from the base metal. While expensive, chrome plating is sometimes competitive in situations where the engine builder wants a hard, wear resistant surface and also wants to build up journals.

Undersizes for domestic crankshafts are .010, .020, and .030in. The metric equivalent sizes are .25, .50, and .75mm. An exception to the above sizes is for crankshafts with pressure rolled fillets, in which case, the manufacturers often limit grinding to the first undersize. While possible to grind some crankshafts to greater undersizes, welding all but eliminates the need for such extremes. It is standard practice to grind all main journals to the same undersize and all crankpins to the same or another undersize. While individual crankpins could be ground to different undersizes, it is generally not done. Many shops prefer grinding main journals last to help ensure that the centerline is straight when finished.

Crankshaft grinders are specialized machine tools with unique adaptations for crankshafts (see Fig.13-83). They are expensive and are often found only in production engine rebuilding shops or specialty grinding shops. The operator is a skilled specialist that must simultaneously hold taper; roundness, size, fillet radius, stroke length, and surface finish specifications within limits.

Fig.13-83 A Berco crankshaft grinder

Set-up begins with mounting the shaft in chucks or between centers (see Figs. 13-84 and 85). In chucks, the shaft will "buck" and grind out-of-round unless straight within limits. The same "bucking" occurs when grinding between centers that are not true in the shaft. While it seems illogical for centers to be off center, consider that original equipment grinders do not use the centers; the centers were used only for roughing operations before grinding. True damaged or inaccurate centers in a lathe prior to grinding between centers (see Fig.13-86).

Fig.13-84 A crankshaft set up in chuck

Fig.13-85 A crankshaft set up between centers
(Storm Vulcan)

Fig.13-86 Recutting crankshaft centers in a lathe

The corners of the grinding wheel require dressing to produce the specified radius with a smooth finish (see Fig.13-87). To do this properly requires a radius dressing attachment (see Fig.13-88) although grinders sometimes form the radius off-hand using a dressing stick. These radii also require touching up as the wheel face wears or is dressed. This particular operation has major effect on the strength of the shaft and its importance cannot be under estimated. Many technicians believe that grinding beyond the first undersize reduces strength, when in fact many failures are traceable to sharp fillet radii, not the undersize.

Fig.13-87 A proper fillet radius is smooth and round

variations in the crankshaft show up at this stage. If not close enough to center for grinding, the machinist makes fine adjustments to the angular position without changing stroke lengths. Consider that half a degree error in positioning shows up as approximately .020in (.50mm) run out on the dial indicator. If angular adjustments only do not reduce run out to less than the amount to be ground from the shaft, a change in stroke length becomes necessary.

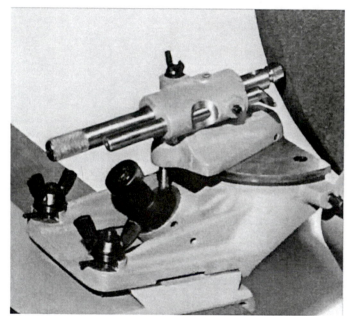

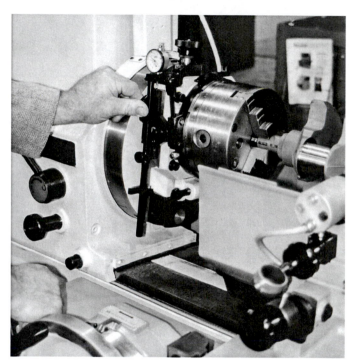

Fig.13-89 Checking stroke offset in a Berco grinder

Fig.13-88 A Berco radius dressing attachment (Courtesy Petersen Machine Tools)

The crankshaft is offset equally for stroke length at each end and the first crankpin indicated into position on the centerline of rotation (see Figs.13-89 and 90). The actual amount of main bearing journal offset from center is half the stroke length. After setting the grinding gauge to the finished size, the first crankpin is ground (see Figs. 13-91 and 92). The crankshaft rotates or "indexes" to each consecutive crankpin position at the grinder work heads (see Fig.13-93). While indexing speeds up positioning, the location must be rechecked with an indicator because production

Fig.13-90 Indicating a crankpin into position (Berco)

Fig.13-91 The grinding gauge is set to size and clamps around the journal

Fig.13-92 Grinding a crankpin

Fig.13-93 Indexing work heads for crankpins

The set-up for grinding main bearing journals requires positioning on the centerline of rotation. The set-up also must ensure that timing gears or sprockets and oil seals at the ends of the shaft are on center. Because of these requirements, positioning is actually from the gear or sprocket location in front, and the rear main seal surface at the rear of the shaft (see Figs. 13-94 and 95).

Fig.13-94 Mains set up from gear or sprocket locations at the front of the crankshaft

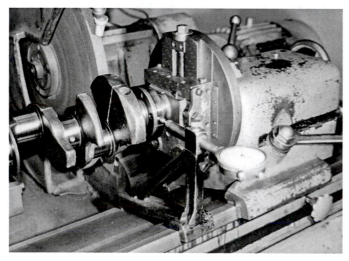

Fig.13-95 Mains set from the rear main seal location at the rear of the crankshaft

As in other machining operations affecting oil clearance, some engine builders prefer double checking shaft specifications prior to grinding. Calculate oil clearance as follows:

Housing bore	2.5000in
Bearing thickness	-.0615
Bearing thickness	-.0615
Oil clearance	-.0020
Shaft diameter	2.3750in

Oil holes require chamfering to prevent sharp edges from gouging the bearings (see Fig.13-96). The machinist chamfers the hole with a rotary grinder and mounted stone before polishing (see Fig.13-97).

Fig.13-96 An oil hole chamfer

Fig.13-97 Chamfering an oil hole with a die grinder

The final surface finish quality necessary for normal bearing service requires polishing after grinding. Grinding leaves microscopic burrs best removed with a fine-grit polishing belt (see Fig.13-98). Polishing in both directions of rotation is recommended to get all burrs, especially on nodular iron shafts. The rear main seal and thrust surfaces are also polished. Polishing is also recommended for engine overhauls even when the crankshaft is not reground. It removes minor scoring and marks caused by measuring and handling. Metal removal is less than .0002in (.005mm) on the diameter as excessive polishing causes hourglass and out-of-round journals.

Fig.13-99 Grinding a flywheel in a Kwik-Way grinder

If heat checking is present, magnaflux inspect the flywheel and look for cracking. While most heat checking is superficial, some may develop into major cracks that penetrate the flywheel. These cracks cannot be ground away. If cracks show on the reverse side of the flywheel, replace it.

It is important that the clutch surface remain parallel to the reverse side of the flywheel bolting to the crankshaft. Keep flywheel run-out under .005in (.12mm) TIR (see Fig.13-100). Correcting run-out sometimes requires resurfacing both sides of the flywheel.

Fig.13-98 Polishing a crankshaft

RESURFACING FLYWHEELS AND REPLACING RING GEARS

Resurfacing the flywheel on engines equipped with standard transmissions is frequently part of servicing the clutch during engine overhaul or rebuilding. Resurfacing restores flatness, removes minor heat checking caused by clutch slippage, and restores smooth clutch operation. Flywheel grinders are designed for simple set up and quick turn-around of grinding jobs (see Fig.13-99).

Fig.13-100 Checking flywheel run-out with an indicator

Recessed, or stepped, clutch surfaces require machining in two steps. The amount required for cleanup of the recessed clutch surface must also be removed from where the pressure plate attaches (see Fig.13-101). This two-step procedure maintains normal pressure for the clutch pressure plate. Specifications are not always available so it is best to measure these dimensions before and after grinding.

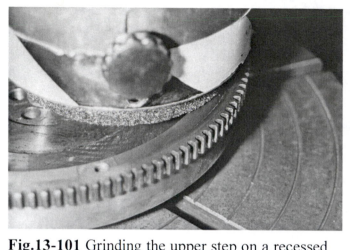

Fig.13-101 Grinding the upper step on a recessed flywheel

Fig.13-102 Tightening the puller collet over the pin

Resurfacing the upper step often requires removing the dowel pins used to locate the pressure plate. The tool used for removing these pins, or pins in block decks, is a collet type puller. The correct sized collet is place over the pin and closed tightly around it using a tapered sleeve and slide hammer (see Fig.13-102). The slide hammer action is then reversed and the pin removed (see Fig.13-103).

Fig.13-103 Removing the pin with the slide hammer

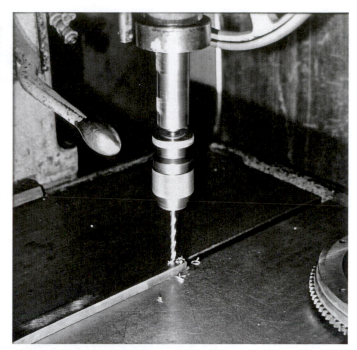

Fig.13-104 Drilling the hole to locate the pin

Should the puller fail to remove the pins, drill holes behind the pins so that they may be driven out with a pin punch. Keep in mind that many of these pins are hardened and cannot be drilled out. To locate holes directly behind these pins, first drill a hole matching the pin diameter into a steel plate clamped to the drill press table (see Fig.13-104). Next, turn the flywheel upside down; insert one of the pins into the drilled hole, and drill through the backside of the flywheel to the pin (see Fig.13-105). Be sure to drill all holes equally spaced around the flywheel to maintain balance.

Fig.13-105 Drilling behind the pin while positioned in the locating hole

This also is a good time to inspect starter ring gear teeth for wear. Replacement of the ring gear is simple and fast. To remove the old ring gear, drive a chisel between gear teeth to expand it and drive it off with a punch. To install the new ring gear, heat it with a torch until uniformly hot around its circumference and then quickly drop the ring gear into position on the flywheel (see Fig.13-106). Wear welding gloves for this procedure and be sure the ring gear is properly seated in place by tapping at several points with a punch and hammer.

Fig.13-107 Checking oil pump end clearance

Fig.13-106 Placing a heated ring gear into assembly around a flywheel

Fig.13-108 Sanding a pump cover to remove wear

OVERHAULING OIL PUMPS

Although the oil pump is probably the single most important part relating to engine longevity, it gets surprisingly little attention. Although engines frequently lose at least some oil pressure after break-in, technicians often attribute this wrongly to bearing clearance opening up and think of it as normal. In the author's experience, an extra 20 minutes preparing the oil pump keeps it working at peak efficiency for the life of the engine.

Begin by checking the parts of the pump against specified service limits. Check both gear and rotor type pumps for end clearance of the gears in the pump housing (see Fig.13-107). If end clearance exceeds .003in (.08mm), sand or lap the pump body to reduce clearance. The end covers for both pumps wear but are easily resurfaced on a sander (see Fig.13-108).

While there is a specified .005in (.12mm) limit of clearance between the gear teeth and the housing, it is equally effective to visually check the housing for scoring (see Fig.13-109). Only gears wobbling on their shafts, or solids going through the pump, cause this scoring. Scrap gear type pumps if there is major scoring or if the gears have contacted the housing.

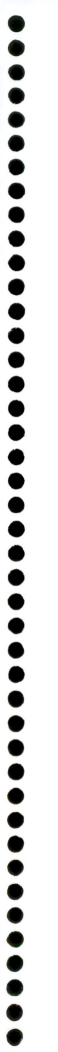

Fig.13-109 Checking clearance between gears and the pump housing

Check rotor type pumps for clearance between inside and outside rotors with a feeler gauge (see Fig.13-110). The service limit is approximately .014in (.36mm). Clearance between the outer rotor and the pump housing should not exceed .010in (.25mm) (see Fig.13-111).

Fig.13-110 Checking clearance between pump rotors

Fig.13-111 Checking rotor-to-pump housing clearance

Check the oil pump drives for both types of pumps for wear especially the hexagonal intermediate shafts between the distributor and the oil pump. Watch for visible rounding of the ends of these drives that leads to failure (see Fig.13-112).

Fig.13-112 Checking for rounded hex oil pump drive shafts

The reason many pumps lose some of their efficiency soon after installation is because sharp edges on gears or rotors cut into the pump housings and pump covers and increase end clearances. To prevent this, deburr or break all sharp edges on gears and rotors. This is no more than a two minute job using a polishing belt (see Fig.13-113).

Fig.13-114 Inspecting the pressure regulator valve and passage

Similar inspection is required for crank driven "rotor" or "gear-rotor" oil pumps but clearances are different. Outer rotor to housing clearance is approximately .012in (.3mm) and the limit between inner and outer rotor is approximately .008in (.2mm) (see Figs. 13-115 and 116). End clearance is comparable to other pumps at .003in (.08mm) although correcting clearance may require replacing the pump cover or rotors. Measure rotor thickness and check against specifications to be sure. For assembly, check the direction of inner rotor installation and torque the cover to specifications.

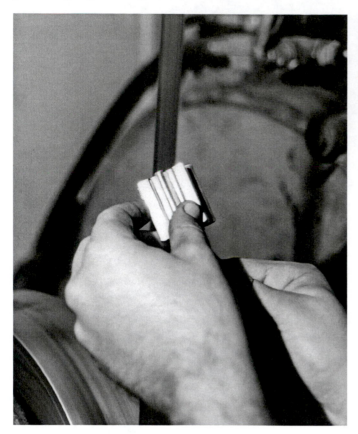

Fig.13-113 Using a polishing belt to break sharp edges on a pump gear

If needed, oil pump repair kits are available with gears or rotors, gaskets or seals, and pressure relief springs. Order oil pump driveshafts or intermediate shafts separately. In preparation for final assembly, thoroughly clean and deburr all parts including the pressure regulator passage and valve (see Fig.13-114).

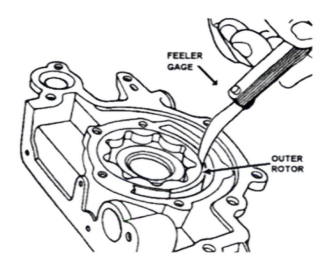

Fig.13-115 Checking rotor to housing clearance in a gear rotor pump (Chrysler)

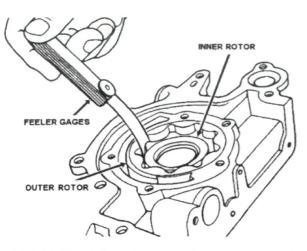

Fig.13-116 Checking clearance between rotors in a gear rotor pump (Chrysler)

Fig.13-117 Press fitting an oil pump screen into assembly with the pump body

Oil pump pickup screens attach to the oil pump, or a passage in the block leading to the pump, with all kinds of devices. Some use pipe threads that must be sealed, some attach with screws and require a gasket, some clamp into place and require an O-ring, and others press into the pump body. For those that press-fit, remove oil from the pick-up and bore, apply a few drops of Loc-Tite sealer and press the pick-up into the pump body. Be sure to use the correct tool (see Fig.13-117). Remember that should the pump pull air into the inlet side, the oil becomes aerated and the engine fails. Also check the position of the pick-up in the oil pan and make sure that it is within a 1/4in (6-mm) of the bottom. A pick-up too far from the bottom pan is equivalent to running the oil level low

SUMMARY

For overhaul, wear of cylinder walls and piston ring grooves are limited. Machinists must hone the minimum from cylinders and thoroughly clean ring grooves. Honing too much increases piston clearance potentially causing engine noise and rapid ring wear.

When honing for oversize pistons and rings, surface finishes must match piston ring requirements. To meet all of these requirements, machinists must have knowledge of honing processes and piston ring materials.

Boring must be done so that the cylinders remain in proper alignment with the crankshaft centerline. Sleeving also requires boring to precise diameters so that the sleeves seal the water jackets. In addition to operating boring machines, machinists must know how to properly position cylinder blocks and boring bars so that cylinder alignment is correct.

Line boring corrects housing bore diameters and alignment. Less obvious is that these operations move the crankshaft and change concentricity with seals front and rear. Machinists need to know how to evaluate the effect of these changes and how to center timing cover and rear main seals.

Errors in fitting piston pins lead to engine failure and noise complaints. In addition to operating boring and honing machines, measuring skill and knowledge of comparative measuring tools are essential in achieving the specified clearances.

Measuring skill and knowledge of comparative measuring tools are also essential in machining connecting rod hosing bores; specifications for bore diameters and roundness are within very close limits. As in pin fitting, both boring and honing processes are used.

The assembly and aligning of pistons and connecting rods is critical to piston, pin, ring, and rod bearing longevity. While full-floating piston pins enable assembly by hand, the installation of the retaining rings requires care. Press-fit assemblies require knowledge of pin presses and press adapters and of rod heaters. Rod and piston alignment must be within close limits and machinists should be familiar with a variety of equipment for the purpose.

The average machinist need not develop skill in regrinding camshafts and lifters or crankshafts because much of this work is sublet to specialists. However, machinists and technicians do benefit from a general knowledge of processes, job requirements, and points of inspection. In the assembly stage, these people have the last opportunity for quality control and must catch defects that cause failures.

Chapter 13

RECONDITIONING ENGINE BLOCK COMPONENTS

Review Questions

1. Machinist A says that glaze breaker hones produce the finish required for ring seating. Machinist B says that these hones reduce taper and improve roundness of cylinders. Who is right?

   a. A only
   b. B only
   c. Both A and B
   d. Neither A or B

2. If the cylinder is in good condition but piston clearance exceeds specifications, fit can be improved by
   a. installing new rings
   b. resizing the piston by knurling
   c. honing the cylinder
   d. removing the ridge

3. Replace pistons if ring side clearance exceeds
   a. .002in
   b. .004in
   c. .006in
   d. specified wear limits or double new clearance

4. Machinist A says that in production, cylinders are bored square with the main centerline or the pan rails. Machinist B says that for portable boring bars to position cylinders correctly, decks and mains must be parallel. Who is right?

   a. A only
   b. B only
   c. Both A and B
   d. Neither A or B

5. Machinist A says chamfer to prevent preignition. Machinist B says chamfers reduce ring breakage on assembly. Who is right?

   a. A only
   b. B only
   c. Both A and B
   d. Neither A or B

6. Machinist A says bore cylinders 1/2in past the bottom end of cylinders to allow for honing. Machinist B says honing stones must pass the bottom end of cylinders to produce straight cylinder bores. Who is right?

   a. A only
   b. B only
   c. Both A and B
   d. Neither A or B

7. Machinist A says that chrome or cast iron rings require cylinder finishes in the 10 to 15 RA range. Machinist B says that moly rings require approximately 30 RA cylinder finishes. Who is right?

   a. A only
   b. B only
   c. Both A and B
   d. Neither A or B

8. The angle of crosshatch in honing machines is controlled by
   a. cylinder length
   b. spindle RPM
   c. stroking rate
   d. a combination of spindle speed and stroking rate

9. To shrink fit a 4-inch diameter cylinder sleeve in an iron block requires a minimum temperature difference of _____ degrees F.
   a. 68
   b. 125
   c. 175
   d. 225

10. Cylinder sleeves for passenger car engines are available in _____in wall thickness(s).
    a. 1/16
    b. 3/32
    c. 1/8
    d. 3/32 or 1/8

11. Maintaining equal clearance in each main
    bearing
    a. prevents oil pressure loss
    b. prevents knocking
    c. maximizes bearing life
    d. minimizes crank distortion and breakage

12. Machinist A says that crankshaft and camshaft
    centerlines move together half the amount
    bores diameters are reduced for line honing.
    Machinist B says that these centerlines move
    apart after line honing. Who is right?

    a. A only   c. Both A and B
    b. B only   d. Neither A or B

13. Machinist A says that a thrust main cap cut
    at an angle increases crankshaft endplay.
    Machinist B says grind main caps in the same
    plane as the original parting face. Who is right?

    a. A only   c. Both A and B
    b. B only   d. Neither A or B

14. Machinist A says that correcting crankcase
    alignment with multiple housing bore diameters
    requires line boring. Machinist B says that
    when welding bore saddles or replacing a main
    cap, line honing is required.   Who is right?

    a. A only   c. Both A and B
    b. B only   d. Neither A or B

15. Machinist A says that cutting main caps
    for line boring or honing can distort main
    seal locations. Machinist B says that
    these operations can affect crankshaft seal
    concentricity at both ends. Who is right?

    a. A only   c. Both A and B
    b. B only   d. Neither A or B

16. Machinist A says that piston pins in press fit
    assemblies have .0005in clearance in pistons.
    Machinist B says that full floating piston pins
    have .001in interference fit in rods. Who is right?

    a. A only   c. Both A and B
    b. B only   d. Neither A or B

17. Machinist A says fit pistons with excessive pin
    clearance for oversized pins or replace them.
    Machinist B says bush and resize these pistons.
    Who is right?

    a. A only   c. Both A and B
    b. B only   d. Neither A or B

18. Machinist A says that cracked rods are resized
    to fit oversize diameters. Machinist B says
    that resizing should restore roundness within
    .0003in and diameter within specified limits.
    Who is right?

    a. A only   c. Both A and B
    b. B only   d. Neither A or B

19. Machinist A says prepare for resizing rods
    by removing .002in from each parting face.
    Machinist B says remove more if needed for
    cleanup. Who is right?

    a. A only   c. Both A and B
    b. B only   d. Neither A or B

20. Rod aligning must follow resizing rod housing
    bores unless
    a. straightened before resizing
    b. honed two at a time
    c. honed in a power stroking hone
    d. bored in alignment with pin bores

21. A connecting rod housing bore is reduced in diameter .004in by grinding the cap and rod. The center distance is
    a. unchanged
    b. lengthened .004in
    c. shortened .004in
    d. shortened .002in

22. Machinist A says that when press-fitting piston pins through rods, heat the small end of the rod to 425 degrees F. Machinist B says press pins into place from one side only. Who is right?

    a. A only   c. Both A and B
    b. B only   d. Neither A or B

23. When press fitting piston pins through connecting rods,
    a. heat the piston pin first
    b. use .002in interference
    c. press from one side
    d. press back and forth until the pin is centered

24. Machinist A says remove bend and twist in rods to within .001 in 6in. Machinist B says correct twist before bend. Who is right?

    a. A only   c. Both A and B
    b. B only   d. Neither A or B

25. Machinist A says that hardened crankshafts are reclaimed by gas atmosphere hardening. Machinist B says that they are reclaimed by soaking in cyanide salts. Who is right?

    a. A only   c. Both A and B
    b. B only   d. Neither A or B

26. Damaged crankpins or journals are welded before grinding so that
    a. cracks may be repaired
    b. fillets may be restored
    c. stroke length may be corrected
    d. shafts may be ground to minimize undersize or kept standard

27. Machinist A says that excessive polishing causes problems with crankshaft journal alignment. Machinist B says that it causes problems with the shape and roundness of journals. Who is right?

    a. A only   c. Both A and B
    b. B only   d. Neither A or B

28. Machinist A says that crankshaft thrust wear is corrected by welding and grinding to specifications. Machinist B says that grinding for oversize thrust bearings corrects this problem. Who is right?

    a. A only   c. Both A and B
    b. B only   d. Neither A or B

29. Machinist A says that before installing a reground crankshaft, chamfer oil holes. Machinist B says polish journals. Who is right?

    a. A only   c. Both A and B
    b. B only   d. Neither A or B

30. Machinist A says that to minimize run-out, reface flywheels parallel to the original crankshaft flange. Machinist B says grind the two steps in recessed flywheels to maintain clutch pressures. Who is right?

    a. A only   c. Both A and B
    b. B only   d. Neither A or B

31. Machinist A says remove flywheel ring gears with a torch. Machinist B says install them with a punch and hammer. Who is right?

    a. A only   c. Both A and B
    b. B only   d. Neither A or B

32. When clearance between oil pump rotors and the pump housing is excessive, the pump should be
    a. fitted with new rotors
    b. fitted with oversize rotors
    c. remachined
    d. replaced

33. Machinist A says correct excessive oil pump gear or rotor end clearance during overhaul. Machinists B says remove all sharp edges to reduce pump wear. Who is right?

    a. A only   c. Both A and B
    b. B only   d. Neither A or B

34. Machinist A says that should the oil pump pick up pull air, the oil becomes aerated. Machinist B says that this causes bearing failure. Who is right?

    a. A only   c. Both A and B
    b. B only   d. Neither A or B

35. Machinist A says that the oil pump pick up must be within one-inch of the bottom of the pan. Machinist B says that it must be within 1/4 inch. Who is right?

    a. A only   c. Both A and B
    b. B only   d. Neither A or B

FOR ADDITIONAL STUDY

1. What cylinder wall surface finish is required for chrome rings? What finish is needed for moly rings?

2. How are the surface finishes above obtained in honing?

3. How should cylinders be aligned relative to the crankshaft centerline?

4. How is this alignment maintained when boring cylinders? What about honing cylinders?

5. List two possible points where maximum piston diameter can be measured?

6. How is a cylinder repair sleeve retained in the block?

7. How much, and which direction, does the crankshaft centerline move following line honing? What about line boring?

8. For a passenger car engine with full-floating piston pins, how much piston pin clearance is needed in the connecting rod? What about in the piston?

9. For a passenger car engine with press-fit piston pins, how much piston pin interference is needed in the connecting rod? How much clearance in the piston?

10. A connecting rod housing bore is 2.2500 inches in diameter and the rod bearings are .0615in. thick. For .0020in. clearance, what should the crankpin diameter be?

11. How does the crank grinder position the shaft so that the rear main oil seal and timing sprocket diameters remain concentric with main bearing journals?

12. List five points of inspection for crankshafts?

13. When correcting bend and twist in connecting rods, which is corrected first? What difference does it make?

14. In replacing flywheel ring gears, how is the old gear removed and how is the new gear installed?

15. When overhauling an oil pump which specifications can be restored? How are pump longevity and efficiency improved?

Chapter 14

RESURFACING CYLINDER HEADS AND BLOCKS

Upon completion of this chapter, you will be able to:
- Compare milling and grinding processes as used for resurfacing castings and list the advantages of each.
- Describe intake manifold alignment problems caused by resurfacing.
- List the steps required to correct intake manifold alignment.
- Determine the correct V-block ratios for resurfacing the intake sides of cylinder heads or intake manifolds and the tops of engine blocks.
- Outline the differences in resurfacing procedures for overhead camshaft and pushrod cylinder heads.
- List special precautions that apply to resurfacing diesel cylinder heads.
- List special requirements that apply to resurfacing air-cooled cylinder heads.

INTRODUCTION

The primary reason for resurfacing decks is to ensure head gasket sealing. Surfaces on cylinder heads and blocks become warped because of overheating or improperly tightened head bolts. Surfaces also burn and erode from the flow of gases through leaking gaskets. Resurfacing restores flatness and the quality of surface finish required for optimum gasket sealing.

For rebuilding, resurfacing is done to restore decks to new specifications. In overhauls or valve grinding, surfacing is done only as required although most technicians prefer resurfacing at least the cylinder heads. With the cylinder head deck surfaces restored, head gaskets are more forgiving of minor warpage in block decks. Always inspect for flatness, dents or scratches, corrosion around water passages, or burning that might lead to gasket failures. Expect that half the cylinder heads inspected require resurfacing for some reason.

It is occasionally necessary to resurface intake or exhaust sides of cylinder heads and manifolds. Intake manifolds sometimes distort because of the exhaust gas passages that heat plenum areas and supply exhaust gas for recirculation into the intake airflow (EGR). The exhaust sides of cylinder heads and exhaust manifolds sometimes warp and erode from exhaust leaks. Positive gasket sealing is impossible without first correcting these problems.

If a head gasket failure has occurred, be especially careful to check surfaces. Resurface both heads on V-block engines even if only one head gasket failed since both heads were subjected to the same operating conditions that led to failure. Machining both heads also keeps compression equal on both sides.

COMPARING RESURFACING MACHINES

Surfacing machines remove metal by sanding, grinding or milling. Some are suited to quick, routine resurfacing of cylinder heads and others resurface both engine blocks and cylinder heads. With the introduction of multi-layered steel gaskets, surface finish specifications were reduced to 20 Ra microinches or sometimes less. These machines will not produce such finishes unless run properly.

Large belt sanders are ideally suited to cleaning up minor surface imperfections on cylinder heads (see Fig.14-1). Because they require no setup or clamping, they are faster than grinding or milling machines. Sanders are also handy for less precise, quick cleanup of exhaust manifolds and other odd surfacing jobs. It is however more difficult to maintain precision and alignment when removing greater amounts of material.

Fig.14-1 Surfacing intake side of a cylinder head on a sander (KwikWay)

Grinding machines produce relatively smooth surface finishes and can cut through hard spots in castings such as welds, seat inserts, or diesel pre-combustion chambers (see Figs. 14-2). Because the casting and the grinding wheel tend to deflect away from each other during grinding, shallow cuts and extra passes are required to obtain flatness. These final passes are made without any additional increases in the depth of cut and are referred to as "sparking out."

Most grinding wheels use an abrasive composition intended for iron although aluminum is sometimes ground with these same abrasives by coating the face of the wheel with soap. To grind aluminum, take shallow passes, dress frequently, and keep the wheel conditioned with the soap. The shallow cuts and the need to spark out make grinding slower than sanding or milling.

With sufficient rigidity in the machine spindle and setup, milling requires only one deep cut to clean up warpage and one shallow cut to improve flatness and surface finish (see Figs. 14-3). There are two basic types of milling cutters now used for resurfacing. The first utilizes a face mill cutter assembly with 10 or more single point carbide cutters (see Fig.14-4). Because of the number of cutters, and the carbide tool material, cutting speeds and feeds are high. The second type of cutter assembly typically uses one or possibly two CBN cutter inserts that run at very high cutting speeds of 3,000 feet per minute but, because there are fewer cutters, feed rates are reduced (see Fig.14-5). For these cutter assemblies, tool manufacturers recommend CBN for surfacing iron castings and PCD for aluminum. In any milling operation, watch for hard spots in castings that can dull cutters.

Fig.14-2 Grinding a cylinder head deck (Berco)

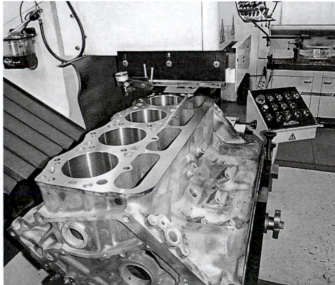

Fig.14-3 Milling a block deck

Fig.14-4 A face mill cutter assembly (Sunnen)

Fig.14-5 A CBN milling attachment (Berco)

The introduction of multi-layer steel (MLS) cylinder head gaskets for aluminum engines placed new emphasis on surface finish quality. Unless coated, these MLS gaskets require a surface finish of 20 Ra microinches or better for proper sealing. For coated MLS head gaskets and other engine gaskets, surfaces finishes up to 30 Ra microinches are satisfactory. With care, very fine finishes can be obtained grinding or milling. Grinding wheels must be carefully dressed, preferably trued with a diamond, and soap also helps. Milling cutters must be properly sharpened and lubricated with a light oil to keep aluminum from adhering to cutting edges, effectively dulling them. In both cases, light cuts and slow feed rates are necessary.

GENERAL PRECAUTIONS

As a rule, remove as little as possible to restore flatness. However, inspecting with a straightedge and a feeler gauge has limited accuracy and does not always show how much is required for clean up. For example, a .004in (.13mm) feeler gauge might just slip under the straightedge, but .008in (.02mm) or stock removal can be required for clean up.

The most obvious complication caused by excessive surfacing is increased compression resulting from the reduced clearance volumes. Increases in compression ratios are slight in most cases and therefore cause no problems. In extreme cases, depending upon engine controls, increases in compression can require using a high octane fuel or retarding the ignition timing. Neither alternative is desirable since high-octane fuel increases operating cost and retarded ignition timing reduces efficiency.

Certain engines have proportionately greater changes in clearance volumes when resurfaced and therefore have greater increases in compression ratios. The greatest changes occur when resurfacing cylinder blocks with large bore diameters or cylinder heads with combustion chambers covering the full cylinder bore. Engines with high compression ratios, such as performance engines or diesels, experience the greatest changes.

In some engines, valve-to-piston clearance is small and excessive resurfacing of a cylinder head can cause interference between valves and pistons (see Fig.14-6). The amount of piston-to-valve clearance has less to do with compression ratio than upon the positioning of valves in the combustion chamber. Do not assume that low compression or low performance engines have more clearance and are therefore forgiving.

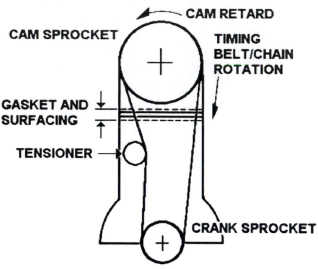

Fig.14-6 Interference between valves and pistons caused by excess resurfacing

removal or head thickness because of limited chain tensioner travel and because of the loss of performance caused by late intake valve closing.

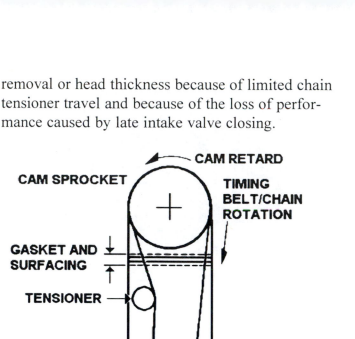

Fig.14-7 Retarded valve timing caused by changes in crank and cam centerlines

In pushrod engines with nonadjustable rocker arms, there is another change to consider. While resurfacing brings the cylinder head and block closer together, the pushrod lengths remain the same. The result is that the pushrods hold valves open because the head is closer to the block. This is a situation similar to having valve stems or pushrods that are too long. The result is loss of compression caused by open valves or possibly bent valves.

Resurfacing causes valve timing changes in overhead camshaft engines because of timing chain or belt slack created as the camshaft and crankshaft move closer together (see Fig.14-7). Typically, crankshaft rotation takes up slack on the driving side of a timing chain and a chain tensioner takes up the slack on the opposite side. Valve timing retards because the camshaft remains stationary until crankshaft rotation takes up the chain slack. Especially important to engine performance, the intake valve closes late and compression drops. To be exact, valve timing retards one degree for every .020in (.50mm) removed. A one-degree change is not much by itself, but it is significant when added to the chain stretch found in overhead cam engines. Keep surfacing within specified limits for stock

Another problem is the change in intake manifold alignment on V-block engines. On these engines, resurfacing not only causes heads to move toward the block, but also causes them to move closer together. Unfortunately, the intake manifold remains just as wide as before but must still fit between the cylinder heads. The result is that manifold bolt holes and intake ports move out of alignment (see Fig.14-8). The most obvious problem is the potential for vacuum leaks. Less obvious is the oil pullover into intake ports resulting from internal vacuum leaks into intake ports.

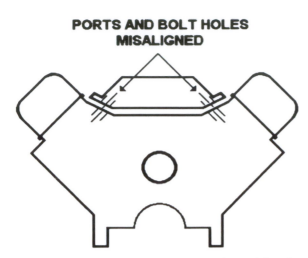

PORTS AND BOLT HOLES MISALIGNED

Fig.14-8 Misaligned V-block heads and intake after resurfacing

Keep in mind that this may not be the first time the heads or blocks were resurfaced and that any additional resurfacing adds to the total. For OHC cam engines, specifications for minimum head thickness are generally available. Expect the maximum surfacing allowed to be in the .010 to .025in range (.25-.60mm). Using thicker head gaskets compensates for half this amount and these should be specified whenever available.

It is obvious that machinists have to keep precise track of amounts removed. In milling this isn't too much of a problem because the cutter will remove the same amount as the depth of cut. Because of wheel wear in grinding machines, the amount removed is less than the depth of cut. If possible, measure deck height or head thickness before and after surfacing. The problem is that sometimes there are no convenient points to measure from. In such cases, install a set screw in a head or intake manifold bolt hole and measure from the surface to the set screw with a depth mike before and after grinding. A "Nylok" or similar self locking screw might be necessary to keep it from turning. If the threads are not square with surface, place a steel ball bearing on top of the set screw (see Fig.14-9).

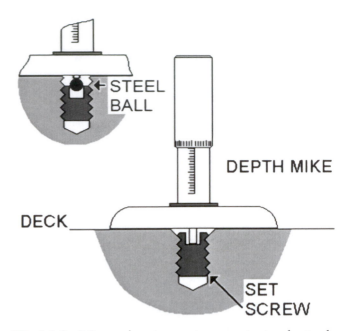

Fig.14-9 Measuring to a set screw to track stock removal

Surfacing heads leaves sharp edges and burrs around the edges of combustion chambers and, unless removed, they become hot and cause pre-ignition. Break these sharp edges with a die grinder or emery cloth (see Fig.14-10).

Fig.14-10 Breaking sharp edges around combustion chambers

Unless head bolt holes are counterbored, surfacing block decks brings head bolt threads closer to the decks. Unless chamfered, these threads pull up into the head gasket and interfere with clamping and sealing. Chamfer the top thread to match the outside diameter of the head bolts using a countersink (see Fig.14-11). This is also a good time to reinstall dowel pins pulled for resurfacing as they help protect the deck while moving the block through other shop operations and assembly.

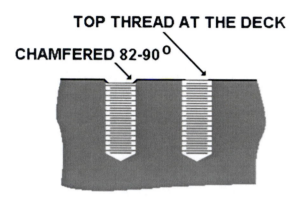

Fig.14-11 Chamfering head bolt holes

CORRECTING V-BLOCK INTAKE MANIFOLD ALIGNMENT

Some cylinder heads warp so badly that saving them requires more resurfacing than machinists and technicians like to do. Also, heads and blocks for performance engines are sometimes resurfaced greater amounts to reduce combustion chamber volumes and deck clearance to minimum specifications. To avoid problems, even in high performance engines, raise compression by changing pistons if available.

In extremes, manifold alignment is corrected only by multiple machining operations. To visualize the problem, study an engine assembly before resurfacing and note that the manifold bolts and intake ports are in alignment. Note also that there is a certain amount of crush, or gasket clearance, between the ends of the manifold and the top of the block (see Fig.14-12). To assembly properly it is

necessary to restore the alignment of the bolts and ports and the gasket crush after resurfacing heads and blocks.

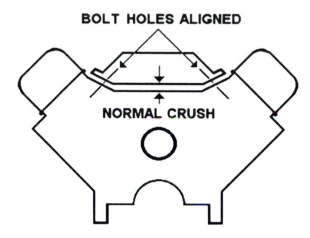

Fig.14-12 Normal alignment of V-block, heads, and intake manifold

First, restore the alignment of bolts and ports by resurfacing the intake sides of the heads to increase the distance between them to match the size of the manifold (see Fig.14-13). Note however that this operation reduces gasket crush at the ends of the manifold because it causes the manifold to sit lower on the engine (see Fig.14-14). The second operation, resurfacing the top of the block beneath the manifold, restores the crush (see Fig.14-15). This step is often omitted by using silicone sealer, instead of gaskets, under the ends of the manifold during assembly.

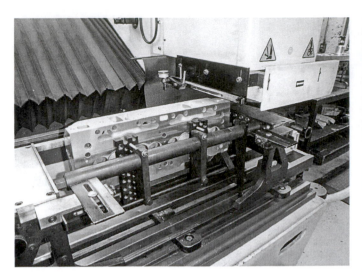

Fig.14-13 Resurfacing the intake sides of heads

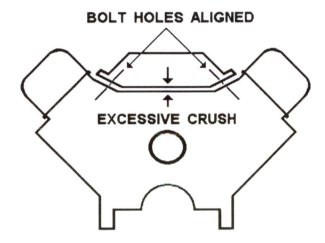

Fig.14-14 Although port and bolt alignment is corrected, there is excess crush on the end gasket

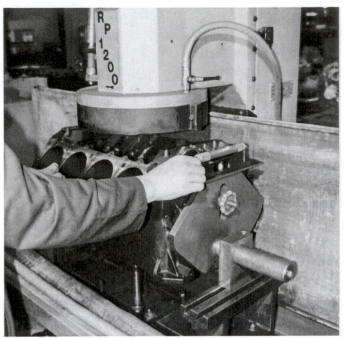

Fig.14-15 Setting up to machine the top of the block for gasket clearance

If the distributor installs through the top of the block, surfacing lowers it into the block. Should the distributor drive the oil pump through an intermediate shaft, the shaft can bind between the pump and the distributor. Also, distributor and camshaft drive gears sometimes wear because lowering the point of gear tooth contact changes it from a rolling to a wiping action (see Fig.14-16).

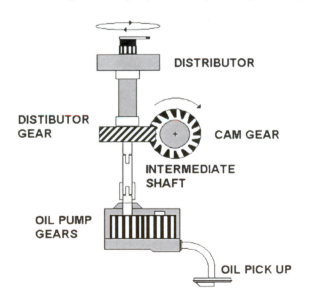

Fig.14-16 The relationship between the camshaft, distributor, and oil pump drives

Each different engine uses a particular set of stock removal ratios to determine how much correction to make in each resurfacing step. For a 350 Chevrolet for example, the total amount removed from block and cylinder head deck surfaces is multiplied by 1.2 for correcting the intake sides and multiplied by 1.7 for the top surface of the block. For example, if .010in is removed from deck surfaces, calculate removal from the other sides as follows:

Intake Sides 1.2 x .010 = .012in removal
Top of Block 1.7 x .010 = .017in removal

The ratios vary according to the angle between cylinder banks for the block and the included angle from the deck of the cylinder head to the intake side (see Fig.14-17). While the included angles of cylinder heads are numerous, there are two common angles for engine blocks, 60 and 90 degrees (see Fig.14-18). Study the combinations of angles laid out in the following table.

Fig.14-17 Measuring the angle between the deck and intake side with a protractor

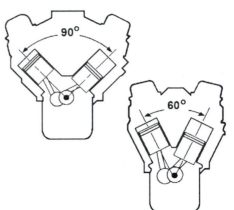

Fig.14-18 V-blocks with 60 and 90-degree angles between cylinders

An option to the machining steps above includes surfacing the sides of the intake manifold instead of the sides of the cylinder heads (see Fig.14-19). Calculations for the amount of resurfacing are the same, but the machinist has the option of selecting the easiest or quickest setup for the particular engine.

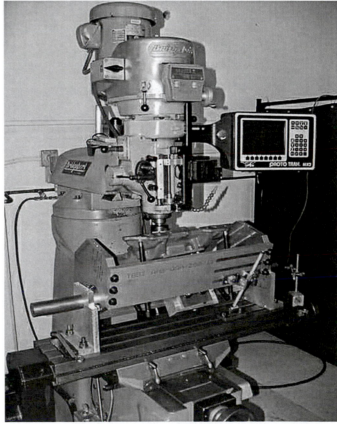

Fig.14-19 Surfacing the intake sides of a manifold using Fowler Machine tooling

90-DEGREE V-BLOCKS		
Included Head Angle	Intake Side Ratio	Top of Block Ratio
95	0.9	1.3
90	1.0	1.4
80	1.2	1.7
75	1.4	2.0
70	1.7	2.3
65	2.2	2.8
60-DEGREE V-BLOCKS		
90	0.6	1.2
80	0.7	1.3
75	0.75	1.4
70	0.8	1.5

DETERMINING V-BLOCK RATIOS

Graphic solutions for these ratios help us visualize the alignment problem. A reconstruction of the problem requires only a few lines and measuring the sides the triangle formed yields all ratios. Using a 350 cubic-inch Chevrolet as an example, begin by making a drawing showing the intersection of the deck of the block, the top of the block, and the intake side of the cylinder head (see Fig.14-20). Take the angles directly from the engine.

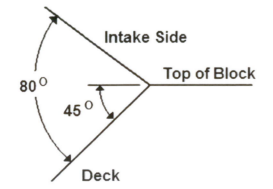

Fig.14-20 Draw the intersection of deck, top of block, and intake side of head

The next step is to establish a reference point on the intake side of the cylinder head. Imagine this point as the center of a manifold bolt hole (see Fig.14-21).

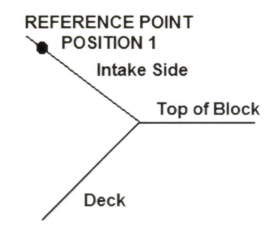

Fig.14-21 Establish reference point position 1

Next, consider that after resurfacing block decks or heads, this point moves. To be exact, it moves at a 90-degree angle to the deck of the block (see Fig.14-22).

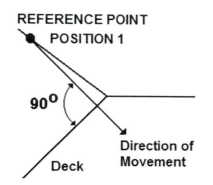

Fig.14-22 The direction of movement from position 1

For purposes of determining ratios, imagine removing 1 inch by resurfacing. Measure along the direction of movement 1in from the reference point and draw another point (see Fig.14-23).

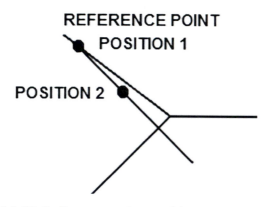

Fig.14-23 Reference point position 2

Position two of the reference point shows that the cylinder head has moved toward the engine block. The reference point must be moved outward again so that the manifold will fit between the cylinder heads. Resurfacing the intake side of the head moves the reference point outward along a line 90 degrees to the intake side of the head (see Fig.14-24).

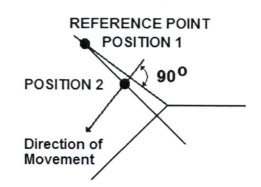

Fig.14-24 The direction of movement from position 2

Resurfacing the intake side must move the reference point outward until it is directly under its original position. Draw a line downward from the original position of the reference point at a 90-degree angle to the top of the block (see Fig.14-25).

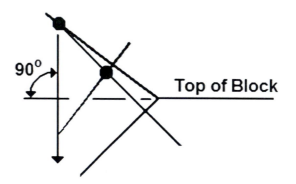

Fig.14-25 A vertical line closing the triangle

Reference point position three is at the intersection of the last two lines drawn. This point completes the triangle required to Figure out ratios (see Fig.14-26)

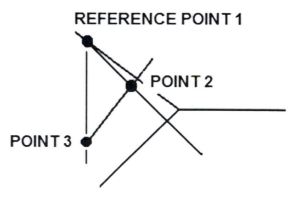

Fig.14-26 Reference point position 3

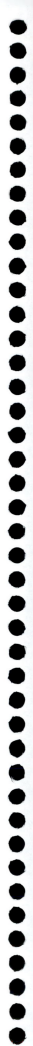

The next step is to measure the sides of the triangle. The measured distances will be equivalent to the stock removal ratios (see Fig.14-27). Here, the intake side ratio is 1.2:1 and the top of block ratio is 1.7:1. If 1 inch were actually removed from the face of a cylinder head, the reference point would move from position one to position two. To correct manifold bolt and port alignment, the intake side of the head must be machined until the reference point moves from position two to position three. To provide clearance under the manifold for gaskets, the top of the block must be machined an amount equal to the distance between reference point positions one and position three.

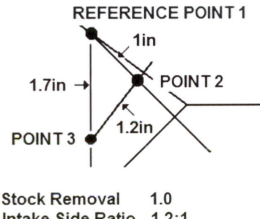

Stock Removal 1.0
Intake Side Ratio 1.2:1
Top of Block Ratio 1.7: 1

Fig.14-27 Measure the three sides of the triangle to determine ratios

This method of determining ratios works for any V-block engine. The only basic information needed is the angle from the top of the block to the deck side of the block and the included angle of the cylinder head. As important as deriving the ratios, is that this exercise helps us visualize the changes in alignment caused by resurfacing.

RESURFACING OVERHEAD CAM CYLINDER HEADS

Check for free rotation of the camshaft before resurfacing any overhead cam head. If the head is warped on the topside, straighten it first to correct

cam alignment and to minimize surfacing. Overhead camshaft centerlines are corrected on some cylinder heads by resurfacing the cam side. These are heads with removal cam bearing housings with no part of the cam bearing bores in the head. To prepare, remove valve guides, dowel pins, and adjusting screws from the topside and position the head casting upside down in the surfacing machine (see Fig.14-28). Take care to maintain parallel alignment between both sides of these heads by checking position in the setup with a depth mike or vernier depth gage (see Fig.14-29). If available, place shims under the bearing supports on reassembly to raise the camshaft centerline and thereby compensate for surfacing both sides of these heads (see Fig.14-30).

Fig.14-28 Resurfacing the cam side of an OHC head

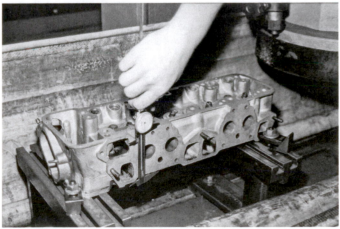

Fig.14-29 Check parallelism by comparing thicknesses at each end of the head

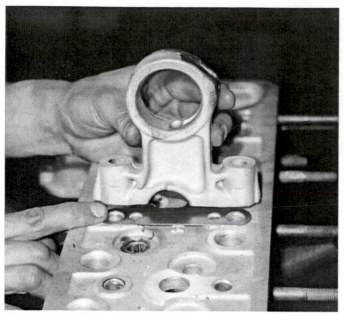

Fig.14-30 Shimming bearing supports on an OHC head

With chain driven overhead camshafts, attach timing covers to the cylinder block when resurfacing. This keeps the timing cover flush with the deck of the block and prevents interference with the clamping of the cylinder head gasket (see Fig.14-31).

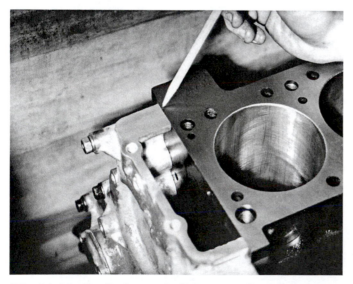

Fig.14-31 Surfacing a timing cover flush with the block on a chain-driven OHC engine

RESURFACING DIESEL CYLINDER HEADS

Diesel cylinder heads present special problems in resurfacing. First, it may be desirable to resurface these heads with the precombustion chambers in place to keep them flush with the cylinder head surface (see Fig.14-32). If possible, remove, clean, and reinstall precombustion chambers before resurfacing.

Fig.14-32 A diesel precombustion chamber ground flush

These precombustion chambers are alloy steel and are difficult to machine. Grinders work well but expect extreme wheel wear. To keep track of stock removal, measure head thickness before and after grinding.

Because of the extremely high compression ratios in these engines, keep track of valve recession in the combustion chamber to prevent piston-to-valve interference. If uncertain, measure from the deck to the valve on tear down and measure again after grinding valves and seats and resurfacing (see Fig.14-33). Grind valve seats slightly deeper to restore valve recession and then correct installed spring and stem heights on the topside of the head.

Fig.14-33 Checking valve recession after surfacing heads

RESURFACING AIR-COOLED CYLINDER HEADS

Air-cooled cylinder heads, primarily Volkswagen and Porsche, require resurfacing procedures that not all machine shops are equipped to perform. With high mileage, cylinder barrels embed into the sealing surfaces around the chambers of these cylinder heads and, unless corrected, the damage leads to seal failures (see Fig.14-34).

Fig.14-34 Check for damaged sealing surfaces on VW-Porsche heads

Repair requires machining in two or possibly three steps. First, reface the seal surface using a facing head (see Fig.14-35). In a facing head, the single-pointed tool feeds on a spiral with the rotation of the chuck. Should the new seal surface extend too far into the combustion chamber, the chamber contour must be remachined to narrow the sealing area (see Fig.14-36). In every case, surface the deck to restore the depth to the seal surface or interference between the head and the barrels occurs on assembly (see Fig.14-37). On individual cylinder heads such as those in six cylinder Porsches, head thickness must also be kept equal (see Fig.14-38).

Fig.14-35 Resurfacing the sealing surface with a facing head

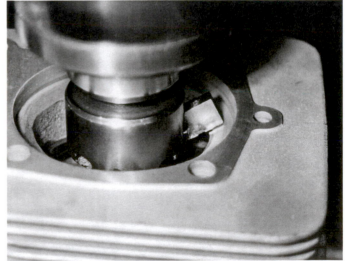

Fig.14-36 Reshaping the chamber contour after facing the sealing area

Fig.14-37 Resurfacing to correct the distance to the sealing surface

Fig.14-38 Head thickness must be uniform on these individual heads

Watch for differences in cylinder lengths and crankcase dimensions. For precise assembly, it is sometimes necessary to shim or face crankcases to equalize total deck heights and to keep cylinder heads parallel in assembly to the crankcase.

SUMMARY

Cylinder head and block decks are resurfaced primarily to ensure head gasket sealing. However, gaskets have changed. There are coated gaskets and multi-layered steel just to name two popular types and each has particular surface finish requirements. Machinists must be capable of producing the required finishes using either milling or grinding processes on iron or aluminum castings.

It is best practice to machine deck surfaces the minimum amount necessary to achieve clean-up. Should the amount required be excessive on V-block engines, machinists must calculate the corrections required and machine the intake sides of cylinder heads so that that the intake manifold will again fit. In overhead cam engines, excessive surfacing increases chain slack and changes valve timing.

There are a number of special requirements that machinists must deal with when surfacing timing covers on overhead cam blocks, precombustion chambers in diesel cylinder heads, aircooled heads, or aluminum verses iron castings. They must also be aware of changes in compression, valve-to-piston clearance, and valve timing changes caused by resurfacing. Customers are not always aware of special problems or changes in the engine and shops must therefore be prepared to explain them.

Chapter 14

RESURFACING CYLINDER HEADS AND BLOCKS

Review Questions

1. Machinist A says that milling machines have faster stock removal rates than grinders. Machinist B says that grinders are better suited to machining surfaces with hard spots. Who is right?

   a. A only
   b. B only
   c. Both A and B
   d. Neither A or B

2. Unless grinding wheels spark out, the surface is
   a. not flat
   b. rough
   c. smooth
   d. overheated

3. With .006in warpage according to a straight edge and thickness gauges, .006in surfacing will provide _____ percent cleanup.
   a. 25
   b. 50
   c. 75
   d. less than 100%

4. Machinist A says that if valve lash is nonadjustable, excessive surfacing of pushrod heads causes valves to remain open. Machinist B says that excessive surfacing of any engine decreases piston to valve clearance. Who is right?

   a. A only
   b. B only
   c. Both A and B
   d. Neither A or B

5. Machinist A says that for resurfacing any given amount, compression ratios increase more when surfacing engines with large displacements. Machinist B says that compression ratios increase more with large bore or high compression engines. Who is right?

   a. A only
   b. B only
   c. Both A and B
   d. Neither A or B

6. Machinist A says that excessive surfacing of OHC heads causes chain binding. Machinist B says that in pushrod engines, it causes incorrect rocker geometry. Who is right?

   a. A only
   b. B only
   c. Both A and B
   d. Neither A or B

7. Machinist A says that excessive surfacing of OHC heads causes valve timing to advance. Machinist B says that it increases chain slack. Who is right?

   a. A only
   b. B only
   c. Both A and B
   d. Neither A or B

8. Machinist A says that limits for surfacing OHC heads are specified as the maximum amount of stock removal. Machinist B says that limits are specified as minimum head thickness. Who is right?

   a. A only
   b. B only
   c. Both A and B
   d. Neither A or B

9. Machinist A says that V-block intake manifold distortion causes internal vacuum leaks. Machinist B says that such leaks cause high oil consumption. Who is right?

   a. A only
   b. B only
   c. Both A and B
   d. Neither A or B

10. Machinist A says chamfer head bolt holes after surfacing block decks. Machinist B says that counterbored head bolt holes in decks need not be chamfered. Who is right?

    a. A only
    b. B only
    c. Both A and B
    d. Neither A or B

11. Machinist A says that after resurfacing heads, breaking the sharp edges around the combustion chamber eliminates sources of preignition. Machinist B says that eliminating sources of pre-ignition requires polishing combustion chambers. Who is right?

    a. A only
    b. B only
    c. Both A and B
    d. Neither A or B

12. Machinist A says that after surfacing V-block heads, intake manifolds assemble without distortion if sealer is used instead of gaskets under the ends of the manifolds. Machinist B says that unless the intake sides are surfaced, ports and fasteners do not align. Who is right?

    a. A only
    b. B only
    c. Both A and B
    d. Neither A or B

13. Machinist A says that because distributors sit lower in some engines after excessive surfacing, distributor drive gears wear and oil pump drives sometimes bind the pump. Machinist B says shim up distributors after surfacing these engines any significant amount. Who is right?

    a. A only
    b. B only
    c. Both A and B
    d. Neither A or B

14. Ninety degree V-block heads with an 80 degree included angle are surfaced .010in, therefore remove _____ in from the intake sides.
    a. .009
    b. .010
    c. .011
    d. .012

15. Machinist A says that straightening heads can eliminate the need for line boring cam bearings. Machinist B says that line boring is required if cam bores are scuffed or worn oversize. Who is right?

    a. A only
    b. B only
    c. Both A and B
    d. Neither A or B

16. OHC camshaft alignment is corrected by surfacing the topside of cylinder heads with
    a. removable cam bearing caps
    b. removable cam bearing assemblies
    c. integral bearing bores
    d. adjustable cam followers

17. If possible, surface chain driven overhead camshaft engine blocks
    a. to minimum deck clearance
    b. less than .008in
    c. with the timing cover attached
    d. with the timing cover removed

18. If possible, surface diesel cylinder heads
    a. with precombustion chambers in place
    b. with precombustion chambers removed
    c. less than .008in
    d. in milling machines

19. Machinist A says that because of precombustion chambers, grinding wheel wear is high. Machinist B says that to know the amount removed, measure head thickness before and after surfacing. Who is right?

    a. A only
    b. B only
    c. Both A and B
    d. Neither A or B

20. Machinist A says that the recessed seal area around chambers require surfacing in VW-Porsche air-cooled heads. Machinist B says correct the depth from the deck to this seal area to prevent interference in assembly. Who is right?

a. A only
b. B only
c. Both A and B
d. Neither A or B

FOR ADDITIONAL STUDY

1. What surface finish is required for cylinder head gaskets?

2. What can you do to improve surface finishes when milling? What about when grinding?

3. A pair of V-block cylinder heads has been surfaced enough to cause manifold alignment problems. What other machining steps are required to correct the alignment?

4. An overhead cam cylinder head needs surfacing. How should the deck be positioned for machining?

5. A chain driven overhead cam cylinder block needs surfacing. What special precautions should be taken?

6. A diesel cylinder head must be ground with the stainless steel precombustion chambers in place? How is the grinding wheel going to perform?

7. How can machinists keep track of stock removal when surfacing heads?

8. An overhead cam cylinder head has been surfaced .020in. How will this effect valve timing?

9. What happens to head bolt holes after surfacing engine blocks? What can you do to prevent problems?

Chapter 15

ENGINE BALANCING

Upon completion of this chapter, you will be able to:
- Define reciprocating and rotating weight.
- List the sequence of steps in equalizing reciprocating and rotating weights of connecting rods.
- Identify internally and externally balanced crankshaft assemblies.
- Select the appropriate bob weight formulas for various V-block crankshafts.
- Calculate bob weights.
- List the sequence of steps in balancing complete crankshaft assemblies including connecting rods, pistons and pins, dampers, flywheels and clutches or flexplates.
- Explain the procedure for balancing torque converters.
- Describe special cases requiring heavy metal.
- Evaluate exchange crankshafts or changes in rotating and reciprocating weight and determine if balancing is required.

INTRODUCTION

Unbalanced forces multiply as engine speed increases and cause not only vibration, but also a crankshaft wobble that damages main bearings (see Fig.15-1). Because of the vibration and potential for damage, engine builders balance all performance engines and, where needed, passenger car engines.

RPM	Oz.	Lbs	Gms	Kgs
500	7.3	.46	207	.207
1000	19	1.2	539	.539
2000	117	7.3	3317	3.317
3000	263	10.2	7456	7.456
4000	464	29.0	13154	13.154
5000	720	45.0	20412	20.412

Values based on a one ounce weight placed one inch from center.

Fig.15-1   As rotational speed increases, the effect of force increases

Passenger car engines are balanced in production but replacement and exchange parts used in rebuilding create out-of-balance conditions. Recommend balancing to customers when engine parts are not entirely original such as when using replacement pistons with different weights or exchange connecting rods or crankshafts.   In performance engines, changes in balance are even more significant and balancing is a necessary part of engine building.

While balancing generally improves smoothness in any engine, some engines cannot be made perfectly smooth because of their inherent design. For example, V6 engines have "rocking couples" caused by unequal reciprocating forces in opposing cylinder banks that also alternate side-to-side. This rocking couple is greater in 90-degree engines than 60-degree engines and is eliminated only by running a balance shaft. The balance shaft is weighted on opposite sides at opposite ends and runs at crankshaft RPM (see Fig.15-2).

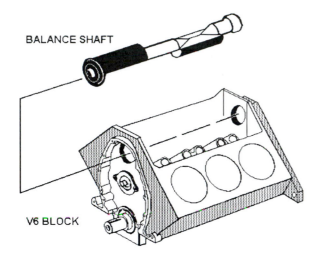

Fig.15-2 V6 balance shaft weights on opposite sides at opposite ends

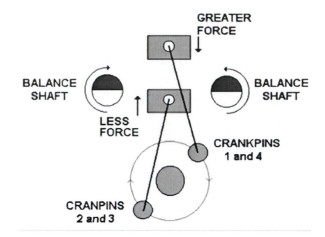

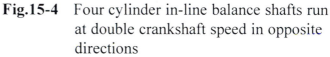

Fig.15-4 Four cylinder in-line balance shafts run at double crankshaft speed in opposite directions

Four cylinder in-line engines also have an inherent balance problem because the forces of pistons traveling downward and those traveling upward do not fully cancel each other. This is because the acceleration of the two pistons passing top center exceeds that of the two passing bottom center (see Fig.15-3). The difference in forces causes vertical motion in the engine mounts. Then, a half revolution later, two pistons approach top center and two others approach bottom center, and the differences in forces reverse direction. The problem is completely eliminated only by installing two balance shafts running at double crankshaft speed in opposite directions (see Fig.15-4).

To quantify these forces, understand that pistons reach maximum velocity when the connecting rod is 90 degrees to the crankshaft, typically between 70 and 75 degrees ATDC (see Fig.15-5). The actual force generated by piston travel is calculated by multiplying the rate of acceleration times the reciprocating weight. While weights are constant, the rates of acceleration vary according to crankshaft position and rod angle. Depending on rod ratio, peak acceleration occurs somewhere between TDC and the point of peak velocity.

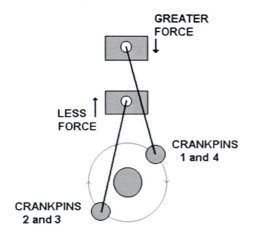

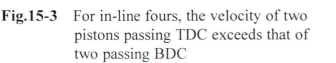

Fig.15-3 For in-line fours, the velocity of two pistons passing TDC exceeds that of two passing BDC

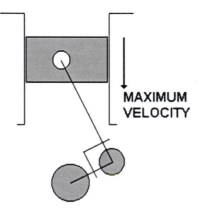

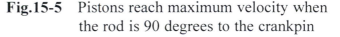

Fig.15-5 Pistons reach maximum velocity when the rod is 90 degrees to the crankpin

Unless unbalanced forces are sufficiently reduced or canceled, the manufacturers must design sophisticated mounting systems to isolate the passenger compartment from vertical engine motion.

WEIGHING PISTONS AND CONNECTING RODS

All weight that travels up and down in the cylinder is reciprocating weight. This includes pistons, piston rings, piston pins, pin retainers for full-floating pins, and the pin ends of connecting rods. Equalizing these weights helps reduce unbalanced reciprocating forces in the cylinders.

Rotating weight travels in a circle with the crankpin and includes the housing bore end of the rod and the rod bearing. Equalizing these weights helps reduce unbalanced rotating forces in the crankshaft. Record both reciprocating and rotating weights of connecting rods before removing weight. Do this by weighing one end, reversing the setup, and then weighing the opposite end (see Fig.15-6 and 7). The weight at either end is also found by subtracting the weight of the opposite end from the total weight.

Fig.15-7   Weighing the rotating ends of connecting rods

Record the weights of all pistons in a set (FIG.15-8). The pins can be weighed separately but it is convenient to include their weight with that of the pistons. Mark the weights directly on top of each piston to prevent mixing up pistons and weights while making corrections.

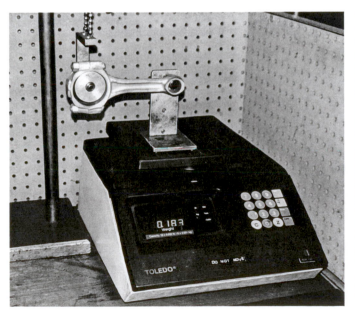

Fig.15-6   Weighing the reciprocating ends of connecting rods

Fig.15-8   Weighing the pistons with pins

Because variations in the weights of piston rings are negligible, it is necessary to record the weight of rings for one piston only. If used, also record the weight of piston pin retainers for one piston. Variations in rod bearing weights are also negligible and it is necessary to record the weight of one pair of bearing inserts for one rod only.

BALANCING CONNECTING RODS

Before balancing connecting rods, be aware that reducing weight at the housing bore end causes the weight at the pin end to increase slightly. To understand this, study the set-up for weighing rods and note that the balance boss at the housing bore end lifts the pin end. Reducing weight at this end reduces this lifting effect and therefore causes an increase in weight at the pin end. For this reason, balance housing bore ends first and then weigh pin ends before balancing them.

As stated, equalizing weights at the housing bore ends helps reduce unbalanced rotating forces in the crankshaft. Begin by reading the recorded weights and sand or grind the bosses at the rod cap to reduce weights as needed (see Fig.15-9). Reduce heavier weights to within one gram of the lightest weight.

Fig.15-9   Removing rotating weight from connecting rods

Also as stated, equalizing weights at the pin ends helps reduce unbalanced reciprocating forces. Read the recorded weights and sand or grind the bosses at the pin end to remove weight as needed (see Fig.15-10). If there is no balance boss, then sand or grind carefully around the pin end taking care not to weaken the rod by excessively thinning the wall around the pin bore. As with rotating weights, reduce heavier weights to within one gram of the lightest weight.

Fig.15-10   Removing reciprocating weight from connecting rods

Sanding or grinding to remove weight leaves small "cuts" on the surfaces. These are potential points of stress concentration. Polish and bead blast these surfaces after balancing.

BALANCING PISTONS AND PINS

Begin by reading the combined weights of pistons with pins. To remove weight from pistons, set-up the pistons in a lathe and face the balance pads under the pin bosses (see Fig.15-11). If there is no balance pad, or if pins are balanced separately from pistons, set-up the pins in a lathe and chamfer the bores using a carbide lathe tool (see Fig.15-12). In a typical set of pistons, it is possible to remove five grams from piston balance pads and three grams from piston pins.

Fig.15-11 Removing weight from piston balance bosses

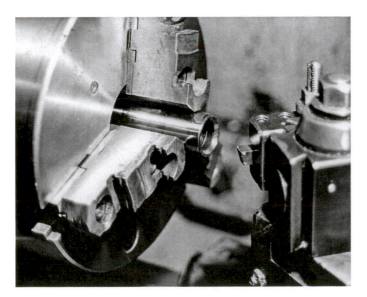

Fig.15-12 Chamfering piston pins to remove weight

Should it be impossible or unsafe to remove enough weight from pistons to bring differences within one gram, selectively assemble pistons to rods. For example, place heavy pistons with pins in assembly with rods having light pin ends. Conversely, place light pistons with pins in assembly with rods having heavy pin ends.

Machinists sometimes selectively assemble pins to pistons in balancing. Do this only if necessary and only after carefully checking pin clearances because the manufacturers sometimes selectively fit pins to obtain specified clearance.

BALANCING CRANKSHAFTS

There is no assurance that a statically balanced crankshaft will run smoothly in the engine. Static balance means only that the weight is evenly distributed around the center of rotation. However, weight is often located on one side of center at one end and the other side of center at the opposite end (see Fig.15-13). When the crankshaft rotates, it wobbles because each end of the crankshaft is unbalanced (see Fig.15-14). If the crankshaft wobbles more than allowed by main bearing oil clearance, bearing damage results.

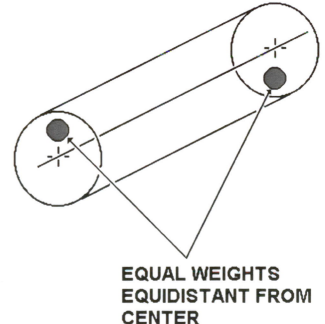

EQUAL WEIGHTS EQUIDISTANT FROM CENTER

Fig.15-13 Because these weights and radii are equal, the shaft is statically balanced

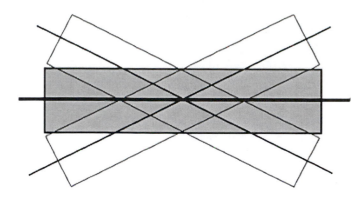

Fig.15-14 A statically balanced shaft wobbles because it is dynamically unbalanced

To eliminate the wobble, crankshafts require balancing in two planes. That is, balance requires detection and correction at both ends of the crankshaft. Of course, correction at one end of the crankshaft affects balance at the other end. To detect unbalanced conditions, the balancer runs the crankshaft at a speed adequate to generate measurable forces. Setting up the balancer requires entering the radius from the center of rotation to the counterweights, the distance between them, and their distance from the bearings supporting the shaft into the control panel (see Fig.15-15). From this data, a computer locates the point of unbalance at each end counterweight and calculates the necessary weight removal or addition.

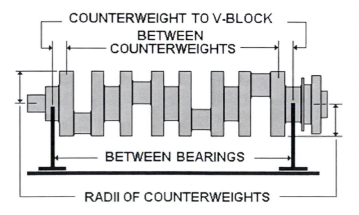

Fig.15-15 Measurements needed to calculate corrections

The amount of correction varies with the input radius and requires adjustment when the radius to the actual point of correction changes. For example, it may not be possible to add weight where the balancer suggests and it becomes necessary to remove weight on the opposite side of the crankshaft. If the radius on the opposite side is different, the amount of correction also changes. For example, a correction of 10 grams 3 inches from center would be equivalent to 15 grams 2 inches from center (see Fig.15-16). While some balancer computers do the calculation once given the new radius, developing on-the-job judgment requires understanding the logic involved.

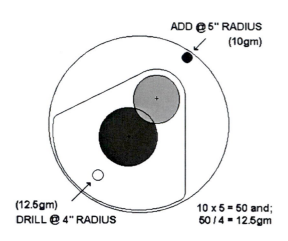

Fig.15-16 The effect of radius on corrections (note the mixed metric-English units)

As suggested, correcting balance requires drilling to remove weight or adding weight by attaching balance weights at the end counterweights. The amount of weight removed by drilling is found in tables such as the one below (see Fig.15-17). The values are based upon densities or grams per cc of iron and steel and the corresponding volumes of drilled holes at varying diameters and depths. Adding weight requires selecting from weights with assorted diameters and thickness and driving them into existing holes and then securing them by welding (see Figs. 15-18 and 19). Do not add weights on top of counterweights because they interfere with the undersides of piston pin bosse

Weights from Drilled Holes							
Hole	----------- Hole Diameter ------------						
Depth	1/4	3/8	1/2	5/8	3/4	7/8	1
1/8	0.8	1.8	3.1	4.9	7.1	9.6	13
1/4	1.6	2.5	6.3	9.8	14	19	25
3/8	2.4	5.3	9.4	15	21	29	38
1/2	3.1	7.0	13	20	28	38	50
5/8	3.9	8.8	16	24	35	48	63
3/4	4.7	11	19	29	42	58	75
7/8	5.5	12	22	34	49	67	88
1	6.3	14	25	39	56	77	100
	----------- Drill Points ------------						
	0.1	0.4	0.9	1.8	3.1	4.8	7.2

Fig.15-18 A selection of weights to correct balance

Fig.15-17 Drilling to correct balance

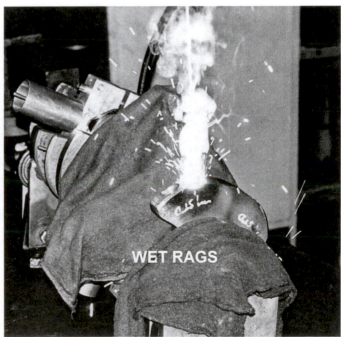

Fig.15-19 Securing weights by welding

Crankshafts for V-block engines present special problems in balancing. The counterweights on these shafts compensate for the crankpin plus the rod rotating weight and up to half of the reciprocating weight. Because counterweight compensation is not total, "bob weights" are added to the crankshaft for balancing (see Fig.15-20). Making up bob weights requires values for rotating and reciprocating weights for each crankpin and therefore all balancing steps on pistons, pins, and connecting rods must be completed first.

Fig.15-20 A crankshaft with bob weights set up for dynamic balancing

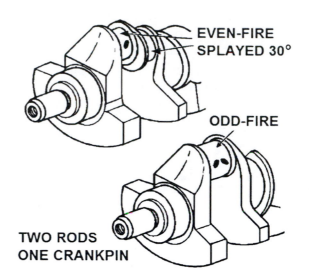

Fig.15-21 An odd-fire V6 crank (bottom) and even-fire crank with splayed crankpins (top)

Ninety-degree V8 and sixty-degree V6 engines use the same bob weight formula, 100 percent rotating weight plus 50 percent reciprocating weight. Ninety-degree V6 engines with three crankpins and two connecting rods per crankpin like a V8 also use the same bob weight calculations. These "odd-firing" engines fire in 90 and 150-degree intervals of crankshaft rotation, and though balanced, they are not necessarily smooth.

To make the 90-degree V6 engine smoother, Buick designed a crankshaft with six crankpins arranged in three pairs with 30 degrees between rods on each crankpin (see Fig.15-21). These "splayed" crankshafts made the engines even firing in 120-degree intervals. Bob weights for this engine use 100 percent rotating weight plus 36.6 percent reciprocating weight. Other manufacturers now have even firing V6's requiring the same bob-weight calculations.

As noted in the introduction, many 90-degree V6 even-fire engines now use balance shafts. Be sure to check bob weight formulas for these engines as they may require 50 percent reciprocating weight instead of 36.6 percent. Consider also that it is possible to balance the replacement piston assemblies to the original weights and not affect balance of the crankshaft. In earlier models, Chevrolet manufactured 90-degree V6's with a similar "splayed" crankshaft design, but with only 18 degrees between rods on each crankpin. While this engine does not fire evenly, there is a reduction in the amplitude of the rocking couple in the horizontal plane. Bob weights for this engine use 100 percent rotating weight plus 46 percent of reciprocating weight.

Considering the variations in design and differences in calculating bob weight values, careful engine identification and calculations are necessary. Study the following sample bob weight calculations:

Sample Reciprocating and Rotating Weights

Reciprocating Weight
Piston	580gm
Piston Pin	140
Rings	50

```
Reciprocating Rod   180
Total               950gm

Rotating Weight
    Rotating Rod      410gm
    Rod Bearings (set)  50
    Total             460gm
```

Example #1
V6, Two Assemblies per Crankpin, 90-Degree

Bob Weight 100% Rotating; 50% Reciprocating

```
Reciprocating Weight (1)    950gm
Rotating Weight (2)         920
Oil, Carbon                   5
BOB WEIGHTS (3)           1,875gm
```

Example #2
V8, Two Assemblies per Crankpin, 90-Degree

```
Bob Weight 100% Rotating, 50% Reciprocating
Reciprocating Weight (1)    950gm
Rotating Weight (2)         920
Oil, Carbon                   5
BOB WEIGHTS (4)           1,875gm
```

Example #3
V6, Even Fire, Six Crankpins, 90-Degree

Bob Weight 100% Rotating, 36.6% Reciprocating

```
Reciprocating Wt. (36.6%)   348gm
Rotating Weight (1)         460
Oil, Carbon                   5
BOB WEIGHTS (6)            813gm
```

In-line and opposed crankshafts do not require bob weights for balancing. The counterweights and rotating forces in four and six cylinder in-line or opposed crankshafts cancel. Reciprocating forces in in-line six or opposed four and six-cylinder engines also cancel. Reciprocating forces in in-line fours do not cancel and the best that can be done is to reduce reciprocating weights, not just equalize them.

Flywheels or flexplates and harmonic balancers should be checked for external balance weights before proceeding with crankshaft balancing. External weights attached to these components are part of the crankshaft balance and if present, the crankshaft, flywheel, and harmonic balancer must be balanced as an assembly (Figs. 15-22 and 23). Because flexplates and harmonic balancers sometimes fail, make all balance corrections at the crankshaft so that replacement of these components has the least possible effect on total balance.

Fig.15-22 An external weight on a flywheel

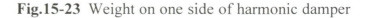

Fig.15-23 Weight on one side of harmonic damper

In a computerized balancer, a crankshaft assembly is balanced when the required correction is six grams or less at a one-inch radius. Another explanation of balance requirements is that the crankshaft wobble cannot exceed the limits of oil clearance, approximately .002in (.05mm), or bearing damage results. If properly balanced by either set of specifications, the wobbling motion is within acceptable limits.

BALANCING FLYWHEELS AND CLUTCHES

If possible, balance the flywheel, clutch, and harmonic balancer by adding them to the already balanced crankshaft while still in the balancing machine. First, assemble the flywheel and harmonic balancer to the crankshaft and balance them. Next, add the clutch to the flywheel and balance it. Be sure to center punch or somehow mark the flywheel, crankshaft flange, and pressure plate for their positions of assembly. Following this practice also minimizes the effect of a replacement clutch or harmonic balancer on total balance.

Balancing replacement clutch pressure plates prevents customer complaints. Test pressure plate balance independent of the crankshaft by adding it to a balanced flywheel mounted on a balance arbor (see Fig.15-24). Make correction by welding a weight onto the pressure plate cover. Because friction disc positions continually change, they are not balanced.

Fig.15-24 Balancing a flywheel and clutch separately on an arbor

BALANCING TORQUE CONVERTERS

Torque converters present special problems in balancing. For example, some Chrysler V8 engines with cast crankshafts are externally balanced at the factory by adding weights to the converter (see Fig.15-25). Unless specialized tooling is on hand, most shops cannot balance converters independently or balance crankshafts with the converters attached. If this is the case, consider balancing rods and pistons to original weights and leaving the original crankshaft balance alone. Always be sure that replacement torque converters are correct for the particular crankshafts.

Fig.15-25 Balance weights on a torque converter

Before balancing converters, completely drain them of fluid. In some converters, this requires drilling and tapping for drain plugs (see Fig.15-26). Because a single drain plug on one side throws the converter out of balance, install two drain plugs equally distant from center on opposite sides. Remove these plugs and drain the fluid after the first few minutes of rotation in the balancer. If any fluid remains in the converter, balance readings will not repeat.

While balancing converters is done, removing all the fluid is difficult and time consuming. The author suggests balancing only those that clearly need it.

Fig.15-27 Tooling to hold the turbine section of a converter on center for balancing

BALANCING WITH HEAVY METAL

Many performance or modified engines require balancing with "heavy metal." This requires drilling holes in the crankshaft counterweights and plugging them with Tungsten alloy plugs that weigh significantly more than the iron or steel they replace (see Fig.15-28).

Fig.15-26 Drain plugs added to a torque converter

A splined shaft similar to the input shaft of the transmission, inserts through the converter hub and into the turbine section of the converter (see Fig.15-27). This tooling supports and centers the turbine and within the converter housing and makes possible the balancing of the total assembly.

Fig.15-28 "Heavy metal" weights installed through counterweights

This method of balancing becomes necessary when changes in the crankshaft assembly are such that bob weight totals are radically different from the original. Such changes commonly result from lengthening the stroke or by installing performance connecting rods and racing pistons.

Changing from external to internal balance also requires heavy metal balancing. Remember that some racecars do not run flywheels or harmonic balancers, and if these were originally part of external balancing, the counterweights must be made heavier to compensate. These weights are expensive and, labor included, are not cost effective for conventional balancing. The chart below details the available sizes of heavy metal weights and the weight difference after installation:

HEAVY METAL PLUGS

Diameter x Length	Net Gain
.437 x .875in.	21gm
.500 x .750	24
.500 x 1.200	35
.750 x 1.200	79
.875 x .500	48
.875 x .700	63
.875 x 1.000	92
.875 x 1.100	101
.875 x 1.200	105
1.000 x .700	83
1.000 x 1.000	127
1.000 x 1.100	133
1.000 x 1.200	145

SUGGESTIONS FOR MINIMUM BALANCING

Full engine balancing is an added expense paid for by the customer. While some are readily sold on the benefits, others are not. Make at least minimum checks for engine balance to protect against unacceptable vibration and customer complaints after rebuilding.

First, compare the relative weights of pistons in a set. Some manufacturers include specified weights in service manuals. Depending on the size of the engine, a spread of 5 to 10 grams in a set is tolerable. Correction to this level means that perhaps only one or two pistons out of a set need weight removed.

On V-block engines, compare the weights of replacement pistons to the original pistons or to the manufacturers' specified weights. Remember that piston weight is a factor in crankshaft design and balance and changes in weight make dynamic crankshaft balancing necessary if the engine is to run smoothly.

Also keep in mind that some engines counter engine motions or rocking couples with "balance shafts." In four cylinder engines, the weighted balance shafts run at double crankshaft speed and in opposite directions. V6's have a single balance shaft weighted on opposite sides and opposite ends running through the lifter passage area at crankshaft speed. In balance shaft engines, the shafts are weighted to compensate for specific reciprocating weights and these weights must remain unchanged when rebuilding. To be safe, balance replacement pistons to match the original pistons.

Also check connecting rod weights, especially if all connecting rods are not from the same original set. Keep total weights within a seven gram spread or the manufacturer's specifications if available.

Check exchange crankshafts for precise interchangeability or balance them. Many crankshafts share dimensions but balance differently. For example, Chevrolet 305 and 350 crankshafts interchange perfectly but balance is radically different for the two applications because of differences in reciprocating weights. For balance reasons, it is better to reuse original crankshafts or balance exchange crankshafts.

It is not unusual for engines to develop vibrations after replacing clutches. Use only the highest quality clutch pressure plates with a guaranteed close level of balance. If in doubt, send clutch pressure plates to a machine shop equipped for balancing.

Following these recommendations requires only a scale weighing in grams. While the engine is not fully balanced, the shop will learn of potential balance problems before assembly and can advise the customer accordingly.

SUMMARY

Reasonable engine balance is a required for smooth operation and for reliability. An unbalanced condition is created when pistons, connecting rods, or exchange crankshafts do not match the original parts they replace. Technicians and machinists should recognize when there is the potential for a balance complaint and properly advise customers regarding balance.

Some engines are subject to secondary motion even when properly balanced. These include V6 and four cylinder inline engines. Only balance shafts eliminate such motions. Those assembling engines should understand the operation of balance shafts and properly time them during assembly.

For balance purposes, weights are separated into reciprocating and rotating groups. In all engines, the weights of components within these weights must be equalized. In V-block engines, replacement parts within these groups must also be equal to the original parts they replace or the crankshaft must be rebalanced. This is also true in four-cylinder in-line engines equipped with balance shafts.

Machinists must balance pistons and connecting rods, calculate and assemble bob weights, and balance crankshafts. They must also balance clutches and flywheels as part of engine balancing or as separate jobs. In doing this work, they need to identify those engines that are externally balanced and balance crankshaft assemblies with vibration dampers and flywheels or flexplates attached.

Chapter 15

ENGINE BALANCING

Review Questions

1. Machinist A says that vibration caused by an unbalance condition is greatest at idle. Machinist B says that unbalance forces increase with engine RPM. Who is right?

   a. A only
   b. B only
   c. Both A and B
   d. Neither A and B

2. Machinist A says that V6 engines have balance "couples." Machinist B says that only 90-degree odd-fire V6 engines have balance "couples." Who is right?

   a. A only
   b. B only
   c. Both A and B
   d. Neither A and B

3. Machinist A says that at approximately 75 degrees after TDC in a four-cylinder engine, pistons 1 and 4 travel faster than pistons 2 and 3. Machinist B says that at approximately 75 degrees before TDC, pistons 2 and 3 travel faster than pistons 1 and 4. Who is right?

   a. A only
   b. B only
   c. Both A and B
   d. Neither A and B

4. Machinist A says that rod bearing housings are reciprocating weight. Machinist B says that the pin end of rods and pistons with rings are rotating weight. Who is right?

   a. A only
   b. B only
   c. Both A and B
   d. Neither A and B

5. Machinist A says no matter the engine type, equalize reciprocating weights. Machinist B says always equalize rotating weights. Who is right?

   a. A only
   b. B only
   c. Both A and B
   d. Neither A and B

6. Machinist A says that balancing a V8 crankshaft requires bob weights that include all rotating and half of reciprocating weight. Machinist B says that this is correct for V8 engines but not for all V6 engines. Who is right?

   a. A only
   b. B only
   c. Both A and B
   d. Neither A and B

7. Machinist A says that to preserve the original equipment balance in V-block engines, replacement pistons must equal one another in weight. Machinist B says that they must match the original weights. Who is right?

   a. A only
   b. B only
   c. Both A and B
   d. Neither A and B

8. Machinist A says that otherwise interchangeable V-block crankshafts are originally balanced to varying specifications. Machinist B says that in-line pistons and rods are rebalanced without affecting crankshaft balance. Who is right?

   a. A only
   b. B only
   c. Both A and B
   d. Neither A and B

9. When testing crankshaft balance electronically, correction must be less than
   a. 1gm at a 3in radius
   b. 6gm at a 1in radius
   c. .003in
   d. oil clearance

10. Externally balanced crankshafts are tested with the flywheel and harmonic balancer
    a. separate from the crankshaft
    b. with the crankshaft
    c. before the crankshaft
    d. separate from each other

11. A crankshaft has been test balanced with a 10gm weight 3in from center. Permanent correction must be made opposite the weight 2in from center by
    a. removing 15gm
    b. removing 7gm
    c. adding 15gm
    d. adding 7gm

12. Machinist A says that on disassembly, mark pressure plates and flywheels for position on the crankshaft to preserve balance. Machinist B says mark clutch discs for position. Who is right?

    a. A only   c. Both A and B
    b. B only   d. Neither A and B

13. Machinist A says that when adding large amounts of weight to counterweights, weld weights on top of the counterweights. Machinist B says drill and fill holes with heavy metal. Who is right?

    a. A only   c. Both A and B
    b. B only   d. Neither A and B

14. Machinist A says that in balance shaft engines, replacement piston weights must match the original. Machinist B says that if they do not match, balancing the crankshaft restores balance. Who is right?

    a. A only   c. Both A and B
    b. B only   d. Neither A and B

15. _____ are not balanced.
    a. Flexplates
    b. Flywheels
    c. Harmonic balancers
    d. Friction discs

FOR ADDITIONAL STUDY

1. How can an unbalanced crankshaft assembly cause engine failure? Under what conditions would such a failure likely occur?

2. How is an unbalance condition created?

3. How does a balance shaft work in a V6 engine?

4. How do balance shafts work in a four-cylinder engine?

5. How do you identify an externally balanced engine?

6. Answer the following questions pertaining to the balancing of an even-firing, 90-degree V6 engine with a balance shaft?
    a. What bob weight formula is used?
    b. How many bob weights are needed?
    c. How much should the replacement pistons weigh?

7. How fast does a balance shaft in a V6 run? In what direction does it run?

8. How fast do balance shafts in a four-cylinder run? In what direction do they run?

Chapter 16

ENGINE ASSEMBLY

Upon completion of this chapter, you will be able to:
- List points in an engine block requiring chamfering or removal of sharp edges prior to assembly.
- Describe the final cleaning measures required for assembly.
- List the sequence of steps and precautions in assembling pushrod cylinder heads.
- Compare pushrod and overhead cam cylinder head assembly procedures.
- Explain how to select drivers and install core plugs.
- Explain how to select camshaft or auxiliary shaft bearing drivers.
- List the precautions in installing camshaft or auxiliary shaft bearings.
- Explain how and when to install oil passage plugs during assembly.
- Describe basic requirements for sealing rotating shafts.
- List the sequence of steps in fitting a rope-type rear main seal.
- List the sequence of steps in fitting a lip-type rear main seal.
- List the clearances to check when installing a crankshaft and main bearings.
- Explain how to use Plastigage to check bearing clearances.
- Explain valve timing is set in assembly and how to check for correct valve timing.
- List the sequence of steps in installing piston rings.
- Explain the procedure and precautions for installing piston and rod assemblies.
- Compare the assembly of pushrod and overhead camshaft cylinder heads to short blocks.
- Compare mechanical and hydraulic valve adjusting procedures.
- List precautions in installing oil pumps and pick-up screens.

- List precautions in installing timing cover, valve cover, oil pan, and intake and exhaust manifold gaskets.
- Rewrite or adapt an engine assembly checklist to suit specific engines using service manuals or other references.
- List the tests and inspections performed in engine run-in stands.
- List the checks made when installing flywheel and clutches, flexplates, and bellhousings.

INTRODUCTION

Assembly is the single most critical process in engine service. All the painstaking attention to inspection and machining of component parts serves no purpose unless assembly receives equally pains-taking care. Failure to clean or check the fit of a part, or tighten a nut or capscrew, can lead to major engine damage. Remember also that the technician or machinist assembling the engine has the last opportunity, and the responsibility, for inspection.

Because most repair shops sublet machining, it is especially effective to have the technician assembling the engine recheck the fits and clearances. This is the final opportunity to control the quality of machined and replacement parts. Be assured that machine shops much prefer correcting problems before they cause engine failure.

CLEANING AND DEBURRING FOR ASSEMBLY

While accumulated sludge and carbon are removed from engine parts in cleaning, machining also produces contaminants such as metal chips and abrasives that are at least as destructive if left in the engine. Be sure to scrub cylinders, valve guides, and oil passages with bore brushes and detergent.

This is also the appropriate time for removing sharp edges or burrs created during machining. Be especially watchful for burrs or sharp edges around crankshaft oil holes, combustion chambers, and both ends of cylinders. Burrs around combustion chambers cause pre-ignition, sharp edges around crankshaft oil holes shave bearings, and sharp edges at the lower end of cylinders shave piston skirts. Chamfer head bolt holes after resurfacing to prevent the top thread from pulling up into the head gasket when tightening head bolts.

Carbon, dirt, and other contaminants in threaded holes make correct torque readings impossible during assembly. Run a tap or steel bore brush through head bolt and main bolt holes. A tap should not remove any significant amount of metal.

Inspect valve lifter bores and cam bearing bores for burrs, corrosion, or varnish buildup that might cause the incorrect fit of replacement parts. Valve lifters must rotate freely in their bores to maintain normal life for both the lifters and camshaft. Cam bearing bores must be clean and deburred so that new bearings press into place without scoring and distorting bearing shells. Deburr these bores with emery cloth or with a drill motor and a ball type hone.

After chamfering, tapping, and deburring, thoroughly scrub the engine block in soap and water or run it through a spray jet tank with detergent solution. Detergents are more effective at loosening honing abrasives and cast iron particles than cleaning solvents. Failure to clean the block thoroughly at this time causes piston ring wear and damage to engine bearings caused by abrasive contaminants in the engine oil.

ASSEMBLING CYLINDER HEADS

As with the block, clean the cylinder head thoroughly before assembly. It is especially important to scrub valve guides. With clearances restored to new specifications, abrasives that remain in valve guides sometimes cause valves to seize after only a few minutes of operation. Abrasives can also combine with lubricants to form a lapping compound that causes severe guide wear. Clean and use an assembly lubricant on valve stems

to prevent galling (see Fig.16-1). Also coat the valve face and seat to prevent corrosion. Lubrication is especially important if the cylinder head is to sit any length of time before installation.

Fig.16-1 A scored valve stem caused by too little clearance, a lack of oil, or dirt

First check for the correct installed heights of valve springs and stems. Next, install a valve, spring shim, valve seal, spring, retainer and valve keepers. Because the square-cut keeper grooves on some valves cut positive seals on installation, place a plastic sleeve over the valve stem to protect them as they slide past the grooves. Remember to cut the plastic sleeve to length just below the keeper grooves so that it can be reused. Also remember not to compress the spring more than required to prevent crushing the valve seal (see Fig.16-2).

Fig.16-2 Adjusting the valve spring compressor to clear seals

On overhead camshaft cylinder heads, check for free rotation of the camshaft first. If the camshaft rotation is free, add rocker arms or cam followers to the assembly and adjust valve lash (see Fig.16-3).

Fig.16-3 Check for free cam rotation in OHC heads prior to final assembly

INSTALLING CORE PLUGS

Water leaks around core plugs (also called expansion plugs, soft plugs, freeze plugs, or Welsh plugs) are a common source of comebacks to the shop. Unless repaired immediately, these leaks lead to overheating and subsequent engine failure. Take the time to install them correctly the first time. First, measure the bore for each core plug. Order core plugs the same size as the bores. The core plugs come with diameters equal to the bore size plus the correct amount for the interference fit. While both deep and shallow cup core plugs are available, use deep cup plugs whenever the depth allows. Core plug kits are packaged for common applications and are available in stainless steel or brass for marine engines.

Clean the bore with emery cloth to remove burrs and scale and coat the outer edges of the plugs with sealer. Some assemblers coat the backsides of plugs to help prevent corrosion. Drive core plugs into place until slightly below the surface of the block and even with the chamfered edge of the bore. While plug manufacturers do not necessarily recommend them, some core plug drivers drive against the outside edge of the plug (see Fig.16-4). If possible, select solid drivers approximately 1/32in (.75mm) smaller than the inside diameter of the core plug (see Fig.16-5). A driver that is too small in diameter distorts the core plug and the interference fit is lost. If a driver fits too closely, the core plug closes up on the driver on installation.

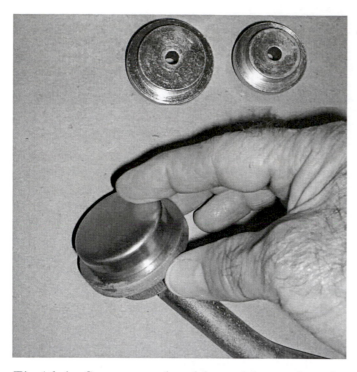

Fig.16-4 Some core plug drivers drive against the outer edge of the plug

Fig.16-5 Drivers fit loosely inside the cup

INSTALLING CAMSHAFT BEARINGS AND CAMSHAFT

Fit new cam bearings carefully to ensure the proper fit of the camshaft through the bearings and normal oiling of the engine. The bearings often vary in diameter according to the bearing position. Lay them out carefully for position before installation. Align cam bearings oil holes with passages in the engine block so that oil feeds through the bearings to other engine parts (see Fig.16-6). Only after checking bearing positions and oil hole alignment, are the bearings ready for installation.

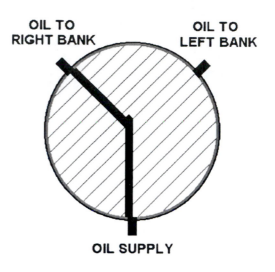

Fig.16-6 Oiling through a cam bearing

Clean and deburr the bearing bores in push-rod engine blocks with a ball-type hone or other tools. Although recommendations vary, the author suggests lubricating the outside surface of the bearings to prevent galling and distortion on installation.

Many shops use solid cam bearing drivers made for each bearing. Each driver has two diameters (see Fig.16-7). The smaller diameter matches the camshaft journal diameter, and the larger diameter slips through the housing bore in the engine block. Camshafts with only one journal diameter require only one bearing driver. If the camshaft journal diameters vary, one driver is required for each bearing. The solid driver fits inside the bearing. An extension enables the driver to reach all bearing positions and a centering cone helps in driving bearings in straight (see Fig.16-8). Chamfer the edge of the bearing facing the driver with a bearing scraper (see Fig.16-9). This prevents bearing material from being "upset" or deformed inward during installation and causing camshaft binding.

Fig.16-7 Match the small diameter of solid cam bearing drivers to cam journals

Fig.16-8 A driver extension and centering cone

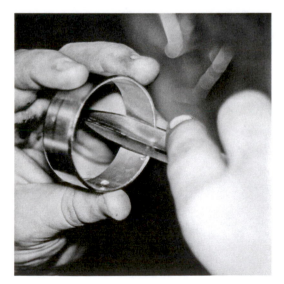

Fig.16-9 Chamfering the bearing prevents upsetting material on installation

Check the alignment of oil holes through bearings after installation by shining a light through oil passages. If the passage through the bearing is not accessible, use an inspection mirror to check for light. Remove any bearing that blocks oil holes and correctly install a new bearing in its place. Because individual bearings are often not available, this often means buying another full set so it pays to perform this job with care and attention the first time.

To be sure that the camshaft is straight, check alignment in V-blocks in the same manner as a crankshaft (see Fig.16-10). Next, check the fit of the camshaft in the new bearings. Turning the block up on end allows the camshaft to be lowered through the bearings with the least amount of interference between the cam lobes and the bearing surfaces (see Fig.16-11). Installing long capscrews in the sprocket or gear holes also helps by creating a "handle" for the camshaft. If checking fit with the block horizontal, take care to prevent scoring the bearing surfaces with the cam lobes on installation. The fit is acceptable if the camshaft can be rotated by hand. Remember to oil cam journals before installation.

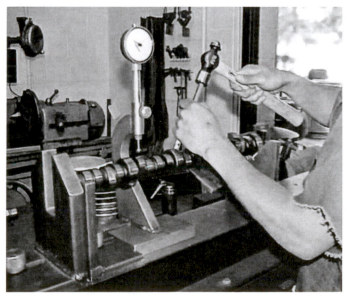

Fig.16-10 Check and straighten the camshaft as needed

Fig.16-11 It is sometimes convenient to install pushrod cams with the block on end

Should the cam bind on installation, check camshaft alignment and straighten it to within .001in (.03mm) TIR. Second, check the bearing fit at each end of the block by turning the cam end-for-end and testing for binding. If the cam is straight, and the bearings at each end have clearance, any binding on installation is in the center bearings. Third, tap the camshaft gently into assembly and then tap it back out of assembly. This marks tight spots in the bearings that can be removed by hand with a bearing scraper. Be sure to clean shavings from oil passages after scraping bearings and before plugging passages. Never allow an engine with camshaft binding go through assembly as bearings lock up on cam journals and spin in their housing bores causing major damage to the block.

Some engines use a thrust plate to limit the endplay of the camshaft. The thrust plate attaches to the front of the engine block behind the timing gear or sprocket. To assure correct endplay, there may be a shoulder on the timing gear, sprocket, or the camshaft. It is also possible to have a spac-

er ring inside the thrust plate between the timing gear or sprocket and front camshaft journal (see Figs. 16-12 and 13). The shoulder or spacer ring is approximately .004in (.10mm) thicker than the thrust plate to allow for end clearance. Rather than attempting to sort out all possible combinations for assembly, check end clearance in assembly. Binding on the thrust plate or extreme end clearance suggests misassembly or incorrectly matched parts.

Fig.16-12 A shoulder on the cam sprocket extends through a thrust plate to set endplay

Fig.16-13 A spacer ring and thrust plate combination for control of endplay

Fig.16-14 Check endplay after pressing on the cam gear

The camshaft timing gear is pressed off and a replacement pressed back on. Be sure to align the thrust plate with the Woodruff key during removal to prevent damage to the thrust plate. Also align both the thrust plate and timing gear with the Woodruff key for reassembly. Heating aluminum timing gears makes assembly easier. Be very careful to install timing sprockets as clean and straight as possible as a timing gear that runs out causes gear noise or sometimes a knock. These noises are difficult to pinpoint and use up considerable time in diagnosis.

Check camshaft endplay with a feeler gauge between the thrust plate and the front journal of the camshaft (see Fig.16-14). Correct excess endplay by replacing the thrust plate.

Coat cam lobes with an anti-scuff lubricant for the final installation. The anti-scuff lubricant protects the camshaft from wear during the first minutes of engine operation. Use oil on cam journals and bearings. Coat the outer edges of the rear cam plug with sealer and install it in the same way as a core plug.

INSTALLING OIL PLUGS

Some oil passages are plugged with tapered pipe plugs. Tapered pipe threads lock and seal very securely with only moderate tightening. The only specific recommendation is to make sure threads are clean and coated with sealer on installation. Core plugs are also used in oil passages. Measure the inside diameter of the bore and order plugs for the measured size. Remember that the new plugs come with the correct interference fit. Select drivers in the same way as for plugs in the water jackets, slightly loose or 1/32in (.75mm) undersize and drive them in slightly below the outside surface of the block. Some machinists assemble using "Loctite" to both seal and help lock oil plugs in place.

Others stake the plugs in place to prevent oil pressure from blowing the plugs out of their bores. This is done neatly with a blunt cold chisel by positioning the chisel across the oil passage, hitting it with a hammer, turning it 90 degrees and hitting it again. This stakes the opening in four places (see Fig.16-15).

Fig.16-15 Oil plugs staked in place

SEALING ROTATING SHAFTS; THE BASICS

Crankshafts, camshafts, auxiliary shafts, and balance shafts all have the same potential for leaking oil. Technicians are concerned about leaks because engines that leak oil eventually run low and cause engine failure. Watch the sealing of crankshaft main seals as failures are too frequent and because it is especially difficult to make repairs after engine installation.

There are at least five basic requirements to seal rotating shafts. First, bearing clearance must not be excessive. For crankshaft rear main seals, this may mean that even the high limits of specifications allow sufficient crankshaft wobble to shorten seal longevity.

Second, seal surfaces on the crankshaft must be concentric with the main journals. If the crankshaft was positioned from worn journals when ground, there may be concentricity errors. When checking crankshaft alignment in V-blocks, check that timing chain sprocket locations are concentric at least within .0015in (.03mm) TIR (see Fig.16-16). Timing gear locations must be even closer or gear noise and wear become a problem. The vibration damper hub also locates on these surfaces and runs out if not concentric with the mains. Also check the rear main seal surface for concentricity within at least .0015in (.03mm) TIR (see Fig.16-17). Should concentricity at these two points be out of specifi-

cations, the only option is regrinding the shaft with proper attention to concentricity.

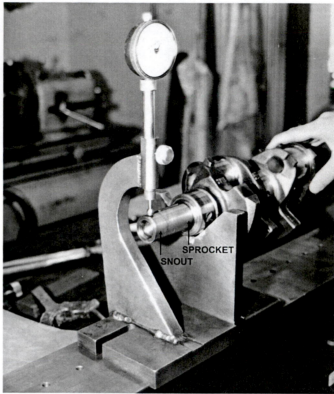

SPROCKET
SNOUT

Fig.16-16 Check run out at snout and timing gear or sprocket location

Fig.16-17 Check rear main seal locations for concentricity on a reground shaft

Third, timing covers and rear main seal locations ideally position within .005in (.12mm) TIR of the crankshaft centerline. Consider the stack up of tolerance for each specification plus the distortion of seal locations caused by grinding main caps for line boring or honing. Even with alignment tools, it may be necessary to drill dowel pin holes in timing covers oversize so that the cover seal can "float" onto center over the damper hubs (see Fig.16-18). In checking rear main seal locations within block castings, the author has found that remachining is necessary to obtain concentricity (see Fig.16-19).

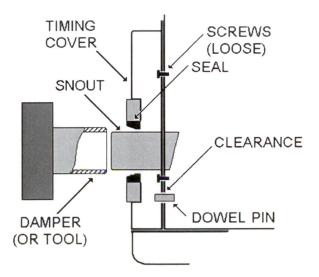

Fig.16-18 Center the timing cover seal over the damper

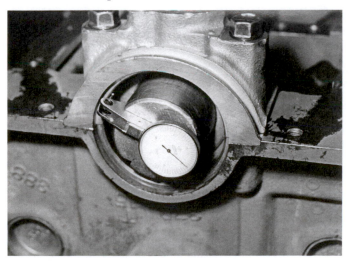

Fig.16-19 Checking concentricity of rear main seal location in the block

Fourth, with the seal partly located in the main cap, poor concentricity is sometimes the result of main caps that do not center on assembly. Of particular concern are those engines with no dowel pins or registers to position caps (see Fig.16-20). In these cases, assemble with grease on bearing surfaces and the caps will hydraulically center themselves over the main journal.

Fig.16-20 Dowels locate rear main caps in many engines

Fifth, check that seal surfaces on crankshafts and vibration dampers are in good condition and polished. Should wear be apparent, install repair sleeves over damper hubs (see Fig.16-21). For rear main seals, install a repair sleeve or possibly grind crankshaft seal surfaces .005 to .010in (.12-.25mm) undersize (see Fig.16-22). Slight changes in the diameter of these surfaces are less critical than the surface finishes. Consider that they are well oversize with repair sleeves and only slightly undersize when reground.

Fig.16-21 A worn damper hub with a sleeve installed

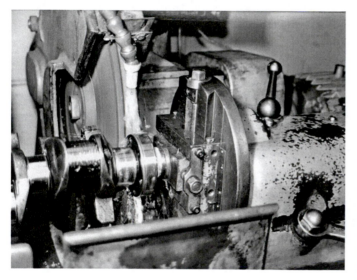

Fig.16-22 Grinding a rear main seal hub

FITTING THE REAR MAIN SEAL

Rear main seals of the rope type must be fitted correctly to prevent oil leaks or binding of the crankshaft. First, these seals are coated with a dry lubricant and if oil soaked, they can swell and bind on the shaft. On assembly, coat them with assembly lube. Fitting the seal begins by installing the rope packing in the groove and shaping it with a driver (see Fig.16-23). The driver is the same diameter as the seal surface of the crankshaft. Last, trim the ends flush with the engine block. Because it is easy to shape the seal in the block, remove the seal half from the block and install it in the rear main cap. Then repeat the procedure for the block

half of the seal and leave it in place when finished. An alternate method of shaping the seal is to roll it into shape using a short length of round stock (see Fig.16-24).

Fig.16-23 Shaping the rear main oil seal

Fig.16-24 Rolling a rear main seal into shape with round stock

Take particular care to prevent the rear main seal from causing bearing misalignment. Rope material jammed between the bearing cap and the engine block causes such misalignment. Check for excess rope material by tightening the bearing cap in place, removing it, and looking for material between the cap and block surfaces. Remove any excess material with a sharp knife.

Lip-type seals are more common in passenger car engines and present minimum problems during assembly. They do not require fitting because they already fit both the block and the shaft. Still, be careful to lubricate the seal with engine oil and to face the seal toward the oil inside the crankcase (see Fig.16-25). Also coat the outside edges with sealer on assembly.

Fig.16-26 Installing side seals in a rear main bearing cap

Fig.16-25 Install the seal lips towards the oil

Other sources of leakage are around the sides or under rear main caps and seal housings. The side seals are sometimes made of an absorbent material that swells in assembly. Because oil causes them to swell, soak them in oil immediately before installation (see Fig.16-26). Some side seals are made of a rubber type material and they tighten in place after wedging a small steel rod behind the seal. To prevent oil seepage between the cap or seal housing and the block, apply a thin coating of gasket sealer across the rear edge of the cap parting line (see Fig.16-27).

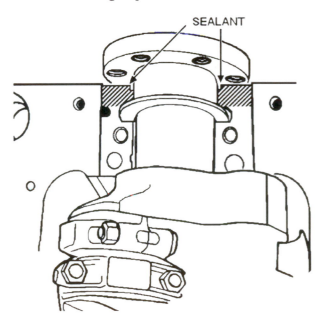

Fig.16-27 Seal under the rear main cap to prevent oil seepage

16-11

INSTALLING THE MAIN BEARINGS AND CRANKSHAFT

Check main bearings for location and position before installation. For example, main bearing sets have an upper half and a lower half. The upper half has an oil hole that aligns with an oil passage in the block. The upper half also has an oil groove while the lower half most often does not. Also take care to place the flanged thrust main bearing in the correct location for the thrust surfaces of the crankshaft (see Fig.16-28).

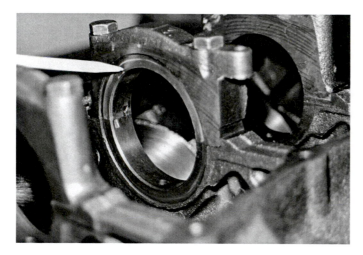

Fig.16-28 Check the position of the thrust bearing in the block and on the shaft

The backsides of bearings and the bores must be absolutely clean and dry. Contamination between the bearing shell and the surfaces of the block or bearing caps causes poor heat transfer and bearing distortion. Place the bearings in position in the block and in the bearing caps and then lubricate them. With the bearings in place and lubricated, lower the crankshaft carefully into place (see Fig.16-29).

Fig.16-29 Lowering a crankshaft into the main bearings

Check oil clearance at this point using Plastigage. Place the plastic strip across the crankshaft journal after wiping oil from the surfaces and tighten the bearing cap to specifications. Then remove the bearing cap and check the clearance against the graduated scale on the Plastigage package (see Fig.16-30). Be aware that the Plastigage used for oil clearance has a limited range of .001 to .003in (.03-.08mm), and when readings are outside these limits, it is best to check clearance by other methods. It helps to warm Plastigage above room temperature before use and be sure to remove the Plastigage when done.

Fig.16-30 Checking bearing clearance with Plastigage

Many machinists prefer checking oil clearance by measuring the bearing inside diameter (see Fig.16-31). With the bearing torqued in place, compare the inside diameter to the shaft outside diameter. The difference in diameters is the clearance.

Fig.16-31 Measuring the installed inside diameter of a bearing

It is good practice to lubricate the main bearing capscrews with engine oil before installation. Do not oil the internal threads because the holes may partially fill with oil, and hydraulic locking will prevent proper tightening of the capscrews. Tightening of the capscrews also should be done in stages of approximately one-third, two-thirds, and full torque. Before tightening the thrust main cap, rap the crankshaft at each end to align the flanges of the thrust main bearing.

Check the crankshaft for free rotation after tightening the bearing caps in place. Keep in mind that a properly fitted rear main seal of the rope type causes some drag in crankshaft rotation. If a problem is suspected, inspect the bearings for wear patterns and recheck the rotation with the seal removed.

Remember that there is an exception to these procedures for those engines having no provision for centering the bearing caps. For engines without dowel pins or registers to locate caps, coat the bearings in the cap with assembly grease to help center

the caps hydraulically. Remember that poorly centered caps cause out-of-round bearing housing bores and distortion or mislocation of rear main seals.

Another exception applies to aluminum engines. In some of these engines, the tightening of main bolts or head bolts must be done in sequence. For example, some aluminum block V8's have head bolts that extend into the main bearing webs and tightening them affects crankcase alignment (see Fig.16-32). If main bolts are tightened before head bolts in these engines, crankcase distortion and crankshaft binding results. If tightened in sequence, assembly progresses without difficulty. In general, check for special assembly procedures when working on any aluminum block engine.

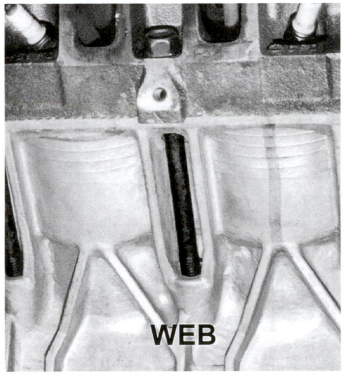

Fig.16-32 Note the location of these head bolts and main webs

Check crankshaft endplay at this time using a dial indicator and pry-bar (see Fig.16-33). Endplay can also be checked with feeler gauges between the thrust face of the crankshaft and the flanged thrust bearing (see Fig.16-34). Remember that thrust cap positioning affects endplay. If tight, recheck with the thrust cap loosened.

Fig.16-33 Checking crankshaft endplay with an indicator

Fig.16-34 Checking crankshaft endplay with a feeler gauge

An alternate method of crankshaft assembly is to "lay-in" the crankshaft. This method is especially valuable if there is suspicion of crankcase or crankshaft misalignment. Begin by installing the crankshaft bearings and crankshaft with a strip of .001in (.02mm) shim stock between the bearing cap and the backside of the bearing shell. Leave rope-type rear main seals out of the assembly and check the crankshaft for free rotation. If the crank-

shaft rotates freely with reduced clearance, both crankcase and crankshaft alignment is acceptable, provided clearance is not excessive. To be sure that clearance is within specifications, remove the shim stock and check clearance with Plastigage.

Another variation from these procedures is the installation of main bearings while the engine is in the chassis. This is done by "rolling" bearings in and out of assembly. First, back off all main bearing capscrews a full turn so that the crankshaft drops. Then remove one main cap and insert the rollout pin into the crankshaft oil hole. The rollout pins are soft aluminum and available in most parts stores. Remove the upper bearing by turning the crankshaft until the rollout pin catches the bearing and forces it to rotate with the crankshaft (see Fig.16-35). Install a new bearing by reversing the procedure. Install the lower bearing in the cap as normal. Repeat the process for each of the other main bearings.

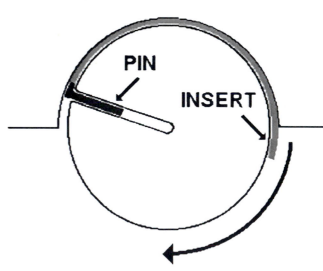

Fig.16-35 "Rolling" out a main bearing in the chassis

SETTING VALVE TIMING

In pushrod engines, the cam drive sprockets and timing chain are installed in alignment to time the engine's valve events. The crankshaft sprocket fits the crankshaft in only one position to assure proper assembly. The camshaft sprocket is twice the diameter of the crankshaft sprocket and fixed in position on the camshaft by a key or a pin. There are timing marks on each sprocket. Valve timing is correct in many applications when the timing marks are on the centerline between the crankshaft and camshaft (see Fig.16-36). Still, check service manuals to be sure of valve timing marks because there are variations. Assemble by sliding the two sprockets and the timing chain into position together. This reduces the twisting or binding of the timing chain during installation.

Fig.16-37 Typical timing marks on timing gears

Fig.16-36 Typical timing marks on timing sprockets

With timing gears, timing marks on the gears typically align on the centerline through the camshaft and the crankshaft (see Fig.16-37). As with timing chains and sprockets, be sure to check service manuals for the correct alignment of timing marks.

To be sure of valve timing, check for equal valve opening at TDC on the exhaust stroke. Equal valve opening shows that valve timing is close to split overlap. If only one tooth off, one valve will be well open and one valve clearly closed, not even close to split overlap.

Some engines have "dual pattern" camshafts with unequal lift and duration for intake and exhaust valves and valve opening is not equal at TDC. For example, if exhaust duration is greater than intake duration, split overlap occurs a few degrees ahead of TDC.

Because assembly by timing marks alone has limited accuracy, engines sometimes do not perform up to their potential. Typically, split overlap occurs in production engines between TDC and 4 degrees before TDC exhaust, that is, both valves are open equally somewhere between these two points. To precisely adjust valve timing requires offset keys or eccentric cam bushings (see Fig.16-38).

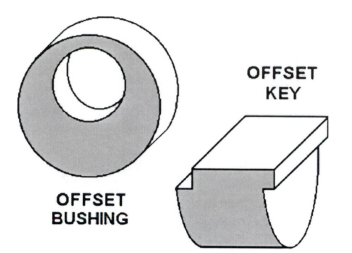

Fig.16-38 Offset keys and bushings for precise valve timing adjustments

Following are some simple procedures to read valve timing in pushrod engines:

1. Install the timing set, timing cover, and damper following manufacturer's recommendations but without gaskets. Use only one or two screws to retain the timing cover.

2. Rotate the crankshaft until both valves are closed. Place an indicator on one pushrod and set it to zero (see Fig.16-39). Next, rotate the engine to TDC exhaust and read the valve opening on the indicator. Repeat these steps on the second valve.

Fig.16-39 Setting an indicator to zero at the pushrod (OHV)

3. Since split overlap occurs when both valves are open the average of these two readings, rotate the crankshaft toward TDC exhaust until either valve is open to the average. Next, read the valve timing at the ignition timing marks.

4. With equal intake and exhaust duration and lift, set valve timing so that split overlap occurs between TDC and 4 degrees advance using an offset key or cam bushing (see Fig.16-40 and 41). For dual pattern camshafts with increased exhaust valve duration, set split overlap to at least 4 degrees advance. While these timing points are advanced, remember that normal chain stretch causes cam timing to retard 2 to 4 crankshaft degrees.

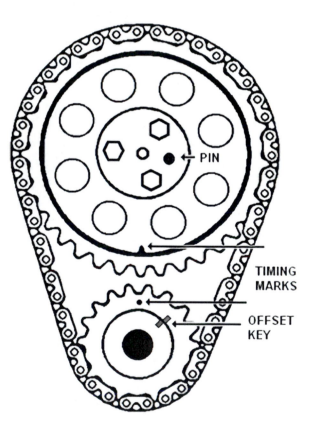

Fig.16-40 Advancing a cam with an offset key

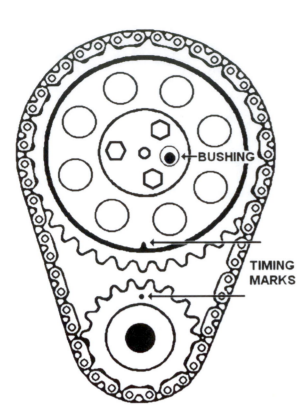

Fig.16-41 Advancing a cam with an eccentric bushing

For overhead cam engines, procedures are slightly different. For one thing, with a complete cylinder head assembly including the camshaft, valves can hit pistons unless both the camshaft and crankshaft are correctly positioned before placing the head on the block. First time both the crankshaft and camshaft according to timing marks and assemble head to the block and then check valve timing more precisely as follows:

5. Using a dial indicator on the intake valve spring retainer, turn the engine in the normal direction of rotation until the intake valve closes. Set the indicator to zero (see Fig.16-42).

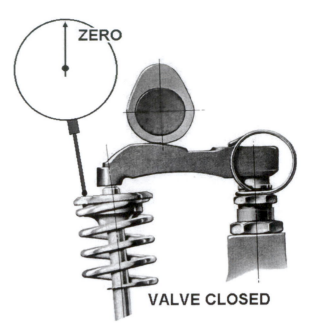

Fig.16-42 Setting the indicator to zero on the spring retainer (OHC)

6. Continue turning the engine forward and record the intake valve opening at TDC exhaust stroke.

7. Move the indicator to the exhaust valve and turn the engine forward until the exhaust valve closes. Set the indicator to "zero."

8. Continue turning the engine forward and record the exhaust valve opening at TDC exhaust stroke.

9. Ideally, the intake valve opening at TDC exhaust will equal exhaust valve opening. If advanced, intake lift at this point will be greater, and if retarded, it will be less.

10. If timing does check out, turn the engine until either valve is open the average of the two readings. Then record the number of degrees from TDC at the timing marks on the damper. If correction is required, remember that camshaft degrees are half of crankshaft degrees.

As with pushrod engines, engine performance is generally very good with timing set split overlap at TDC or a few degrees before. For engines with greater exhaust valve duration or lift, split overlap typically occurs a few degrees before. Remember too that advanced timing on assembly also compensates for normal timing chain stretch during break-in. More precise methods of valve timing are covered in Chapter 17, Preparing Performance Engines.

INSTALLING PISTON RINGS

After oiling the cylinder wall, place each piston ring squarely in the cylinder and check the end gap with a feeler gauge (see Fig.16-43). Check end gap with the piston ring in the lower (unworn) area of worn cylinders. End gap may be checked anywhere in new cylinders. Common practice is to check for minimum end-gap only or approximately .003in (.08mm) per inch of cylinder diameter. However, check specifications for new piston designs as some use hypereutectic alloys and have the top ring moved up on the crown. Minimum end gap on these rings run .005in (.13mm) per inch or more. Without the minimum end gap, the piston rings expand with heat and butt together, causing broken piston rings and scored cylinders. If necessary, increase end gaps by filing (see Fig.16-44).

Fig.16-44 Filing to correct piston ring end gap

While most technicians check minimum end gap only, there is also a specified maximum end gap. Checking maximum end gap catches the installation of rings for a smaller bore diameter. Consider that end gap increases about .003in (.08mm) for each .001in (.03mm) increase in bore diameter and therefore an incorrect ring set has dramatically increased ring gap. In a correct sized cylinder, expect end gap to exceed minimum by up to 50 percent. For example, minimum end gap for a 4-inch (101.6mm) bore would be .012in (.3mm) but a typical ring set commonly checks out with .018in gaps (.45mm).

Piston and ring specifications can vary for the same engine applications leaving open the possibility for mismatching rings with pistons. For this reason, it is a good idea to check ring "back clearance" as well as end gap. Back clearance is the clearance behind the rings when pistons and rings are installed in the cylinder. Check compression ring back clearance by reversing them and inserting them into the piston ring groove. They should not extend outside the groove. Three piece oil control rings are assembled as a set, reversed and slipped into their groove. They should be recessed into the groove with approximately .015in (.4mm) clearance.

Fig.16-43 Checking piston ring end gap

It is essential to read piston ring installation instructions before installing the rings on the pistons. Installing a piston ring upside down or in the wrong ring groove can easily result in high oil consumption. A typical ring installation procedure begins with oil control rings. First install the oil control ring expander with the ends over the piston pin bore. Do not trim the ends of the expander or allow them to overlap as either condition reduces oil control ring tension and causes oil consumption. Next, spiral the lower steel rail into place with the end-gap approximately 2 inches (5cm) to the left of the expander ends. Last, spiral the upper steel rail into place with the end gap approximately 2 inches to the right of the expander ends. This procedure assures that all the end gaps are offset.

Shown in Figure 16-45 are examples of correct compression ring installation. Installing compression rings with an expanding tool helps control the ring expansion so that they open just enough to slip over the piston crown (see Fig.16-46). This helps prevent breaking the rings on installation. Install the lower compression ring first and the top compression ring last. Once installed, rotate the piston rings so that all end gaps are offset. Oil the piston rings and ring grooves and check that rings slide freely in their grooves.

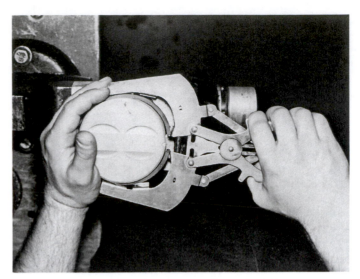

Fig.16-46 Using a ring expanding tool for installing rings

INSTALLING PISTON AND CONNECTING ROD ASSEMBLIES

First lay out piston and connecting rod assemblies in order of installation. The piston usually has a notch or other indicator that must point toward the front of the engine. Also check oil spurt holes, rod numbers, offsets or other references on connecting rods for proper direction.

Now snap each half of the connecting rod bearings into place. Check that bearing tangs fit with clearance in their lock grooves and oil the bearing surfaces. Next, oil the cylinder and rotate the crankshaft throw to BDC. Place rubber tubing over each rod bolt and clamp a ring compressor over the piston rings (see Fig.16-47). If available, a tapered sleeve is preferred to a ring compressor but these fit one bore diameter only (see Fig.16-48). Now push or butt the piston and connecting rod into assembly with a soft-faced hammer or a hammer handle. Guide the connecting rod carefully over the crankshaft throw to prevent nicking the surface. Remove the rubber tubing and tighten the rod cap and bearing in place.

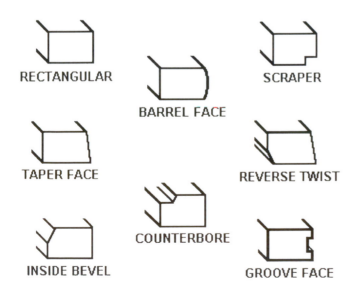

RECTANGULAR

BARREL FACE

SCRAPER

TAPER FACE

REVERSE TWIST

INSIDE BEVEL

COUNTERBORE

GROOVE FACE

Fig.16-45 Examples of correct ring installation

Fig.16-47 Using a ring compressor to install piston and rod assemblies

Fig.16-48 A tapered sleeve for use in place of a ring compressor

Next, check rod bearing oil clearance as with the main bearings. It is best practice to use new self-locking connecting rod nuts to prevent them from loosening. If new lock nuts are unavailable,

remove oil from threads and use an anaerobic adhesive such as Loctite on the threads.

Once tightened into position, check the side clearance between connecting rods with feeler gauges (see Fig.16-49). Specifications range from approximately .006in (.15mm) for single rods to .010in (.10mm) for rods paired together on one crankpin. If side clearance is too tight, lightly sand or draw-file the side faces of the connecting rods. Excessive side clearance causes increased oil throw-off onto cylinder walls and causes increased oil consumption. However, correcting this condition often requires replacing the connecting rods or crankshaft and the condition therefore goes uncorrected. Since oil flow through bearings increases dramatically with clearance, be concerned if both bearing clearance and side clearance are high.

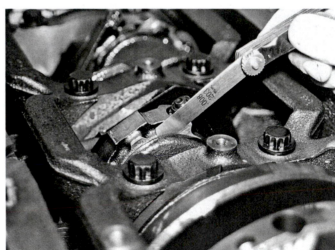

Fig.16-49 Checking rod side clearance

ASSEMBLING CYLINDER HEADS TO ENGINE BLOCKS

Many checks and procedures have been taken up to this point to ensure positive gasket sealing. Surfaces are flat and clean, head bolt threads are clean, and sharp edges around combustion chambers are deburred. Check for the dowel pins in the engine block and replace them if missing (see Fig.16-50). These pins keep the cylinder head and gasket in alignment during assembly.

Fig.16-50 Dowel pins are needed to position gaskets on blocks

Be sure to check pre-combustion chamber height on diesel cylinder heads (see Fig.16-51). Specifications allow pre-combustion chambers to be approximately .002in. (.05mm) above or below the head surface. Also check deck clearance on diesel engines (see Fig.16-52). Remember that with compression ratios over 20:1, a deck clearance change of .010in (.25mm) raises the compression ratio as much as a full point. For some engines, compensation is possible by selecting head gaskets of a different thickness.

Fig.16-51 Check pre-combustion chamber height in a diesel

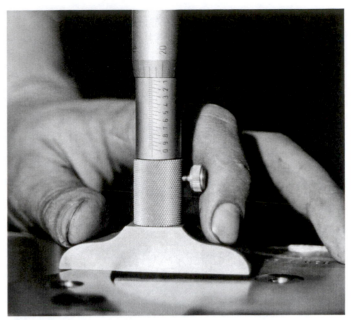

Fig.16-52 Check deck clearance in a diesel

Now check the cylinder head gasket for the correct position for assembly. Some gaskets are marked with a TOP or FRONT to help in correct assembly. On some engine blocks, the dowel pins are offset to different positions at each end of the block so that gaskets fit in only one position. Keep in mind that the passages through head gaskets meter the coolant circulation and improper assembly could cause overheating. Oil passages to the valve train sometimes go through head gaskets, and improper assembly could block these passages.

Check head bolts for length, position, and thread condition. Some head bolts have special configurations such as studs at the top end for mounting accessories. Unless sealer is called for, lubricate the threads of head bolts with engine oil and place them in position. Especially in blind holes, do not oil internal threads because the oil causes hydraulic locking when tightening the head bolts. Use silicone sealer on head bolts that extend into the water jackets. With this sealer, there is minimum change in clamping force at specified torque and the seepage of water around head bolts into the crankcase is prevented. Tighten head bolts in three stages; to one-third, two-thirds, and then specified torque following the correct pattern. Usually, the pattern begins near the center of the head and works alternately left and right around each combustion chamber. Be sure to check service manuals for the exact procedure for each engine.

Of course, torque-to-yield head bolts require tightening to a specified torque value and tightening further by degrees of rotation. Some of the newer torque wrenches read both torque and degrees rotation (see Fig.16-53). Check specifications carefully because some of these head bolts are reusable and some are not. If in doubt, replace them.

Fig.16-53 An electronic torque wrench reading both torque and degrees rotation

For pushrod engines, now install the valve lifters and pushrods. Coat the bases of valve lifters and tips of pushrods with anti-scuff lubricant. Check the fit and rotation of each valve lifter on installation.

INSTALLING ROCKER ARMS

When preparing to assemble rocker arms and shafts, or fasten rocker arm assemblies to the engine, be sure to check rocker arm positioning over valve stems. Many engines use rocker arms that are offset right or left and incorrect assembly causes misalignment with the valve stems (see Fig.16-54).

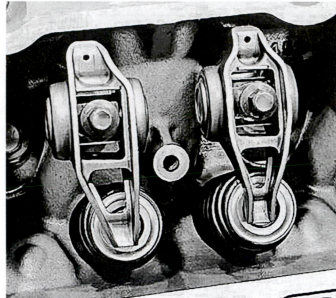

Fig.16-54 Checking positions of offset rocker arms over valve stems

Check the positioning of rocker arm shafts as they must position with oil holes aligned for lubrication. Follow assembly instructions for the particular engine so that rocker arm oil holes position correctly (see Fig.16-55). Make sure that oil plugs are installed in the ends of shafts and lubricate all parts before assembly.

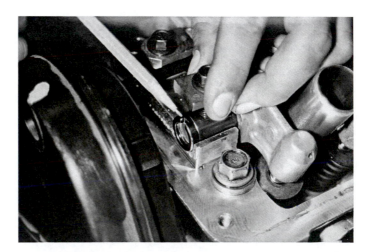

Fig.16-55 Check rocker shaft assembly position to assure proper oiling

ADJUSTING VALVES

Now adjust valve lash to clearance specifications on non-hydraulic valve trains (see Fig.16-56). One method is to turn one cylinder at a time up to TDC and adjust both valves. While this procedure is easy to follow during engine assembly, there are faster methods in service manuals for each particular engine. These methods call for adjusting valves on different cylinders at each crankshaft position and vary according to cylinder firing order.

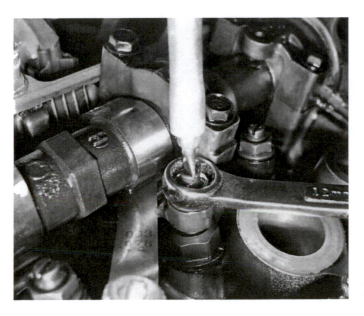

Fig.16-56 Adjusting valve lash or clearance

It sometimes happens that valve lash noise remains after adjustments. When this happens, recheck lash by rotating the engine and look for increased lash at other points around the base circle. If lash is greater at other points in the rotation, adjust lash at those points.

Engines with hydraulic valve lifters and adjustable rocker arms adjust by two basic methods. One is to adjust each rocker arm to zero lash plus a pre-load (1/2 turn for example) while in the valve closed position and then make final adjustments after starting the engine. To limit the mess caused by oil spray, install valve covers with holes cut in the top for access to the adjusting nuts. With the engine running, back off each rocker arm until it rattles, tighten it until quiet, and then very

slowly tighten it one-half to three-quarters of a turn more. Slow tightening allows lifters to bleed down and prevents bending valves. The second way is to collapse each lifter by prying down on the pushrod end of the rocker arm and checking for specified clearance between the valve stem and rocker arm (see Fig.16-57). By any procedure, adjustment is correct when the pushrod seat is depressed into the lifter body approximately .040in. (1mm).

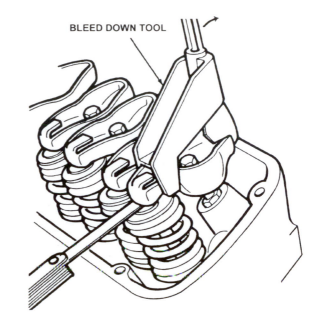

Fig.16-57 Collapsing a hydraulic lifter to check rocker arm adjustment (Ford)

Be aware that non-adjustable rocker arms assemble by tightening to a specified torque. Identify non-adjustable rocker arms by looking for a shoulder on the rocker arm stud. These studs often have a 3/8in diameter on the bottom and a 5/16-24 thread on the top. Remember that installed stem height is critical with non-adjustable rocker arms as it is possible for rocker arms to hold taller valves open.

Do not fill hydraulic valve lifters with oil before assembly as this sometimes causes valve lifters to hold valves open causing them to interfere with pistons. It is safer to circulate oil through the entire engine with a pre-oiler or by driving the oil pump externally after adjusting valves. This way, the pushrod, rocker arm, and valve spring limit plunger travel in the lifters and prevent over filling.

INSTALLING THE OIL PUMP

The engine assembly is nearly complete at this point. A last step is to prepare for the installation of the oil pump. Some technicians check the pump operation and prime it by dipping the pump pickup in oil and turning the pump by hand. Other procedures call for greasing the pump to ensure that it primes when the engine starts. Check service manuals for specific recommendations.

Other checks are required. First, check that the oil pump pickup is within 1/4in (6mm) of the bottom of the oil pan (see Fig.16-58). After checking the position of the pickup, secure or fasten the pickup tube to the pump body. In some cases, the pickup bolts to the pump body and the only particular concern is with gasket sealing. In other cases, the pickup tube presses into the pump body.

it into place. As mentioned earlier, if brazing the pick-up tube to the pump body, first remove all internal parts.

Check the position and attachment of the pickup tube before installing the pump. If the oil pump pickup should fall out, oil pressure is lost. Should the pickup fit loosely, it pulls in air and engine lubrication suffers. If it is too high, oil washes away from the pick-up on braking, acceleration, and cornering.

Some engines use an intermediate shaft to connect the oil pump to the distributor (see Fig.16-59). Some shafts and retaining clips or sleeves install with the pump from the lower side of the engine. Failure to catch this detail requires removing the oil pan from the assembled engine to install the intermediate shaft.

Fig.16-58 Comparing height of the pick up above the pan rails to oil pan depth

If press fitting is required, use the proper installation tool, clean the bore, apply a few drops of anaerobic sealer to the pickup tube, and press

Fig.16-59 An oil pump drive with a sleeve around the drive

Depending on the engine, oil pumps install in the crankcase, in the timing cover, or, as with gear-rotor pumps, over the snout of the crankshaft. Pumps are driven by the camshaft, or by chains off of the crankshaft, or directly by the crankshaft. Check the service references for each specific engine.

PRE-OILING THE ENGINE

Pre-oiling internal parts prevents the scuffing of bearings, cylinders, and camshafts. A common cause of engine damage, especially to bearings, is scuffing caused by dry starts. To pre-oil using the engine's own oil pump, first install the oil pan and fill it with oil. In pushrod engines, install valve lifters to block oil passages through lifter bores. Then drive the oil pump by hand with a speed handle (see Fig.16-60). Rotation by hand provides 20 PSI or more oil pressure. With drill motors, use caution and run the drill slowly to prevent damaging the pump internally. Of course, pre-oiling by hand is limited to camshaft driven pumps.

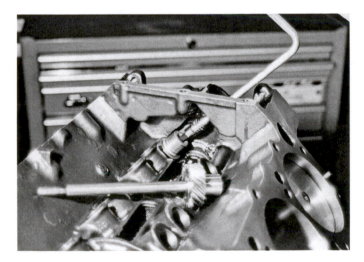

Fig.16-60 Pre-oiling the engine by hand

Remember that the direction of rotation is the same as for the distributor. Rotate the engine to several positions to be sure oil circulates to all engine parts. This procedure also leaves the oil pump primed so that the engine has immediate oil pressure on start up.

If available, connect a "pre-lubricator," or pressurized oil tank, to the main engine oil passage (see Fig.16-61). Crankshaft driven pumps can only be pre-oiled in this way. The location for the oil pressure switch is a good place for this connection. Be sure to fill the oil filter before pressuring the system. Unless the filter is filled first, it is possible to trap air in the system that can prevent the pump from priming on start up. Should this occur, loosen the filter, allow air to escape and retighten the filter.

Fig.16-61 A Goodson PL40 pre-lubricator

In any engine, it is best practice to assemble the oil pump with assembly lube to speed priming. Even when primed, it is safe practice on installation to disable the ignition or fuel injection and crank the engine with spark plugs removed until oil pressure builds and circulates. Removing spark plugs removes pressure from crankshaft bearings, and disabling ignition or fuel injection prevents fires.

A pre-lubricator is also an excellent diagnostic tool. For example, locate excess oil flow through bearings or leaking oil passages by removing the oil pan watching flow rates. It is also possible to check for oil circulation through the various oil passages.

INSTALLING TIMING COVERS

Timing cover seals are another source of oil leaks. Be sure to apply gasket sealer around the seal case before installing it in the timing cover. Also lubricate the seal with engine oil to prevent damage during the first few minutes of operation. A deep groove sometimes develops on the vibration damper seal surface, making an effective seal very difficult. For many engines, a thin-wall sleeve is available to slip over the vibration damper (see Fig.16-62). If no repair sleeve is available, check to see if the replacement seal locates in the wear groove. If it does, try to reposition the seal to a different depth on installation. Replace the vibration damper if these methods do not work.

Fig.16-63 Oil slingers deflect excess oil from seals

Be sure the timing cover centers over the crankshaft. One method of doing this is to place an adapter through the seal and over the snout of the crankshaft (see Fig.16-64). Leave the fasteners loose during this procedure to allow the cover to shift in location. It is sometimes necessary to drill dowel pinholes in timing covers oversize to gain sufficient movement to allow centering.

Fig.16-62 Installing a damper repair sleeve

Many engines use an oil slinger to deflect excess oil from the timing cover seal (see Fig.16-63). Do not forget to install the slinger between the crankshaft timing gear or sprocket and the timing cover or an oil leak is certain.

Fig.16-64 Centering the timing cover over the crankshaft snout

HINTS ON GASKETS SEALS AND SEALANTS

Several points regarding cylinder head gaskets require review. They must be installed in the correct position and not restrict water or oil passages. Another suggestion is to apply gasket sealer only and assemble between the timing sprocket and front seal to surfaces with corrosion damage and to the junctions of gaskets. Do not over use sealer; keep it out of the interior of the engine and do not use it on coated gaskets.

Do apply silicone sealer to all intersections of gaskets, such as at the corners of oil pan gaskets. Silicone sealers work well in those locations because the sealer readily flows into gaps where gaskets come together. Also, seal the corners of V-block intake manifolds where gaskets, the engine block, and cylinder heads all join together (see Fig.16-65). Remember to remove oil from surfaces before applying silicone sealer or it will not stick to surfaces and leaks result.

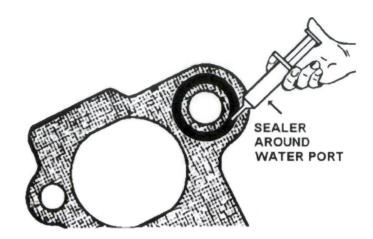

Fig.16-66 Seal around manifold water passages (Courtesy Fel-Pro Inc.)

The end gaskets under V-block intake manifolds sometimes cause difficulty during assembly (see Fig.16-67). The problem is that the gasket sometimes slips out of position when tightening the manifold in place. Use gasket cement to prevent this. Apply the cement to the gasket and to the block surface under the gasket. Allow the cement to dry in the air for a few minutes before installing the gasket. If the cause of difficulty is excessive gasket crush after machining cylinder heads, discard these gaskets and use silicone sealer only.

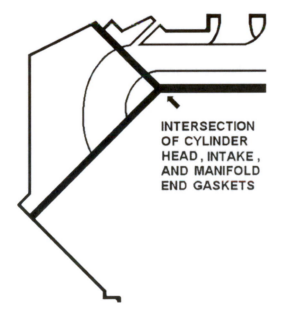

Fig.16-65 Seal the corners of V-block intake manifold gaskets

Silicone sealers also work especially well around water passages. For example, it is good practice to apply a bead of sealer around the water ports between intake manifolds and cylinder heads (see Fig.16-66).

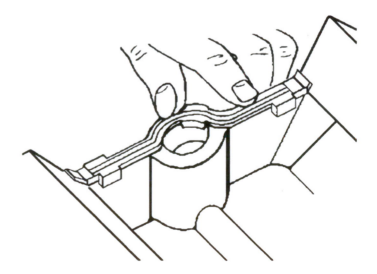

Fig.16-67 Manifold end-gaskets (Courtesy Fel-Pro Inc.)

Straighten all sheet metal before assembly. Watch especially for distortion around the bolt holes in oil pans and valves covers from over tightening. Check the fit of gaskets and alignment with bolt holes. Remember that cork gaskets dry and shrink and it is sometimes necessary to expand them by wetting them with warm water. Be careful not to over tighten bolts on assembly. Snug them lightly to approximately 100 inch-pounds during assembly and retighten them after the engine warms up.

Many newer engines assemble with anaerobic sealers and varying types of silicone sealers instead of some gaskets. Some of these engines also use plastic oil pans or valve covers. Although gasket manufacturers package sets that include gaskets for some of these applications, it is sometimes necessary to assemble with chemical sealants only. For example, dealership technicians seal engines using particular sealants identified by part numbers. Independent garages and machine shops use equivalent sealants or sealant kits provided by gasket manufacturers.

USING AN ASSEMBLY CHECKLIST

To catch the all-important details, follow a basic checklist. Here, sub-assemblies such as cylinder heads and piston and rod assemblies are prepared in advance. The following two checklists are general, one for pushrod engines and another for overhead cam engines. Modify these checklists for specific engines as required.

ASSEMBLY CHECKLIST
Pushrod Engine

Customer _____
Invoice No. _____ Date _____
Engine _____
Assigned to _____

Directions: Check or enter the specification in the space provided.

I. INSPECT
____ Oil passages and threaded holes for cleanliness
____ Cylinder chamfers at top and deburring at bottom
____ Size of cylinders, pistons, rings
____ Size of crankshaft, main bearings, rod bearings
____ Pilot bearing or;
____ Torque converter, flex-plate fit to crankshaft
____ Intake manifold cleanliness, including crossover

II. CYLINDER HEAD ASSEMBLIES
____ Uniform stem heights
____ Solvent check valve sealing

III. ROD AND PISTON ASSEMBLIES
____ Check housing bore diameters
____ Check Pin fits, full-floating only
____ Align rods and pistons
____ Check ring end gap, end gap spacing in assembly

IV. OIL PUMP AND PICKUP
____ Check end-plate wear
____ Check rotor or gear endplay
____ Check rotor to body clearance
____ Deburr rotors or gears
____ Clean, lubricate, install pressure-relief valve
____ Clean or replace pickup screen
____ Assemble pump
____ Attach pick up to pump

V. ASSEMBLY
____ Install core plugs
____ Install cam bearings
____ Check camshaft straightness
____ Check camshaft fit in bearings
____ Blow passages clear
____ Grease camshaft and install
____ Install oil plugs
____ Install camshaft thrust plate
____ Fit, install rear main seal (2 pc)
____ Install main bearings, thrust washers
____ Install crankshaft
____ Check main bearing clearance
____ Position thrust main cap
____ Check crankshaft endplay
____ Torque main caps
____ Install main seal (1 pc)
____ Install cam thrust plate, timing set
____ Check camshaft endplay
____ Install rods and pistons
____ Check rod bearing clearance
____ Check rod side clearance
____ Torque, Loctite rod nuts
____ Install oil pump and drive shaft, crankcase installations
____ Install valve lifters
____ Check for split overlap TDC exhaust
____ Install timing cover
____ Install oil pan
____ Install cylinder heads, torque head bolts
____ Install rocker arms and pushrods
____ Adjust valves

VI. SPECIAL NOTES

_____
_____
_____

ASSEMBLY CHECKLIST
Overhead Cam Engine

Customer _____
Invoice No. _____ Date _____
Engine _____
Assigned to _____

Directions: Check or enter the specification in the space provided.

I. INSPECT
____ Oil passages and threaded holes for cleanliness
____ Cylinder chamfers at top and deburring at bottom
____ Size of cylinders, pistons, rings
____ Size of crankshaft, main bearings, rod bearings
____ Pilot bearing or;
____ Torque converter, flex-plate fit to crankshaft
____ Intake manifold cleanliness

II. CYLINDER HEAD ASSEMBLIES
____ Solvent check valve sealing
____ Check, adjust valve lash

III. ROD AND PISTON ASSEMBLIES
____ Check housing bore diameters
____ Check Pin fits, full-floating only
____ Align rods and pistons
____ Check ring end gap, end gap spacing in assembly

IV. OIL PUMPAND PICKUP
____ Check end-plate wear
____ Check rotor or gear endplay
____ Check rotor to body clearance
____ Deburr rotors or gears
____ Clean, lubricate, install pressure-relief valve
____ Clean or replace pickup screen
____ Assemble pump
____ Attach pick up to pump

V. ASSEMBLY
____ Install core plugs
____ Install auxiliary shaft or balance shaft bearings, as needed
____ Install auxiliary shaft or balance shafts
____ Blow oil passages clear
____ Grease camshaft and install
____ Install oil plugs
____ Install main bearings, thrust washers
____ Install crankshaft
____ Check main bearing clearance
____ Position thrust main cap
____ Check crankshaft endplay
____ Torque main caps
____ Fit, install rear main seal
____ Install rods and pistons
____ Check rod bearing clearance
____ Check rod side clearance
____ Torque, Loctite rod nuts
____ Install oil pump and drive shaft or chain, crankcase installations
____ Set crankshaft and camshaft to TDC
____ Install cylinder head assembly, torque head bolts
____ Time auxiliary shaft or balance shafts
____ Install timing chain, tensioners, guides, oil slinger
____ Check for split overlap TDC exhaust
____ Install timing cover and seal
____ Install oil pump, timing cover installations
____ Check oil pump pick-up position in pan, attach to pump
____ Install oil pan
____ Install timing belt, adjust tensioner

VI. SPECIAL NOTES

_____

_____

_____

TESTING IN A RUN-IN STAND

Many engine builders test assembled engines by driving them with electric motors. The advantage to this method of dynamic testing is that, unlike dynamometer testing, the engine does not require fuel, ignition, cooling, and exhaust systems. Possible tests include; oil pressure and circulation, compression, guide sealing, main and timing cover sealing, engine noises, and lifter rotation in pushrod engines.

Except for chain driven overhead camshaft engines, it is convenient to test the assembled short block first (see Fig.16-68). The short block mounts in the stand and an oil line delivering filtered oil to the engine connects to the block for pre-oiling. If necessary, install the distributor after pre-oiling and test oil pressure using the engine's own oil pump. The oil line from the test stand also connects to a pressure gauge.

Fig.16-68 Testing a short block in a run-in stand

In testing the short block, visually check cylinder walls for scoring or excess oil and listen for engine noises under a completely unloaded condition. Keep in mind that some noises, such as from piston pins, are loudest while under no load. Also keep in mind that if the oil is not fully up to temperature, the oil pressure in testing is approximately double the pressure in a warm engine.

Should there be unacceptable noise, cylinder problems, or oil pressure problems, disassemble and inspect the problem components. Test engines without the pan gaskets installed and check for dirt or metallic particles in the oil pan before final assembly. While bearing material in the oil may not affect test results, it definitely suggests shortened engine life.

Next, assemble the cylinder head to the short block in the engine test stand and resume testing. Check valve adjustments and test compression. Keep in mind that the rings are not seated and the engine must turn over sufficient revolutions to reach maximum compression. If tested at cranking speed, test results approximate specified compression. During testing of the long block, use valve covers cut away to create access to valve adjusters and to make possible visual checks of valve train oiling (see Fig.16-69). Check valve guide sealing by looking for excess oil in each valve port. Also visually check pushrod engines for valve lifter rotation. Again, should problems be found, make the necessary corrections. Because the engine has not reached operating temperature, reuse cylinder head gasket should disassembly be required.

Fig.16-69 Checking the valve train during testing

ASSEMBLING FLYWHEELS AND FLEXPLATES

Resurface flywheels as necessary, assemble in proper position on crankshaft flanges, and torque to specifications. Flywheel bolts sometimes extend through the crankshaft flange into the crankcase and require sealing to prevent oil leaks. These are special fasteners so do not replace them with off-the-shelf hardware.

To check flywheel and crankshaft flange condition, mount an indicator to the engine block and read face run-out near at the outer edge of the flywheel (see Fig.16-70). Be sure to keep pressure in one direction while rotating the engine or crankshaft endplay might cause errors in the reading. The TIR should be within .005in (13mm). If not within specifications, first check run-out on the crankshaft flange and then parallelism between the front and backside of the flywheel. Correction may require grinding the crankshaft flange square or surfacing the backside of the flywheel parallel with the front. Other than torquing procedures, automatic transmission flexplates do not require these checks.

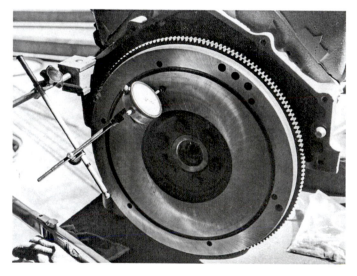

Fig.16-70 Checking flywheel run-out

Install the clutch pilot bearing at this time. For some crankshafts, there are differences between automatic and manual transmission crankshafts. This is particularly important for exchange crankshafts because provision for a pilot bearing is not always made in automatic transmission crankshafts. For some domestic engines, the crankshafts interchange but dimensions for the pilot bearings are different for converting automatic transmission crankshafts to manual. To prepare for installation, clean and deburr the crankshaft bore including the breaking of any sharp edges. Chamfer the inside and outside corners of the bushing to ease installation and to prevent deforming the hole on installation. Lubricate the bearing and drive it in using a driver that fits flat against the face of the bushing (see Fig.16-71).

Fig.16-71 Installing a clutch pilot bearing in the crankshaft

If the pilot bearing is the anti-friction type, pack the ball bearings with grease, coat the outside diameter with Loctite, and install it using a driver against the outer bearing race. Use caution because driving against the inner bearing race damages the bearing.

To assemble clutches to flywheels, insert a clutch alignment tool through the clutch hub into the pilot bearing in the crankshaft (see Fig.16-72). Be sure the clutch plate hub faces the correct direction and that it clears the flywheel when the friction material is flush against the flywheel face. Check the clutch pressure plate for marks that show the correct position on the flywheel and torque to specifications. Clutch pressure plates use special high strength fasteners that cannot be replaced with standard hardware. Tighten these bolts in steps going around the bolt circle to limit distortion of the clutch housing.

Fig.16-72 Aligning the clutch plate to the pilot bearing

For automatic transmission torque converters, be sure to install them all the way into the front pump of the transmission before attempting to connect the transmission to the engine. Do this by rotating the converter while lightly pushing it back into the transmission. The engagement of all members can be felt as it "clicks" into place. After attaching the bellhousing, it should be necessary to pull the torque converter forward to the flexplate. If not, the converter is not yet fully engaged into the transmission. Failure to get this step right destroys the front pump in the transmission.

For both automatic and manual transmissions, some engines use a plate between the bellhousing and the engine block. This plate positions the starter motor and the plate thickness is part of proper transmission spacing.

ATTACHING BELLHOUSINGS

In discussions of sealing rotating shafts, and line boring or honing, it was pointed out that the crankshaft centerline moves. Transmissions must assemble to engines with their input shafts and crankshaft on the same centerline. While heavy-duty engines call for checking this relationship, it is rare for technicians to check passenger car engine and transmission assemblies. Test by mounting a dial indicator to the crankshaft flange and sweeping the bore of the transmission bellhousing with the indicator (see Fig.16-73). The TIR should not exceed .010in (.25mm) but readings often exceed .025in (.6mm).

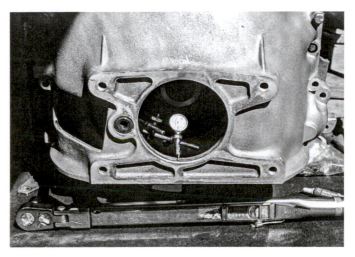

Fig.16-73 Indicating in a flywheel bellhousing

Offset dowel pins restore this relationship to specifications by adjusting the bellhousing on center (see Fig.16-74). Without offset dowel pins, remove the pins, adjust the bellhousing onto center, and then ream the holes for oversized pins. At a minimum, failure to correct this relationship causes transmission input shaft and clutch pilot bearing wear and noise.

Fig.16-74 Offset dowel pins for positioning the bellhousing

SUMMARY

Assembly is a critical stage in engine building. If not done properly, the engine will fail. If parts are incorrect for the application, dirty, or not within specifications, the engine will fail. The assembler must know correct engine assembly procedures and catch any problems that can lead to failure at this stage. The engine may be 99 percent correct but unless the other percentage point is made right, guess what, the engine will fail.

Prior to actual engine assembly, subassemblies are prepared. These include cylinder heads, piston and connecting assemblies including piston rings, and oil pumps. Cylinder heads must be assembled with careful attention to valve seals and installed heights and, for overhead cam heads, valve lash adjustments. Rods and pistons are assembled to each other with attention to cylinder number and direction and ring installation. Oil pumps are inspected, deburred, cleaned, lubricated and assembled. Inspect all parts for cleanliness and damage picked up in handling.

Technicians and machinists must be able to measure engine parts and check clearances using standard measuring tools and Plastigage. They must be familiar with any special tools such as valve spring compressors, cam bearing drivers, and ring compressors. Skill and knowledge is needed

to properly fit cam bearings, install oil plugs and core plugs, install piston and rod assemblies, and timing components.

For both quality control and productivity, assemblers need to know the proper sequence of steps and points of inspection for assembly. The Assembly Checklists in this chapter are a good start but these must be customized to suit particular engine applications. Unless steps are in sequence and inspections done at critical points, time will be lost disassembling parts of the engine to check clearances or, worse yet, assumptions will be made that "it's OK".

Technicians and machinists need to learn different methods for tightening fasteners. Some fasteners are "torque-to-yield" and are tightened by degrees of rotation and some are not reusable. Other fasteners are tightened in stages with a torque wrench. There is also a need to understand uses for thread lubricants or sealers and their effect on clamping force.

Considerable attention must be given to details. Consider the time lost repairing minor oil or water leaks in engines? Even with internal engine components properly assembled, small oil or water leaks will bring the engine back to the shop. At worst, the engine will be run out of oil or water and fail. At best, the leaks will be repaired at the shop's expense. Carefully check those rear main seals, timing cover gaskets and seals, and valve cover and oil pan gaskets.

Testing in a run-in stand is one way of catching a number of problems such as low compression, defective valve guide sealing, low oil pressure, and some oil leaks. Some problems, however, can slip past testing. For example, an engine assembled with dirt in oil passages is not necessarily going to show problems in testing but the bearings will fail prematurely as a result of the dirt. There is no substitute for clean work habits and thorough inspection prior to assembly.

Chapter 16

ENGINE ASSEMBLY

Review Questions

1. Technician A says check for chamfering on the top end of cylinders. Technician B says check for sharp edges on the bottom ends. Who is right?

   a. A only
   b. B only
   c. Both A and B
   d. Neither A or B

2. Prior to assembly, _____ head bolt holes.
   a. clean and chamfer
   b. lubricate
   c. Helicoil
   d. tap

3. Technician A says place a sleeve over the valve stem to protect positive seals on installation. Technician B says adjust the valve spring compressor stroke to protect stem seals on installation. Who is right?

   a. A only
   b. B only
   c. Both A and B
   d. Neither A or B

4. Technician A says compress springs below installed height specifications for keeper installation. Technician B says that if compressed too far, the valve seal is crushed. Who is right?

   a. A only
   b. B only
   c. Both A and B
   d. Neither A or B

5. Technician A says to check for free camshaft rotation in pushrod engines before assembling heads. Technician B says to check for free camshaft rotation in overhead cam engines before assembling heads. Who is right?

   a. A only
   b. B only
   c. Both A and B
   d. Neither A or B

6. A core hole in an engine block measures 1 1/2in. The core plug diameter for this hole is _____ in.
   a. under 1 1/2
   b. exactly 1 1/2
   c. over 1 1/2
   d. 1 9/16

7. Technician A says drive core plugs against the outer edge. Technician B says drive core plugs with a driver that fits with clearance inside the cup. Who is right?

   a. A only
   b. B only
   c. Both A and B
   d. Neither A or B

8. Technician A says that oil holes in cam bearings are for cam journal lubrication. Technician B says that, depending on the engine, they also oil lifters or rockers. Who is right?

   a. A only
   b. B only
   c. Both A and B
   d. Neither A or B

9. Technician A says to prevent bearing distortion on installation, deburr bores and lubricate bearings before driving them into place. Technician B says that the driver must match the cam journal diameter. Who is right?

   a. A only
   b. B only
   c. Both A and B
   d. Neither A or B

10. Machinist A says correct camshaft binding in new bearings by grinding or polishing journals to fit. Machinist B says straighten the camshaft and hand scrape tights spots in bearings. Who is right?

    a. A only   c. Both A and B
    b. B only   d. Neither A or B

11. Technician A says that a camshaft thrust plate controls endplay. Technician B says that some engines do not use thrust plates. Who is right?

    a. A only   c. Both A and B
    b. B only   d. Neither A or B

12. Technician A says to drive timing gears into assembly with a hammer. Technician B says that timing gears that run out or have damaged gear teeth make noise or knock. Who is right?

    a. A only   c. Both A and B
    b. B only   d. Neither A or B

13. Technician A says grease cam journals for assembly. Technician B says oil cam lobes for assembly. Who is right?

    a. A only   c. Both A and B
    b. B only   d. Neither A or B

14. Technician A says stake pipe plugs in oil passages in place on installation. Technician B says lubricate oil passage core plugs on installation. Who is right?

    a. A only   c. Both A and B
    b. B only   d. Neither A or B

15. Technician A says that for any rotating shaft to seal, seals and sealing surfaces must be concentric to the shaft. Technician B says that bearing clearance cannot be excessive. Who is right?

    a. A only   c. Both A and B
    b. B only   d. Neither A or B

16. Machinist A says that cutting main caps for line boring or honing moves the crankshaft centerline. Machinist B says that oil seals front and rear may require centering after line boring or honing. Who is right?

    a. A only   c. Both A and B
    b. B only   d. Neither A or B

17. Technician A says that to seal timing covers, fit damper hubs with a repair sleeve or replace them. Technician B says center the timing cover and seal over the damper hub. Who is right?

    a. A only   c. Both A and B
    b. B only   d. Neither A or B

18. Machinist A says tools for fitting rope type rear main seals match the main bearing journal diameter. Machinist B says fit them with round stock. Who is right?

    a. A only   c. Both A and B
    b. B only   d. Neither A or B

19. Machinist A says that cotton rope type rear main seals are coated with a dry lubricant. Machinist B says soak these seals in oil prior to assembly. Who is right?

    a. A only   c. Both A and B
    b. B only   d. Neither A or B

20. Machinist A says install lip type seals with silicone sealer on the seal lip. Machinist B says install these seals with the outside case lubricated. Who is right?

    a. A only   c. Both A and B
    b. B only   d. Neither A or B

21. Technician A says that upper and lower main bearing inserts are the same and install in any position. Technician B says check that the backsides of bearing inserts and housing bores are clean and dry for assembly. Who is right?

    a. A only
    b. B only
    c. Both A and B
    d. Neither A or B

22. Technician A says lubricate bearings for Plastigage readings. Technician B says that it helps to warm Plastigage before use. Who is right?

    a. A only
    b. B only
    c. Both A and B
    d. Neither A or B

23. Other than Plastigage, check bearing clearance
    a. by comparing the inside diameter of the installed bearing to the shaft diameter
    b. with a dial indicator
    c with a dial bore gauge
    d. with shim stock

24. Unless positioned with dowel pins, align main bearing thrust flanges in assembly by
    a. tapping the crankshaft forward and backward before torquing the thrust main cap
    b. grinding the thrust main cap
    c. assembling bearings over dowels
    d. machining thrust bearing flanges

25. Technician A says that "Laying in a crankshaft" refers to using shim stock to check crankshaft and crankcase alignment. Technician B says that it is an alternate method of checking bearing clearance. Who is right?

    a. A only
    b. B only
    c. Both A and B
    d. Neither A or B

26. Set valve timing during engine assembly by
    a. measuring intake valve opening at TDC intake stroke
    b. measuring exhaust valve opening at TDC exhaust stroke
    c. making sure that both valves close at TDC compression
    d. lining up camshaft and crankshaft timing marks

27. Should there be any doubt regarding valve timing, check
    a. timing marks on the damper
    b. part numbers
    c. overlap at TDC exhaust
    d. valve to piston clearance

28. Technician A says that in pushrod engines, install cylinder heads with the camshaft turned to any position. Technician B says that in OHC engines, align the camshaft and crankshaft to timing marks before installing cylinder heads to avoid bending valves. Who is right?

    a. A only
    b. B only
    c. Both A and B
    d. Neither A or B

29. Technician A says that at split overlap, both valves are open equally. Technician B says that with "dual pattern" cams, lifts are not equal at TDC. Who is right?

    a. A only
    b. B only
    c. Both A and B
    d. Neither A or B

30. Technician A says that except for some hypereutectic pistons, minimum ring end gap is .003in per inch of bore diameter. Technician B says expect up to 50 percent more than minimum in typical ring sets. Who is right?

    a. A only
    b. B only
    c. Both A and B
    d. Neither A or B

31. Technician A says that rubber tubing on rod bolts protects _____ when installing piston and rod assemblies.
    a. main journals
    b. bearings
    c. rods
    d. crankpins

32. Check _____ on crankshaft installation.
    a. rod side clearance
    b. sealing of the rear main
    c. end play and main bearing oil clearance
    d. rod bearing oil clearance

33. Machinist A says oil threads in blind head bolt holes. Machinist B says seal head bolt threads that extend into water jackets. Who is right?

    a. A only
    b. B only
    c. Both A and B
    d. Neither A or B

34. Technician A says that head gaskets locate over dowel pins or rings for proper positioning. Technician B says install head gaskets correctly to assure proper oil and coolant circulation. Who is right?

    a. A only
    b. B only
    c. Both A and B
    d. Neither A or B

35. Machinist A says tighten head bolts in sequence with a torque wrench to obtain uniform clamping force. Machinist B says tighten torque-to-yield head bolts by degrees of rotation to assure uniform clamping force. Who is right?

    a. A only
    b. B only
    c. Both A and B
    d. Neither A or B

36. Torque to yield head bolts are
    a. reusable
    b. not reusable
    c. sometimes reusable, sometimes not
    d. replaced with non-torque to yield bolts

37. Machinist A says check rocker arms for right and left offsets in assembly. Machinist B says check oil holes in rocker shafts for direction in assembly. Who is right?

    a. A only
    b. B only
    c. Both A and B
    d. Neither A or B

38. Lash adjustments are made with lifters or followers on camshaft
    a. base circles
    b. opening ramps
    c. closing ramps
    d. lobes

39. Technician A says that adjustable rocker arms in pushrod, hydraulic lifter engines are tightened to zero lash plus another part of a turn. Technician B says that in some of these engines, lash is checked with the lifters bled down. Who is right?

    a. A only
    b. B only
    c. Both A and B
    d. Neither A or B

40. Technician A says tightly seal oil pump pickups where they attach. Technician B says position the pickup screen at least 1 inch from the bottom of the pan. Who is right?

    a. A only
    b. B only
    c. Both A and B
    d. Neither A or B

41. Machinist A says assemble oil pumps with grease to speed priming. Machinist B says that before start up, crank engines with spark plugs removed and ignition disabled until oil pressure builds up. Who is right?

    a. A only
    b. B only
    c. Both A and B
    d. Neither A or B

42. Machinist A says that an oil slinger goes between the timing sprocket and timing cover seal. Machinist B says that rear main oil slingers are integral to the crankshaft. Who is right?

    a. A only   c. Both A and B
    b. B only   d. Neither A or B

43. Machinist A says that intake manifold distortion on assembly causes vacuum leaks in some V-block engines. Machinist B says that manifold distortion in these engines is avoided by installing gaskets under the ends of the manifold. Who is right?

    a. A only   c. Both A and B
    b. B only   d. Neither A or B

44. Technician A says keep bellhousing or transmission case TIR within .010in. Technician B says that if not concentric, replace the bellhousing or transmission case. Who is right?

    a. A only   c. Both A and B
    b. B only   d. Neither A or B

45. Technician A says center clutch friction discs before tightening pressure plates in place. Technician B says check that torque converters fully engage front pumps before bolting automatic transmissions in place. Who is right?

    a. A only   c. Both A and B
    b. B only   d. Neither A or B

FOR ADDITIONAL STUDY

1. What is the purpose of cylinder chamfers? What about the lower edge of cylinders?

2. Why are head bolt holes chamfered? Which ones do not require chamfers?

3. How are valve seals protected on assembly?

4. What causes camshafts to bind in their bearings? What is done to correct the problem?

5. List in sequence the steps for installing cam bearings in a pushrod engine. Include any required checkpoints along the way.

6. How is a cam bearing driver selected?

7. How is a camshaft straightened?

8. What happens if a cam gear has run-out?

9. What are the basic requirements for the sealing of any rotating shaft?

10. What is the common difference between upper and lower main bearing inserts?

11. Why would a connecting rod need to face a particular direction?

12. Why do pistons need to face a particular direction?

13. In which direction is a piston pin offset?

14. What can cause crankshaft endplay to be tight?

15. Aside from lining up timing marks, how can you tell if valve timing is correct?

16. What checks are made for ring assembly and installation?

17. Which threads require sealing? Which threads require lubrication?

18. Describe two ways of making sure that clamping force is uniform.

19. List three ways of priming an oil pump before starting an engine.

20. How are bellhousings centered with the crankshaft?

Chapter 17

PREPARING PERFORMANCE ENGINES

Upon completion of this chapter, you will be able to:

- Define the limits of cylinder pressure with pump gasoline.
- Describe how to improve volumetric efficiency.
- Explain how to increase flow through ports.
- Describe the advantages of larger valves.
- Explain how maximum valve lift is determined.
- Describe the relationship between valve lift and the valve curtain.
- Explain the relationship between airflow into the cylinder and piston velocity.
- Compare static and effective compression ratios.
- Explain how cylinder pressure is maximized within the limits of pump gasoline.
- Calculate effective compression based upon static compression and intake valve duration.
- Select a camshaft based upon effective compression and the percentage of exhaust flow relative to intake flow.
- Select an exhaust system according to cylinder displacement.
- Explain ignition-timing requirements in relationship to speed, load, and the point of maximum cylinder pressure.

INTRODUCTION

Performance engines cover the range from blueprinted production engines to fully prepared race engines. While actual blueprint drawings are not always available, these engines are machined and assembled to original or specialty equipment manufacturer specifications. In today's market, many customers seek improved output without sacrificing economy, reliability, or longevity. These engines go into pickup trucks, motor homes, and passenger cars and often need to idle acceptably and pass exhaust emission testing.

Of course, race engines must produce maximum power and there is little concern for idle

quality or emissions. However, rules sometimes limit the amount of fuel a racecar can carry making fuel efficiency important. After all, an extra pit stop could make the difference between winning and losing.

As for reliability, the saying is "you have to finish to win". This is a concern when using production engine components designed to operate within a passenger car duty cycle. While passenger cars operate a fraction of the time at wide-open throttle and within a limited RPM range, race engines operate most of the time at wide-open throttle and in higher RPM ranges. Because of these differences, passenger car engines require considerable preparation for racing or even high performance use. Some basic measures to improve reliability include:

1. For high RPM applications, sufficient bearing oil clearance to cool bearings.

2. High volume oil pumps as needed to maintain pressure with increased oil flow.

3. Increased oil sump capacity to keep the oil pump from pulling air and causing bearing failures.

4. Windage trays to reduce crankcase turbulence.

5. Increased valve spring pressure to prevent valve float.

6. In pushrod engines, pushrod guide plates or slotted rocker arms to stabilize pushrod valve trains at high speeds.

7. Lightweight piston assemblies and lighter connecting rod small-ends to reduce reciprocating forces.

8. High strength rod bolts to withstand the reciprocating cycles.

9. Precise engine balancing to stabilize crankshaft rotation at high speeds.

IMPROVING EFFICIENCY

Essentially, there are only two ways to improve output. First are modifications that enable cylinders to fill on each intake stroke and thereby improve volumetric efficiency (VE). To raise this efficiency, flow through the intake manifold and ports into cylinders must be increased. A typical engine develops maximum VE near peak torque such as when accelerating at wide open throttle.

Next are modifications that maximize cylinder pressure, at least within the limits of the fuel. This means raising mean effective pressure, or average cylinder pressure, on the power stroke. While raising compression increases mean pressures, keeping pressures below detonation limits is critical. This requires fuel with adequate octane and precise control of combustion through the calibration of fuel mixtures and spark timing.

Modifications to compression, valve trains, intake and exhaust systems, ports, fuel systems, and ignition systems work together to improve VE and to maximize mean cylinder pressure. This requires thoughtful selection and coordination of any changes in these systems.

The pressure-volume curve in Figure 17-1 shows the increase in pressure caused by compression and gas expansion after ignition. With peak pressure limited by the resistance of the fuel to detonation, it is the average or effective pressure under the curve that is important to power output.

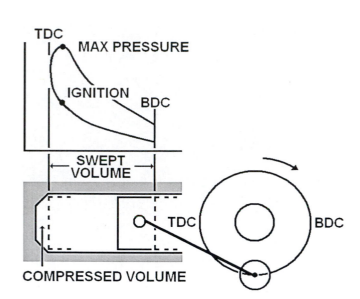

Fig.17-1 Pressure-volume relative to piston travel, crank rotation, and time

Mean effective pressure is calculated based upon torque output as measured in a dynamometer. Because of the required dynamometer test data, the result is called Brake Mean Effective Pressure or BMEP. As mentioned, peak pressure occurs at the point of maximum VE near peak torque. The following is a sample calculation of BMEP for a high performance 5.7-liter Chevrolet engine:

Rated Torque and Horsepower
  360 Pound Feet @ 3,600 RPM
  350 Horsepower @ 5,600 RPM

Horsepower @ Peak Torque
  = TQ x RPM ÷ 5252
  = 360 x 3,600 ÷ 5252
  = 246

BMEP @ Peak Torque
  = HP x 13,000 ÷ Displacement in Liters x RPM
  = 246 x 13,000 ÷ 5.7 x 3,600
  = 156 PSI

BMEP @ Peak Horsepower
  = HP x 13,000 ÷ Displacement in Liters x RPM
  = 350 x 13,000 ÷ 5.7 x 5,600
  = 142 PSI

Pump gasoline is generally adequate for BMEP well above this range but we must consider the resistance of the particular engine to detonation. Specifically, consider the following factors:

1. Aluminum cylinder heads help dissipate the heat from combustion.

2. Cold air induction cools the chamber during overlap and the intake stroke.

3. Short power runs build less heat than running under sustained load.

4. Rich fuel mixtures under peak loads absorb heat in the combustion chamber.

5. Lower engine operating temperatures help resist detonation (but increase wear and may not be compatible with computer engine management systems).

6. Shifting the torque curve upward even 500 RPM lowers BMEP significantly.

7. Carburetors and distributors require adjustments to fuel mixture and spark timing as altitude and ambient conditions change. Failure to maintain adjustments could lead to detonation.

8. Computer engine management systems with feedback fuel mixture control, knock detection and electronically controlled spark timing, adjust as operating and ambient conditions change.

IMPROVING FLOW THROUGH PORTS

Improving flow through ports requires reducing restrictions in the port and at the valve and eliminating turbulence. This does not call for "hogging" out ports in all directions. In fact, because airflow velocities are higher, the smaller port with adequate flow may well yield better throttle response and overall performance. It is also necessary to keep in mind that with production castings, wall thicknesses limit stock removal and getting carried away could result in very thin casting walls or holes that go into water. If avail-

able, check casting wall thickness with an ultrasonic tester before grinding into them. Remember that there are core shifts to deal with, and just because one port permits considerable grinding, does not mean that all ports will.

The areas of airflow loss through a port, and potential gains, are shown in Figure 17-2. With casting limits in mind, there are certain measures that improve flow without greatly enlarging the port size.

FLOW LOSS	PERCENT
1. Wall friction	3-4
2. Contraction around pushrod	2
3. Short turn	11
4. Expansion behind guide	4
5. Expansion 25 degrees	12
6. Expansion 30 degrees	19
7. Bend to exit valve	17
8. Expansion exiting valve	31

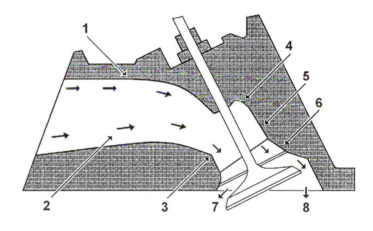

Fig.17-2 Areas of flow loss through a port (Superflow)

Keep in mind that flow potential through low and mid-lift is largely determined not by the port but by the valve and seat contours and therefore the first improvements begin with a good valve and valve seat. A valve seat machined to production specifications is a good start with typical 30, 45 and 60 degree angles. More flow may be found by varying the angles above and below the seat (see Fig. 17-3). If equipped to do so, machining a continuous radius is even better than compound angles. In combination with the seat, flow is sometimes enhanced by reshaping the underside of the valve. However, to determine which of these combined angles and shapes are best requires flow testing.

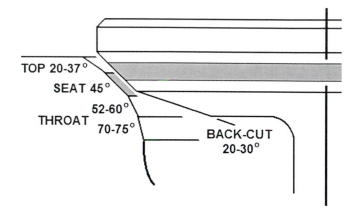

Fig.17-3 Varying angles above and below the seat

The port or throat diameter just below the valve seat should be approximately 85 percent of the valve diameter. This keeps velocity high and helps turn the airflow over the valve seat and under the valve face. The throat angles below the valve seat should remain after working on the port bowl. Be careful in this area as production castings may not allow much metal removal.

For porting, first examine the short-turn radius from the valve seat around to the floor of the port (see Fig.17-4). Beginning approximately 1/4in below the valve seat, shape the bowl to form a smooth, round transition into the port (see Fig.17-5).

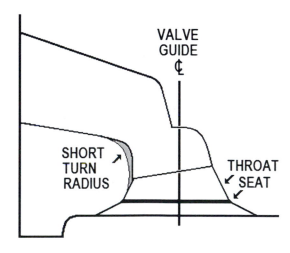

Fig.17-4 Improving the "short turn" radius

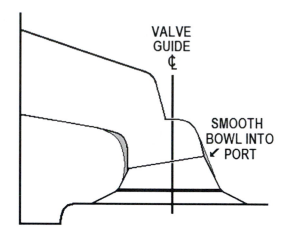

Fig.17-5 Shaping the "bowl" below the valve seat

Next, using the gasket as a template, make a layout of the port shape on the intake side of the head (see Fig.17-6) but before enlarging the port entry to match, compare the layout area to that of the valve and of the intake runner exit from the manifold. Ideally, the port entry would match the valve area and it would also be possible to align the port entry to the intake runner exit. In some cases, the gasket area is larger than necessary limiting the work required in this area.

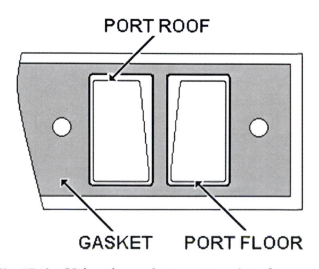

Fig.17-6 Using the gasket as a template for intake port dimensions

Keep in mind that where ports turn, airflow is best around the largest possible radius and it is typically best to enlarge the radius around the roof of a port, not the floor. Therefore, beginning from the layout line on the upper side of the port, continue the roofline to the port bowl (see Fig.17-7).

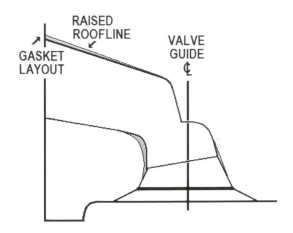

Fig.17-7 Aligning the port roof to the lay-out line

Regarding the other three sides of the port, remember to compare the size of the opening to the runner size in the intake manifold. Manifold runners often taper down at the port entry to boost air-flow velocity. Turbulence can be avoided by more closely matching the exit from the manifold to the port entry however it is important that airflow from the manifold not encounter restrictions at the port (see Fig.17-8). To be safe, leave the manifold runner exit slightly smaller than the gasket opening so that the manifold runner does not overlap the port entry should there be any misalignment in assembly with the cylinder heads.

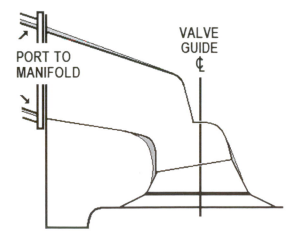

Fig.17-8 Aligning manifold runners with intake ports

The area just below the valve seat is of critical importance. Make sure that bowls continue smoothly downward from the valve seat for 1/8 to 1/4in. From this point, reshape the bowl as necessary to smooth the transition into the port. Remember, when reshaping this area of the bowl, do not remove the throat angle below the valve seat. While carbide burrs are used for this operation, it is possible to shape the bowl using 75-degree "bowl hog" cutters (see Fig.17-9). These cutters are especially helpful when opening up bowls for larger valves.

Fig.17-9 Opening up the bowl for larger valves using a bowl hog

Valve guide bosses that extend into ports and obstruct flow can be reshaped to reduce restriction (see Fig.17-10). If possible, remove valve guide bushings when working in this area and taper the port ends of the new guides. Be conservative however on cutting down the guide length as this shortens longevity.

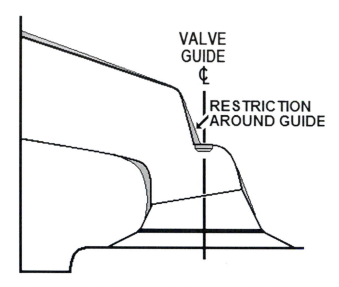

Fig.17-10 Reducing restriction around guide bosses

The measures covered to this point are certain to increase flow but it is sometimes found that while the flow is greater, it stops increasing at relatively low lifts. As mentioned earlier, flow up to mid lift is largely controlled by the valve and valve seat. Above mid lift, the port is the primary factor. When port flow is low at higher valve lift, the port is likely the restriction. For example, in pushrod engines, there is sometimes a restriction going around the pushrod holes but before removing metal, first polish the port walls and test flow as polishing alone may account for a two percent increase. After polishing, remove not more than .010in from around the pushrod side of the port and repeat testing (see Fig.17-11). If there is no increase in flow, stop and look for a restriction elsewhere in the port but continue by removing only small amounts between tests.

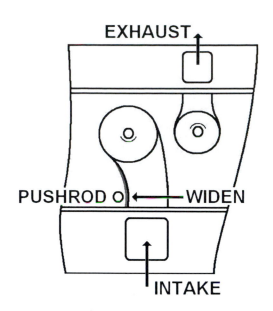

Fig.17-11 Removing material from around a pushrod hole

When working on port runners, keep in mind that airflow becomes turbulent and flow is reduced when a sudden contraction or expansion is encountered. To avoid this, keep cross-sectional areas along the port length uniform, or if tapered, keep the taper smooth and gradual.

Repeat these same procedures for the exhaust ports; smooth the short-turn radii, raise port roofs to layout lines, open the entry into the port bowl below the valve seat to 85 percent of the valve diameter. Because rough surfaces and sharp edges caused turbulence, finish all ports and bowls by smoothing imperfections.

Because some areas are difficult to reach, grinding and polishing requires a variety of tools. Most useful are long-shank carbide burrs, mounted stones, and polishing rolls. Remember to be conservative when grinding on production castings. While there is seldom a problem within 1/4in of the valve seat, the lower sides of the bowl are frequently thin. To check the consistency of the porting without a flow bench, check the most restricted areas and match dimensions port to port. It is also possible to check consistency by CC'ing port volumes.

It should be noted that many production 4-valve per cylinder heads have flow rates so high that porting will not yield the same increases in performance. With these cylinder heads, work is best limited to cleaning up casting imperfections and increased performance is more likely gained through other measures such as compression increases, camshaft changes, and manifold and throttle body changes.

REDUCING RESTRICTION AT THE VALVES

The most obvious restriction to flow through valves is the amount of valve lift and this is one of the primary considerations when selecting a camshaft. While flow testing is the best means of determining the maximum valve lift, there is another way of estimating how much lift is likely to help increase output. Basically, as shown in Figure 17-12, an open valve creates a valve curtain that circles the valve and extends from the face down to the seat. For maximum flow, the area of this curtain should be approximately equal to the valve area.

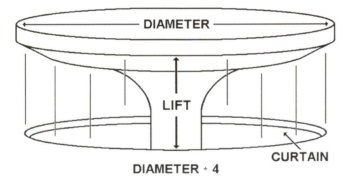

Fig.17-12 Valve area and valve curtain

As shown in the following calculations, valve and curtain areas match when valve lift is equal to one-quarter of the valve diameter:

Valve diameter 2.020in

Valve area	$= Pi \times R \times R$
	$= 3.1416 \times 1.010 \times 1.010$
	$= 3.2047in2$ and;

| Valve lift | $= diameter \div 4$ |
| | $= .505$ and; |

Curtain area	$= circumference \times lift$ or;
	$= Pi \times diameter \times lift$
	$= 3.1416 \times 2.020 \times .505$
	$= 3.2047in2$

Note that lift calculations apply mostly to intake valves and that porting is likely needed to benefit from this amount of lift. Exhaust valves, although smaller, often open equally as far making lift disproportionate to diameter. This is done to get exhaust valves well open by BDC of piston travel to help with "blow-down" and clear cylinders of exhaust gas before pistons reverse direction toward TDC.

Another obvious way of reducing intake restriction is to install oversized intake valves. Larger valves increase flow beginning well below maximum valve lift. This is important because stock heads with stock intake manifolds typically do not benefit from extreme valve lift and improving flow at lower lifts often yields the greatest total flow increase.

Some limitations to consider are the sizes of valve reliefs cut into pistons and the shrouding effect of the combustion chamber (see Figs. 17-13 and 14). Intake valve diameters are generally limited to one-half of the cylinder bore size except for canted valves which can be somewhat larger. Exhaust valves should also be enlarged proportionately. Should the oversized valves require seat inserts, keep in mind that they must be larger than the valves but small enough not to exceed casting wall thickness.

Fig.17-13 Check that intake and exhaust valves clear their respective reliefs in pistons

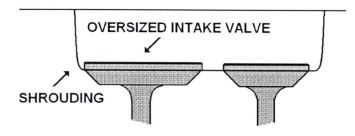

Fig.17-14 "Shrouding" restricts flow around valves

Most valve reliefs in pistons allow slightly oversized valves to clear but major increases in diameter require cutting larger diameter reliefs. Also, extreme increases in valve lift could require machining reliefs deeper into the pistons. Be sure to check these reliefs for clearance prior to assembling pistons to rods as re-cutting reliefs in pistons with the rods attached is more difficult. If re-cutting reliefs, maintain .200in "crown" thickness (see Fig.17-15).

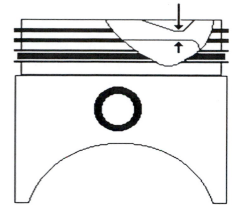

Fig.17-15 Maintain .200in crown thickness when re-cutting valve reliefs in pistons

Although larger intake valves increase flow, valve shrouding in the combustion chamber can restrict flow and should be avoided. Partly because of shrouding, intake valve diameters are limited to approximately one-half of bore diameter. In iron heads, minor shrouding can be reduced using an oversized seat grinding stone with the sides tapered and the corners rounded. Carefully grind down along the wall of the chamber until the corner radius blends with the chamber (see Fig.17-16). In iron or aluminum heads, this operation can be done in guide and seat machines with formed cutters. Doing this work with a die grinder requires care to obtain the same uniformity.

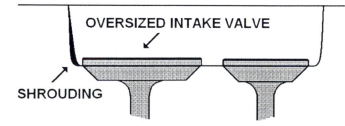

Fig.17-16 Reduce shrouding around intake valves

Remember to check that grinding the combustion chamber wall does not interfere with head gasket sealing (see Fig.17-17). If there is interference, it will be necessary to run less clearance around the valve. Also consider that removing material from the chambers reduces compression. These are good reasons for not going too far oversize with valves.

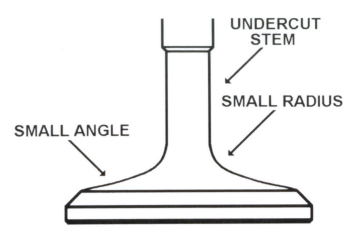

Fig.17-18 A high-flow valve

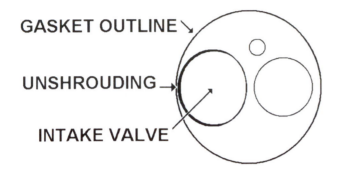

Fig.17-17 Check that chamber wall clears the gasket

Simply replacing production valves with "high-flow" performance valves can significantly increase flow in some engines (see Fig.17-18). Flow restriction is reduced by smooth contours on the backsides of valves, reduced fillet radii and undercut valve stems.

The table below shows the extent to which these valves can increase flow. Note that the increase is proportionately greater at lower valve lifts. Should high-flow valves be unavailable for your particular engine, production valves can be modified to similarly reduce restriction.

Valve Lift	Stock Intake CFM	High Flow Intake CFM	Stock Exhaust CFM	High Flow Exhaust CFM
.150	21.5	35.5	8.0	15.5
.200	45.5	61.0	20.5	25.5
.250	77.0	96.5	37.5	46.0
.300	115.0	127.0	52.5	64.0
.350	142.0	152.5	67.0	79.5
.400	169.5	177.5	81.0	91.5

Remember though that the performance valves are stainless steel or alloys well suited to the heat and stress of performance engines. These valves are also often lighter in weight, a helpful factor in reducing the required spring pressure. To evaluate different shapes, modify a production valve and flow test. Possibilities include undercutting the stem and machining the fillet radius to between 1/4 and 3/8in in an engine lathe (see Fig.17-19). The backside of the valve can also be machined to an angle of between 20 and 30 degrees in a valve grinder (see Fig.17-20). Complete the work by polishing the backside of the valve and fillet radius with "emery cloth".

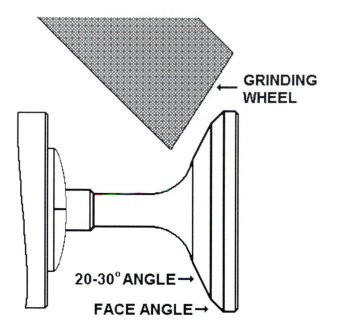

Fig.17-20 Grinding a 20-30 degree angle behind the valve face

On exhaust valves, also polish a small radius or chamfer around the upper edge of the valve head to reduce turbulence in the gas flow around the valve into the exhaust bowel and port (see Fig.17-21).

Fig.17-19 Machining the fillet radius in a lathe

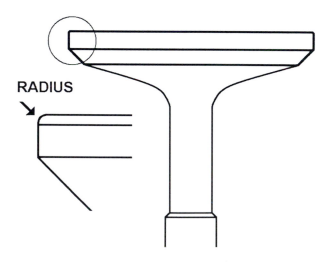

Fig.17-21 The smooth upper edge of a well prepared exhaust valve

As for undercut stems, this is often a means of reducing the weight of valves and the undercutting by itself may or may not increase flow. When port velocities are high, 300 feet per second or more, there is a "flagpole" effect that can occur behind the valve stem. Much as a flag whipping in the wind up on a flagpole, there can be a turbulent condition on the backsides of valve stems that can actually reduce flow. It is best to flow test with different valves before choosing which is best.

Port velocity is however very important and, in this regard, reducing losses in velocity due to expansion are a challenge. To appreciate this, note that when air passes through the port and enters an enlarged space, it expands and slows down, effectively putting on the brakes. To maintain velocity, it may actually help to fill in the excess volumes as opposed to continually looking for ways of enlarging the port. Velocity is important not only within the port but also because it helps get the air past the valve and seat and well into the much enlarged volume within the cylinder.

Of those areas of loss in Figure 17-2, only those cleanup operations that most often produce gains in flow have been included in this discussion. Some improvements are small and some larger but, all together, the flow increase is significant.

FLOW TESTING

The limited measures discussed here produce significant gains however the effectiveness of each step in porting requires verification by testing port flow. This is done in a "flow bench" such as shown in Figure 17-22.

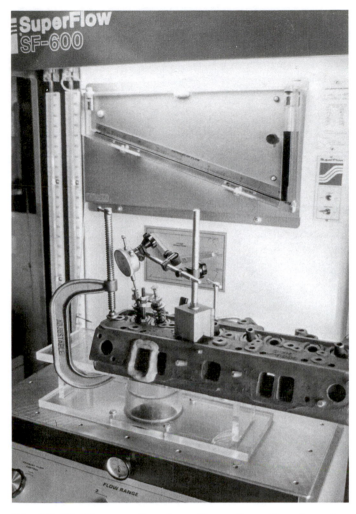

Fig.17-22 Flow testing in a Superflow flow bench

In testing intake flow, note that air is drawn into the cylinder through the intake port. The cylinder adapter is closely matched to the engine's bore diameter and a radius is formed around the port entry to reduce turbulence (see Fig.17-23). For exhaust tests, a mode selection switch is used to reverse flow. A short pipe is sometimes attached to the exhaust port outlet (see Fig.17-24). A means of opening the valve is required and a dial indicator to measure valve opening. Stock or aftermarket valve springs are replaced with very light springs for testing.

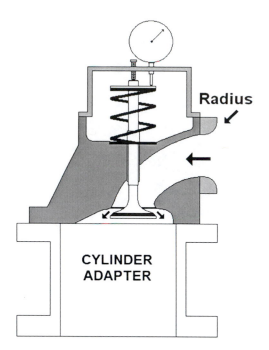

Fig.17-23 Setting up for intake flow testing

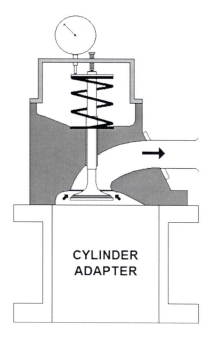

Fig.17-24 Setting up for exhaust flow testing

In manual flow bench operation, combinations of orifice openings in the test cabinet are selected to test at the desired test pressure. The flow control valve is opened to the desired test pressure and flow percentage read on the flow meter. If at least 70 percent flow is not achieved, changes in the orifice selection are necessary (see Fig.17-25).

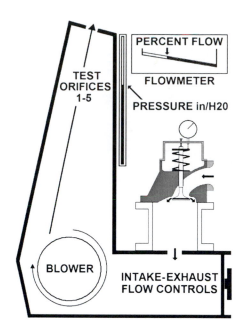

Fig.17-25 Flow test settings, intake

With the percentage from the test, raw CFM is read from a table (see Fig.17-26). Tests are repeated in increments of .025 to .100in up to maximum valve lift or to the point where flow stops increasing. For more accurate CFM, corrections are made for atmospheric pressure and temperature differences between air in the flow bench and atmosphere. Because different orifice selections and test pressures are used at different valve lift points, final CFM numbers are converted to a standard pressure for comparison, typically 28 in/H20. With computer driven flow benches, these tasks are automated.

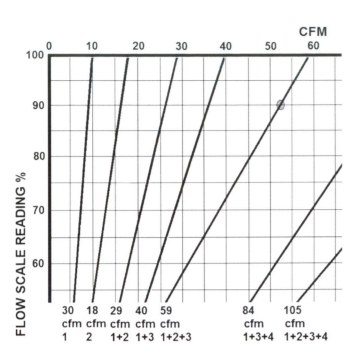

mately .450in valve lift. Third, although exhaust flow increased significantly, it still does not maintain 75 percent of intake flow. From this information, the engine builder learns the amount of valve lift required for peak flow and, assuming minimum flow loss through the intake manifold, that more exhaust duration is needed to achieve 75 percent of intake flow.

Fig.17-26 Flow is 53 raw CFM or 90 percent of 59 CFM with orifices 1, 2 and 3 open

In this Chevrolet 305 cubic inch cylinder head, compare the gains in flow in CFM over the stock casting after cleaning up the ports using the limited measures suggested here. First, peak intake flow increased 14 percent and exhaust flow 34 percent. Aside from peak flows, average flows increased by 14 and 19 percent respectively. Second, both flow rates flattened out at approxi-

	FLOW TEST Chevrolet 305					
					Exhaust Percent Intake	
Valve Lift	--Intake CFM--		--Exhaust CFM--			
	Before	After	Before	After	Before	After
.050	26.2	28.2	21.0	21.2	80.2	75.2
.100	55.3	58.6	40.3	44.8	72.9	76.5
.150	79.4	86.4	58.6	67.6	73.8	78.2
.200	104.0	114.5	77.1	86.7	74.1	75.7
.250	122.3	138.1	93.2	100.6	76.2	72.8
.300	141.9	158.1	102.1	114.6	72.0	72.5
.350	159.5	175.5	105.5	125.7	66.1	71.6
.400	168.3	187.8	107.0	132.7	63.6	70.7
.450	171.8	**201.4**	**108.0**	138.8	62.9	68.9
.500	**174.1**	197.4	108.0	**141.9**	62.0	71.9

Valves are1.840in intake and 1.500in exhaust. CFM is at 28 in/H20.

IMPROVING FLOW THROUGH MANIFOLDS

After doing all this work on the ports, it is now necessary to introduce another reality. That is, production manifolds and many of those legal for street use, are designed for low and mid-range performance. While delivering excellent drivability, they may cause flow losses of ten percent or more. For example, for our typical small block pushrod V8, production and legal street manifolds are "dual plane" and commonly reach peak flow as early as 180 CFM.

So, what do we gain by porting cylinder heads in excess of this limit? To see the gain, we need to think of flow "under the curve" as shown in Figure 17-27. Increasing the flow area under this curve requires taking several measures. Some increase will come from porting, perhaps with an emphasis on flow at low and mid valve lift, some will come from changing or reworking the intake manifold and ultimately, on camshaft selection.

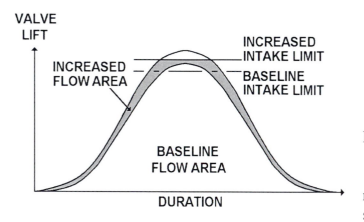

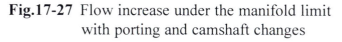

Fig.17-27 Flow increase under the manifold limit with porting and camshaft changes

Aftermarket Intake manifolds also must be selected to meet the engine's flow requirements. First, be sure that the manufacturer of the manifold specifies that it is the correct selection for the displacement and RPM range of the engine. If the vehicle is emission controlled, also be sure that it is approved for street use. For all manifolds, be sure that port runners align with intake ports, and if they do not align, make corrections as required. Also check that entries from the plenum to the port runners have smooth, round corners. Misalignments or sharp edges at these points cause turbulence and inhibit airflow.

For V-block engines, there are single and dual plane manifolds to choose from. In single plane manifolds, all runners to cylinders connect to a common plenum below the carburetor (see Fig.17-28). These runners are typically short and low in restriction. Low and mid-range torque may suffer with these manifolds because of the lack of velocity at low RPM but peak power is very good. A single plane manifold is a good choice for power in the upper half of the RPM range.

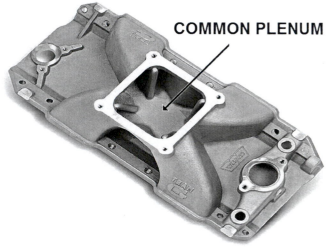

Fig.17-28 A single plane manifold

Dual plane manifolds have long, narrow runners arranged in two sets. For a V8 engine, the sets of runners are for cylinders firing 180 degrees apart. This arrangement isolates runners so those cylinders with closely spaced intake strokes do not rob each other of airflow (see Fig.17-29). Because these longer runners boost velocity, low and mid-range torque is excellent. Dual plane manifolds are a good choice for street driven engines but are generally too restrictive for high RPM ranges.

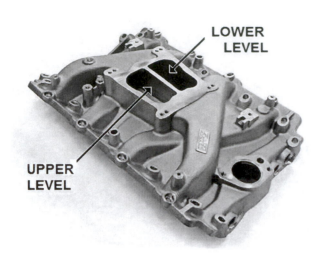

Fig.17-29 A dual plane manifold

Original equipment fuel injection manifolds are typically designed for drivability and low and mid range performance. If the intended use is street driving, match the runner alignment to the intake ports, clean up and polish runners and performance with these stock manifolds is very good. Increasing output at higher RPM may require porting or replacing runners with higher capacity units. Larger throttle valves and performance manifolds for popular applications are available from aftermarket suppliers. In a well-tuned intake system, velocities approach 250 to 300 feet per second at peak power. If the system is too restrictive, velocities increase but flow decreases. If runners are too large for the cylinder displacement, velocities will be low and throttle response will suffer especially at low speeds.

All manifolds for street driving are a compromise between power output and drivability. For example, in flow testing, peak airflow through a typical dual plane manifold will be found to be approximately 10 percent less than through the port only. On the other hand, mid-range torque and low speed throttle response is excellent with these manifolds. Race manifolds will not suffer the same airflow loss but also not have the same drivability in the lower ranges. The same is true for production fuel injection manifolds which are typically tuned for mid-range power.

What can be done to increase flow through production manifolds? Some improvement comes from the runner to port matching already covered. Increase also comes from putting a radius on the entry to the runners from plenum side. To gain access to these areas however may require opening up access to the plenum, for example, by opening up the four holes in a four-barrel manifold to two long slots. Some cylinder-to-cylinder scavenging such as in single plane manifolds is gained by adding an open carburetor spacer or by cutting down the wall separating the two planes (see Fig.17-30). Although desirable, it is very difficult if not impossible, to reach inside and work on runners along their full length, except perhaps by "extrude honing".

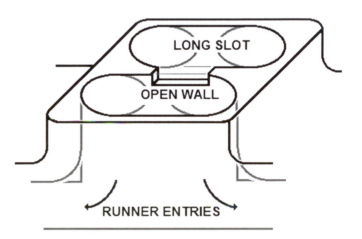

Fig.17-30 Modifications to a production dual plane manifold

All engines benefit from cool inlet air. For wet manifold engines, restricting the flow of exhaust heat from exhaust ports to the floor of the intake manifold benefits the engine in two ways. First by not heating inlet air and secondly by increasing exhaust flow in ports used to heat the intake. Although frequently missed in assembly, many gasket sets include crossover restrictors as specified for particular applications by the manufacturers. Unless the specified restrictors are installed, inlet air is overheated and flow through the effected exhaust ports flow is reduced as much as 25 percent. Keep in mind however, that while a cooler manifold increases inlet air density, completely blocking crossovers reduces fuel vaporization and cuts off heat to carburetor chokes and exhaust gas to exhaust recirculation systems. A good compromise for street engines is to reduce, not eliminate, this heat by using these restrictors (see 17-31).

Fig.17-31 Intake manifold exhaust crossover restrictor

EXTRUDE HONING

The extrude honing process adds another dimension to porting. The process consists of pumping an abrasive with the consistency of putty through ports and manifold runners (see Fig.17-32). Following the principles of hydraulics, when this putty encounters restrictions, velocity increases and so does the cutting action of the abrasive. This process reduces restrictions and leaves runners highly polished thereby significantly increasing flow rates.

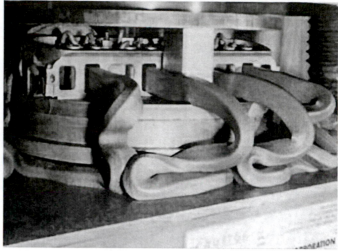

Fig.17-32 Extrude honing intake ports (Courtesy Extrude Honing Corp.)

It is particularly well suited to intake manifold runners that are difficult to rework by other means. By varying the process runner by runner, flow rates can also be equalized. Dry fuel injection manifolds benefit directly from the highly polished finish but, in wet manifolds for carbureted engines, it is necessary to keep fuel vaporized and uniformly distributed. The "as cast" finish in the runners of production manifolds adds turbulence along runner walls and aids vaporization. For street engines with highly polished ports and manifold runners, at least some manifold heat is advisable.

DEALING WITH TUMBLE AND SWIRL

In a 4-valve chamber, tumble describes the rolling action promoted by parallel streams of air passing through two valves into the cylinder on the intake stroke. In a 2-valve chamber, swirl describes the action promoted by the angle of approach to the valve, into the cylinder and around the cylinder wall (sees Fig.17-33). In some engines, swirl is maintained by the shape of the combustion chamber. At the right level, tumble and swirl enhance ignition and combustion by mixing and moving air and fuel around the chamber and across the spark plug with fewer misfires and a faster burn with less spark advance. However, in extremes, these motions can interfere with flame propagation.

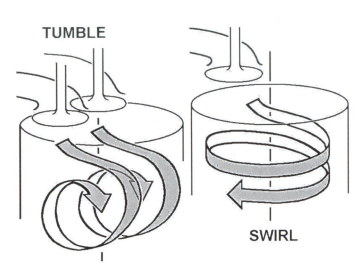

TUMBLE

SWIRL

Fig.17-33 Tumble and swirl motions

If maximum airflow is the primary objective, these actions can be turbulent and inhibit airflow. Cylinder heads with ports and combustion chambers designed to promote swirl may be good for street applications but may not be the best choice for total performance where high flow is the primary objective (see Fig.17-34). Certainly, in making modifications to these cylinder heads, one should keep in mind that mixture motion and combustion could be affected.

Fig.17-34 The heart shape of a high swirl chamber

SYNCHRONIZING VALVE OPENING AND PISTON TRAVEL

The synchronizing of intake valve opening with piston movement at the point of maximum piston velocity greatly improves the efficiency of the intake stroke. Because cylinder pressure is low when the piston reaches maximum velocity, and because only atmospheric pressure forces air into the cylinder, getting the intake valve open through this zone is crucial to improving VE. Maximum piston velocity occurs when the connecting rod is 90 degrees to the crankshaft as shown in Figure 17-35.

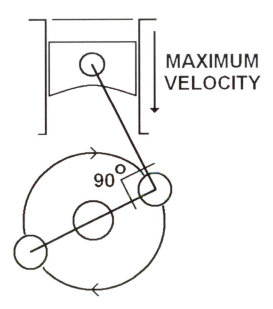

MAXIMUM VELOCITY

90°

Fig.17-35 The point of maximum piston velocity

To get valve opening in synch with piston travel, it helps to select a camshaft that opens valves rapidly. To grasp the importance of this, it is important to understand that even camshafts with the same specified duration and lift are not necessarily equal. The two camshafts in Figure 17-36 have 260 degrees duration and .420in valve lift but one has more flow or area under the curve. The cam with greater area under the curve opens the valve faster and increases flow during the period of high piston velocity.

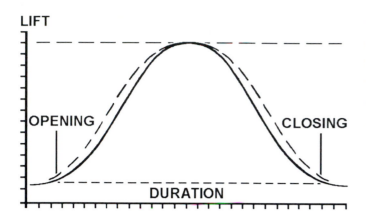

LIFT

OPENING

CLOSING

DURATION

Fig.17-36 Cams with equal lift and duration but different areas under the curve

Opening the valve a few degrees earlier or later affects this synchronization. Changing the rocker arm ratio also changes the valve-opening curve. For example, changing from a 1.5 to a 1.6:1 rocker arm not only increases valve lift, it also creates more area under the curve in synch with piston travel away from TDC. However, higher ratios also increase camshaft loads.

The drawing below shows the relationship between crankshaft stroke, rod length, and the point of maximum piston velocity (see Fig.17-37).

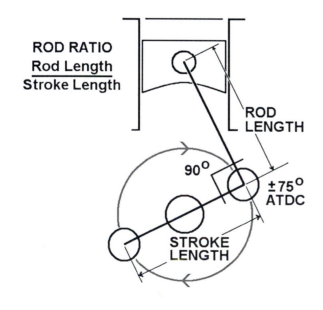

ROD RATIO
Rod Length
Stroke Length

ROD LENGTH

90°

±75°
ATDC

STROKE LENGTH

Fig.17-37 Crankshaft angle at the point of maximum piston velocity

The rod ratio is the relationship of rod length to stroke length. It is calculated by dividing the rod length by the stroke and commonly ranges from 1.5 to 1.9. The crankshaft angle where piston velocity reaches maximum varies according to rod ratio. The table below gives crankshaft angles for maximum piston velocity for a range of rod ratios.

PISTON VELOCITY TABLE

Rod Ratio	Crankshaft Angle ATDC
2.0:1	76 degrees
1.9	75
1.8	74
1.7	74
1.6	73
1.5	72
1.4	70

Rod Length ÷ Stroke Offset = Tangent

Because exhaust gas leaves the cylinder under pressure on the exhaust stroke, the relationship between valve opening and piston travel is viewed differently. Essentially, is it desirable to have as much of the exhaust gas as possible out of the cylinder before the piston passes BDC. This is so that the piston does not compress exhaust gas as it approaches TDC or force exhaust gas back into the intake manifold when the intake valve opens. With these possibilities in mind, get the exhaust valve well open by BDC.

MAXIMIZING CYLINDER PRESSURE

Most technicians and machinists think of compression ratios in terms of advertised or specified numbers. These are static ratios calculated as follows:

(Displacement + Clearance Volume)
Clearance Volume

Note that this calculation uses displacement based upon the specified stroke length beginning at bottom-dead-center. However, when the engine is running, compression does not begin until the intake valve closes 60-degrees or more after bottom-dead center. Displacement with the piston at the intake closing position is much smaller. Similarly, the calculated compression ratio using the displacement at intake closing is much lower.

Engine builders must think in terms of effective compression ratios or those produced with the engine running. With very good VE and 91-octane premium unleaded gasoline, effective compression ratios are limited to approximately 7:1. Engines with stock intake and exhaust systems typically run lower VE and run acceptably with effective compression ratios of 7.5:1 or more. These ratios seem illogical if one thinks in terms of 12.5:1 racing pistons or even 9.0:1 advertised compression ratios in production engines, but if an efficient engine exceeds these numbers, detonation is likely. Static and effective compression ratios can be compared visually in Figure 17-38.

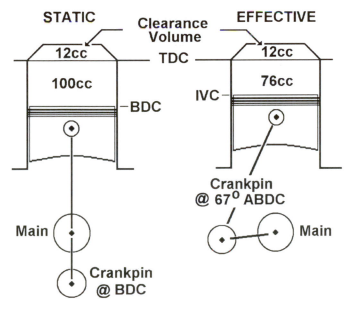

100cc Displacement + 12cc Clearance Volume = 9.33:1
12 cc Clearance Volume

76cc Displacement + 12cc Clearance Volume = 7.33:1
12 cc Clearance Volume

Fig.17-38 Static and effective ratios for the same cylinder

Then why would an engine builder want 12.5:1 pistons? The answer is that with high static compression ratios, intake valves can open earlier and close later and still maintain the desired effective compression. With the increased intake valve duration, VE improves. It is common to see efficient street engines with static compression ratios between 9 and 10:1 so that they can use sufficient intake duration for efficiency. Much higher compression is not always advised for street engines since the required duration and resultant overlap negatively affects vacuum, idle quality, and emissions.

To understand these ideas, it is helpful to gather the required information and walk through some basic calculations. The data below is for a 350 Chevrolet engine:

Bore (.030in over)	4.030in	102.36mm
Stroke	3.480in	88.39mm
Displacement (1 cyl.)	44.39in^3	727.70cc
Gasket thickness	.038in	.97mm
Gasket volume	.49in^3	7.95cc
Deck Clearance	.010in	.25mm
Deck clearance volume	.13in^3	3.34cc
Valve reliefs in piston	.43in^3	7.00cc
Chamber volume	4.64in^3	76.00cc

Note: Divide cubic inches by .061 for cc

Gathering some of this data requires measuring directly because the information is either not in specifications or the engine varies from specifications, low compression being common. To begin, measure deck clearance with a depth mike and "CC" combustion chamber volumes (see Fig.17-39). The following is an example of static compression ratio calculations using the data from the 350 Chevrolet engine above.

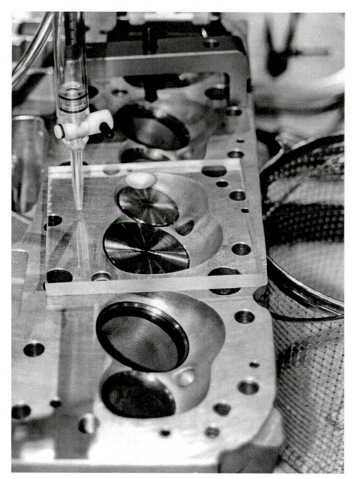

Fig.17-39 Measuring chamber volumes by "CC'ing"

To find the effective compression for this engine, first check camshaft specifications. In this case, the advertised intake duration is 260 degrees and the intake lobe centerline of the camshaft is 112 degrees. With the valve wide open at 112 degrees ATDC on the intake stroke, the intake valve closes 130 degrees later, or 62 degrees ABDC on the compression stroke (see Fig.17-40).

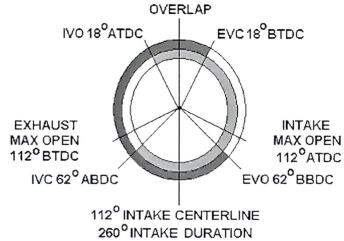

Fig.17-40 A typical valve timing diagram in crankshaft degrees

STATIC COMPRESSION RATIO

Clearance volume is the sum of the volumes:
Deck clearance + valve relief + gasket + chamber or;

$$3.34 + 7.00 + 7.95 + 76.00 = 94.29cc$$

Displacement = 727.70cc

Compression ratio:

(Clearance Volume + Displacement) ÷ Clearance Volume or;

$$(94.29 + 727.70) ÷ 94.29 = 8.72:1$$

The relationship between crankshaft position and swept volume at intake valve closing is shown graphically in Figure 17-41. If the intake closing point and rod ratio are known, it is easy to find the percentage of cylinder volume at intake closing using the following table. With a 5.7 inch rod and a 3.48 inch stroke, the rod ratio is 1.64:1. From the table, the percentage of cylinder volume with a 1.65:1 rod ratio, the nearest to 1.64, and an intake closing point of 62 degrees ABDC, is 79.1. Find the cylinder volume at this point by multiplying the percentage times the cylinder displacement. Using this new information, recalculate the effective compression ratio as follows:

Clearance volume
= 94.29cc
Displacement at 62 degrees ABDC

= 79.1% of 727.70
= 575.6cc
Effective compression
= (94.29 + 575.6) ÷ 94.29
= 7.10:1

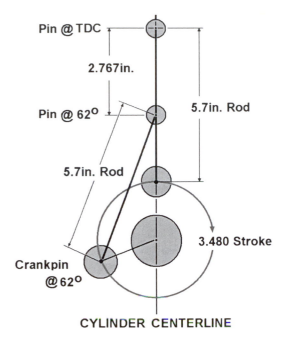

Pin @ TDC

2.767in.

Pin @ 62°

5.7in. Rod

5.7in. Rod

3.480 Stroke

Crankpin
@ 62°

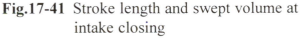

CYLINDER CENTERLINE

Fig.17-41 Stroke length and swept volume at intake closing

Without camshaft specifications, first determine the correct intake closing point based upon the desired effective compression ratio. Cylinder displacement, preferred effective compression ratio and rod ratio are required. Using the same engine as above but with 72 instead of 76cc combustion chambers for higher compression, first calculate the percentage of cylinder displacement at intake closing:

Preferred effective compression ratio = 7:1
Displacement = 727.7cc
Clearance volume = 90.29
Rod ratio = 1.64:1
Displacement @ IVC
= CV (ECR-1)
= 90.29 x 6
= 541.7cc
Percentage @ IVC

= 541.7 ÷ 727.7
= 74.4%

To determine the correct intake valve closing point for 7:1 effective compression, go the following table. Find the nearest percentage to 74.4 in the 1.65 rod ratio column, the closest to 1.64, and then read the intake closing point in the left column. In this case, the nearest percentage is 74.3 and the intake valve closing point is 69 degrees ABDC.

CYLINDER VOLUME AT INTAKE VALVE CLOSING
Percent by Rod Ratio

ABDC	1.90	1.85	1.80	1.75	1.70	1.65	1.60	1.55	1.50
45	88.4	88.5	88.6	88.7	88.8	88.9	89.0	89.1	89.3
46	87.9	88.0	88.1	88.2	88.3	88.4	88.5	88.7	88.8
47	87.3	87.5	87.6	87.7	87.8	87.9	88.0	88.2	88.3
48	86.8	87.0	87.1	87.2	87.3	87.4	87.5	87.7	87.8
49	86.3	86.4	86.5	86.6	86.8	86.9	87.0	87.2	87.3
50	85.8	85.9	86.0	86.1	86.2	86.3	86.5	86.6	86.8
51	85.2	85.3	85.4	85.5	85.7	85.8	86.0	86.1	86.2
52	84.6	84.8	84.9	85.0	85.1	85.2	85.4	85.5	85.7
53	84.1	84.2	84.3	84.4	84.5	84.6	84.8	85.0	85.2
54	83.5	83.6	83.7	83.8	84.0	84.1	84.3	84.4	84.6
55	82.9	83.0	83.1	83.2	83.4	83.5	83.7	83.8	84.0
56	82.3	82.4	82.5	82.6	82.8	82.9	83.1	83.3	83.4
57	81.6	81.8	81.9	82.0	82.2	82.3	82.5	82.7	82.8
58	81.0	81.1	81.3	81.4	81.5	81.7	81.9	82.1	82.2
59	80.4	80.5	80.6	80.8	80.9	81.1	81.3	81.4	81.6
60	79.7	79.8	80.0	80.1	80.3	80.4	80.6	80.8	81.0
61	79.0	79.2	79.3	79.5	79.6	79.8	80.0	80.2	80.4
62	78.4	78.5	78.7	78.8	79.0	79.1	79.3	79.5	79.7
63	77.7	77.8	78.0	78.1	78.3	78.5	78.7	78.9	79.1
64	77.0	77.1	77.3	77.5	77.6	77.8	78.0	78.2	78.4
65	76.3	76.5	76.6	76.8	76.9	77.1	77.3	77.5	77.7
66	75.6	75.7	75.9	76.1	76.2	76.4	76.6	76.8	77.1
67	74.9	75.0	75.2	75.4	75.5	75.7	75.9	76.1	76.4
68	74.2	74.3	74.5	74.6	74.8	75.0	75.2	75.4	75.7
69	73.4	73.6	73.7	73.9	74.1	74.3	74.5	74.7	75.0
70	72.7	72.8	73.0	73.2	73.4	73.6	73.8	74.0	74.2
71	71.9	72.1	72.3	72.4	72.6	72.8	73.0	73.3	73.5
72	71.2	71.3	71.5	71.7	71.9	72.1	72.3	72.5	72.7
73	70.4	70.6	70.7	70.9	71.1	71.3	71.5	71.8	72.0
74	69.6	69.8	70.0	70.2	70.3	70.6	70.8	71.0	71.3
75	68.9	69.0	69.2	69.4	69.6	69.8	70.0	70.2	70.5
76	68.1	68.2	68.4	68.6	68.8	69.0	69.2	69.5	69.7
77	67.3	67.4	67.6	67.8	68.0	68.2	68.4	68.7	68.9
78	66.5	66.6	66.8	67.0	67.2	67.4	67.7	67.9	68.2
79	65.7	65.8	66.0	66.2	66.4	66.6	66.9	67.1	67.4
80	64.8	65.0	65.2	65.4	65.6	65.8	66.0	66.3	66.6
81	64.0	64.2	64.4	64.6	64.8	65.0	65.2	65.5	65.7
82	63.2	63.4	63.5	63.7	63.9	64.2	64.4	64.7	64.9
83	62.4	62.5	62.7	62.9	63.1	63.3	63.6	63.8	64.1
84	61.5	61.7	61.9	62.1	62.3	62.5	62.7	63.0	63.3
85	60.7	60.8	61.0	61.2	61.4	61.6	61.9	62.2	62.2
86	59.8	60.0	60.2	60.4	60.6	60.8	61.0	61.3	61.6
87	59.0	59.1	59.3	59.5	59.7	60.0	60.2	60.5	60.7
88	58.1	58.3	58.5	58.7	58.9	59.1	59.3	59.6	59.9
89	57.2	57.4	57.6	57.8	58.0	58.2	58.5	58.7	59.0
90	56.4	56.5	56.8	56.9	57.1	57.4	57.6	57.8	58.1

Note: Calculated using .050in. pin offset

To maximize cylinder pressure, first machine the engine to achieve the maximum specified static compression. In production engines, this requires reducing deck clearance and clearance volumes to minimum specifications. Next, select an intake valve duration that produces an effective ratio close to 7.0:1. If the exact intake duration is unavailable, other durations and valve timing points can be made to work. For example, if effective compression is high, use longer duration and adjust the effective ratio by advancing the camshaft to the desired intake closing point. In the engine above, advancing the cam 4-crankshaft degrees increases the effective compression ratio by .3 points.

In general, don't just select parts from a catalog but instead build performance engines according to a plan. What will static compression be? What intake valve duration is correct for this compression? The following formulas help the engine builder figure out required information while still in the planning stages.

USEFUL FORMULAS

1. For compression ratios:

   $CR = (D + CV) \div CV$
   Where D is displacement and CV is clearance volume

2. For clearance volume at a predetermined static compression ratio:

   $CV = D \div (CR-1)$
   Where CR is the static compression ratio

3. For the displacement necessary for a predetermined effective compression ratio:

   $D = CV (CR-1)$
   Where CR is the effective compression ratio

4. For the stroke length at a particular displacement:

   $S = D \div A$
   Where A is the cross-sectional area of the cylinder

5. For the cross-sectional area of a cylinder:

   $A = 3.1416 \times R \times R$
   Where Pi is 3.1416 and R is the cylinder radius

6. For cubic inch displacement:

   $D = A \times S$
   Where S is the stroke length

7. To convert cubic inches to cubic centimeters:

   $CC = CID \div .061$ and the reverse;
   $CID = CC \times .061$
   Where CID is the cubic inch displacement

Sample Problems

1. What must the clearance volume be for 10:1 static compression? The engine is a V8 with a displacement of 355 cubic inches. For clearance volume:

   $CV = D \div (CR-1)$
   D in in^3 = 355 ÷ 8 = 44.38in^3 per cylinder
   D in cc = 44.38 ÷ .061 = 727cc per cylinder
   Clearance Volume = 727 ÷ 9 = 81cc

2. What is the stroke length at intake closing and 7:1 effective compression? The clearance volume is 81cc, displacement 355 cubic inches, and bore 4.030in. For displacement at 7:1 compression:

   $D = CV (CR-1)$
   = 81 (7-1)
   = 81 x 6
   = 486cc

   For the stroke at 7:1 compression:

   $S = D \div A$
   D = 486 x .061 = 29.65in^3
   A = 3.1416 x 2.015 x 2.015 = 12.76in^2
   S = 29.65 ÷ 12.76 = 2.323in

SELECTING A CAMSHAFT

While principles pertaining to piston velocity and effective compression have been covered, a further discussion of valve events is necessary to fill-out this area of knowledge. Summarized below are the four key valve events and their effect on engine performance.

1. Intake Valve Closing (IVC): The point of intake closing determines the effective compression ratio. If closed too early, cylinder pressure will be excessive and the engine will be subject to detonation under load. If closed too late, cylinder pressure will be low and there is a reversion of pressure back into the intake manifold especially at low RPM. In all engines, closing the valve well past BDC improves VE because cylinder pressure remains low through this range of piston travel.

2. Exhaust Valve Opening (EVO): The point of exhaust opening marks the end of the power stroke. If opened too early, pressure from the expanding gases is cut short with a resultant loss of output and fuel efficiency. If opened late, exhaust gas pressure in the cylinder drops and evacuation on blowdown is less efficient. By opening the valve at just the right time, maximum output is extracted from the expanding gases and the "blow-down" of exhaust gas exiting the cylinder promotes efficiency of the exhaust stroke.

3. Intake Valve Opening (IVO): The point of intake opening marks the end of the exhaust stroke and the beginning of the "valve overlap period." If opened early, exhaust pressure dilutes the incoming intake charge. If opened late, the effective length of the intake stroke is reduced and efficiency reduced. Ideally, the intake opening occurs when cylinder and manifold pressures are equal.

4. Exhaust Valve Closing (EVC): The exhaust closing occurs after the start of the intake stroke and marks the end of the "valve overlap period." If closed too early, exhaust gas is

trapped in the cylinder and the rise in pressure forces exhaust into the intake manifold. This referred to first as "pressure reversion." If closed too late, vacuum is applied to the exhaust port reducing vacuum and diluting the incoming air with scavenged gases.

Some kind of trade-off is nearly always involved in selecting camshafts. For example, long duration improves VE but the length of the overlap period increases with duration. Long overlap periods create oscillations in manifold pressure along with exhaust gas reversions at low RPM. These conditions cause the loping idle associated with performance camshafts. Still, as can be seen in the following table, stroke cycles at high RPM are in milliseconds and increased duration is necessary for good VE. Therefore, it is possible extend the RPM range of the engine and increase power dramatically by increasing duration but the compromise is that the increased overlap can reduce idle quality and low speed performance.

Duration in Milliseconds					
Degrees	---------- RPM in Thousands ----------				
Duration	5	6	7	8	9	10
240	8.00	6.67	5.71	5.00	4.44	4.00
250	8.33	6.94	5.95	5.21	4.63	4.17
260	8.67	7.22	6.19	5.42	4.81	4.33
270	9.00	7.50	6.43	5.63	5.00	4.50
280	9.33	7.78	6.67	5.83	5.19	4.67
290	9.67	8.06	6.90	6.04	5.37	4.83
300	10.0	8.33	7.14	6.25	5.56	5.00

Degrees per Second =
   Rpm x 360 Degrees ÷ 60 Seconds and;
Duration in Milliseconds =
   Degrees Duration ÷ Degrees per Second

The effects of long durations and increasing overlap do however vary depending on exhaust system backpressure. With a good street exhaust system, back pressure is likely 2-PSI or more at peak power. With open exhaust, backpressure drops to nearly zero. Without backpressure, duration and overlap can be increased without the negative effects of pressure reversions during overlap (although low end torque may suffer). It is important in camshaft selection to keep this in mind. With street exhaust systems, less duration or wider lobe centers reduce overlap and pressure reversions and thereby improve idle quality.

The length of duration and the timing of the four valve events are typically supplied with the camshaft. However, the usefulness of the duration specified varies depending upon the method of measurement. Many performance camshaft suppliers specify advertised duration and duration measured at .050in lifter rise. The nearest thing to an industry standard is SAE duration measured at .006in lifter rise.

Determining effective compression requires knowing precise duration seat-to-seat and finding these points may require following lifter rise with a dial indicator and locating the end of the opening ramp and the sta1rt of the closing ramp (see Fig.17-42). Further complicating this task is the valve train "compliance" in running engines caused by cam bearing clearance, lash, component deflection, and losses normal to the operation of hydraulic lifters or lash compensators.

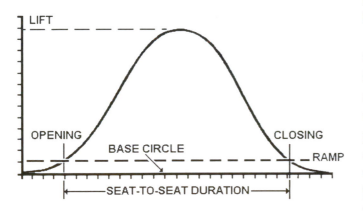

Fig.17-42 A valve lift curve with ramps

The best way of obtaining good camshaft information is to use available computer software, instrumentation and fixtures for analysis (see Fig.17-43). With two rotations of the camshaft, a computer generated report yields a full lift curve, area under the curve, lobe centers, and the length of ramps. Also generated are graphs of acceleration, velocity, and jerk; information needed for the fine tuning of the entire valve train including spring requirements.

Fig.17-43 A cam set up for analysis

Short of using software and hardware to generate an accurate valve opening and closing curve, or checking the valve events in assembly, advertised duration is the best choice for estimating the timing of valve events. Although the ramps may vary slightly from SAE specifications at .006in lifter rise, specifications at .050in lifter rise are no help in determining effective compression.

In addition to the four basic valve events, there two additional points to check on valve opening curves. At the point of peak piston velocity, 75 degrees ATDC on average, we would like intake valve lift to be where there is at least 70 percent of peak intake flow. This ensures good intake efficiency from this point through peak lift. Second, we would like to see exhaust valve lift at BDC to be where there is 70 percent of peak exhaust flow. This enhances exhaust "blow-down". Flow data is required and lift at these lift points can be checked in assembly or against a cam plot.

In general, prior to selecting a camshaft, realistically limit the parameters of engine operation. For example, if the engine and transmission are computer controlled, why build the engine to exceed the OE redline? Also, if the engine must pass an exhaust emission test, or idle in gear, limit overlap to maintain vacuum and acceptable idle quality. Following is a series of considerations that if taken in sequence, help in camshaft selection and timing.

1. Machine deck clearances and combustion chamber volumes to minimum specifications and calculate or measure the static compression ratio.

2. For pump gas, select intake duration and timing to produce an effective compression ratio close to 7.0:1. If VE is limited, a higher ratio is acceptable.

3. Test cylinder head flow and compare intake and exhaust flow rates. In normally aspirated engines, if exhaust flow does not drop below 75 percent of intake flow, exhaust duration equal to intake duration is satisfactory. If exhaust flow drops below 75 percent of intake flow, up to 10 additional degrees exhaust duration might be beneficial. Keep in mind however that because restrictive intake manifolds or undersize throttle bores reduce total intake flow below port capacity, low exhaust port flow can still produce 75 percent of intake flow. If more exhaust flow is needed, improved exhaust port flow is preferable to adding exhaust duration because of the effect of the duration on overlap and manifold vacuum.

4. Based upon flow test results, valve lift need not exceed the point of maximum port flow although higher lift will extend the time period at maximum flow. Keep in mind that many production cylinder heads especially when equipped with production manifolds do not benefit from valve lift much over 1/4 valve diameter and that piston-to-valve clearance may limit valve lift.

5. For single cams, time the camshaft according to the intake lobe centerline. That is; if the specified centerline is 110 degrees ATDC, this is the point when the intake valve reaches maximum lift. Check the point of intake closing to be sure that effective compression is acceptable. If effective compression is low, consider advancing the cam up to 4 crankshaft degrees. If effective compression is high, consider increasing intake duration. With dual cams, the exhaust cam will also need to be timed according to the centerline.

6. If maintaining manifold vacuum is important, such as with computer controls, keep in mind that excess exhaust duration significantly reduces vacuum. Consider wider lobe centers to reduce overlap and maintain vacuum.

It may not be possible to meet all valve lift and timing objectives perfectly, however, the closer to optimum, the greater the increase in output for any engine. Finally, there is a lot of sometimes confusing jargon that applies to camshaft grinding. The following glossary of terms should assist in understanding this jargon and interpreting camshaft specifications:

1. **DURATION** - The valve open period in crankshaft degrees. Up to a point, longer periods open are associated with improved volumetric efficiency.

2. **LIFT** - Cam lift or valve lift measured in inches or millimeters. Up to the limits of port flow, greater lift is also associated with improved volumetric efficiency.

3. **OVERLAP** - The period at the end of the exhaust stroke when both valves are open. With split overlap, both intake and exhaust valves are open equally at TDC exhaust stroke. Overlap helps initiate intake flow especially at high RPM.

4. **BASE CIRCLE** - The portion of the cam lobe concentric to the center of rotation and, when in contact the lifter or follower, the valve is closed. Because valves are on their seats

during this period, this is also when valve cooling is most efficient.

5. **RAMPS** - Opening ramps "preload" or take up valve train slack before opening the valve and extend from the base circle to where valve opening actually begins. Closing ramps extend from the ends of the closing flanks to the base circle and reduce "bounce" or shock on valve closing.

6. **FLANKS** - The areas on cam lobes where valve opening or closing rates are greatest.

7. **NOSE** - The valve slows to a stop on opening and slowly begins closing as it travels across the nose.

8. **LOBE CENTERLINE** - The centerline of the lobe frequently used to locate wide-open points in crankshaft degrees.

9. **LOBE CENTERS** - The separation between intake and exhaust lobe centerlines in camshaft degrees sometimes referred to as the lobe separation angle. Increases in the angle reduce overlap and decreases in the angle increase overlap.

10. **LOBE TAPER** - The taper ground into some camshaft lobes to promote lifter rotation and reduce camshaft wear.

11. **SYMMETRICAL CAM LOBES** – Lobes ground the same on opening and closing sides of lobe centerlines.

12. **ASYMMETRICAL CAM LOBES** – Opening and closing sides of lobes are ground to different contours.

13. **DUAL PATTERN GRINDS** - Intake and exhaust lobes are ground to different duration and/or lift specifications.

14. **VELOCITY** – Rate of change in lift in inches per degree of camshaft rotation inches/degree). Velocity is limited by lifter diameter or follower size.

15. **ACCELERATION** – Rate of change in velocity in inches per degree of camshaft rotation (inches/degree squared). Values are both positive and negative and are important in determining required spring pressures.

16. **JERK** – Rate of change in acceleration in inches per degree of camshaft rotation (inches/degree cubed). High jerk amplifies spring surge.

As a tip for those retro-fitting a roller lifter cam in place of a flat tappet cam, keep in mind that the roller cam is steel, not nodular iron. This requires installing a compatible distributor drive gear, typically bronze or "melonized" hardened steel, to minimize wear.

SELECTING CAMSHAFTS FOR FORCED INDUCTION

With forced induction, whether by exhaust driven turbocharger or a mechanically driven supercharger, camshaft duration must be selected according to final compression ratios including boost. Compression ratios increase with boost and, unless duration and the intake valve closing point are adjusted, pressures become excessive and detonation results. The high VE in forced induction engines makes detonation an even more serious concern. As in normally aspirated engines running on 91-octane pump gasoline, it is recommended that effective compression based upon the intake valve closing point be limited to approximately 7.0:1.

The increase in compression is equal to boost PSI divided by atmospheric PSI and the final compression ratio is equal to the static compression ratio plus this increase. In other words:

Compression Increase
= Boost PSI ÷ Atmospheric PSI and;

Final Compression Ratio
= Static Ratio x (Boost Increase + 1)

Sample Problem

What is the final compression ratio of an engine with 7.5:1 static compression and 8-PSI boost? Calculate as follows:

8 PSI Boost ÷ 14.7 Atmospheric PSI = .544 and;
7.5 (.544 + 1) = 7.5 x 1.544 = 11.6 Compression Ratio

For street driven engines, with some exception for those with pent roof combustion chambers, final compression ratios above 12:1 require excessive duration and are generally not recommended. Depending upon the efficiency of intercoolers in reducing inlet air temperatures, a final ratio as low as 10:1 could be the practical limit before detonation becomes a problem. To use the table below for final compression ratios, the static ratio and the amount of boost must be known.

COMPRESSION RATIOS WITH BOOST

Static Ratio	\-------------Boost PSI ---------------				
	4	6	8	10	12
6.5	8.3	9.2	10.0	10.9	11.8
7.0	8.9	9.9	10.8	11.8	12.7
7.5	9.5	10.6	11.6	12.6	13.6
8.0	10.2	11.3	12.4	13.4	14.5
8.5	10.8	12.0	13.1	14.3	15.4
9.0	11.4	12.7	13.9	15.1	16.3
9.5	12.1	13.4	14.7	16.0	17.3

Additional exhaust flow is necessary in forced induction engines. With volumetric efficiencies well in excess of 100 percent, it is necessary to purge the cylinder of proportionately more exhaust gas. In addition to increasing exhaust valve size and porting, additional exhaust duration is typically called for. Basically, if a normally aspirated engine requires 75 percent exhaust flow relative to intake flow, an engine with forced intake requires 80 percent or more.

With such long intake and exhaust durations, there is a tendency to dilute the intake charge with exhaust gas during overlap. Unless addressed through camshaft design, the problem is compounded in turbocharged engines where exhaust backpressures can double under load and increase pressure reversions into the intake and the cylinder during overlap.

With wet manifold engines, fuel is also forced out the exhaust during overlap elevating the already high exhaust temperatures. For these reasons, it is not unusual to find lobe centers increased over those for normally aspirated engines to reduce overlap.

Valve springs also deserve special attention in engines with forced induction. Consider what happens behind a 2-inch diameter intake valve in an engine with 8-PSI boost. The area of the valve is approximately 3.1 square inches, and with 8-PSI boost, there is a 24-pound force attempting to force the valve open. As a result, intake valve spring pressures must be increased to compensate.

MATCHING INTAKE AIRFLOW TO THE ENGINE

So far, the engine has been dealt with as an air pump and modified to better fill the cylinders and thereby improve volumetric efficiency. To realize maximum performance, the intake and exhaust systems must have tuning and flow capacity matching the requirements of the engine.

Specifically, carburetors and fuel injection systems must be capable of adequate airflow measured in cubic feet per minute or CFM. Without forced induction, only exceptional production or highly modified engines are likely to exceed 100 percent volumetric efficiency. Since we are discussing four-stroke cycle engines, this means filling the cylinders every two revolutions of the crankshaft. CFM requirements at 100 percent VE are calculated as follows:

RPM x Displacement ÷ 3456 = CFM

Where displacement is in cubic inches and; 3456 is two cubic feet of air in cubic inches.

In applying these calculations to a particular engine, consider that we are calculating the quantity of air required to fill the cylinder without considering that we are also filling the combustion chamber. Also, some inlet air escapes through the exhaust valve during overlap. All things considered, expect that a race or performance engine typically requires 25 percent or more intake airflow over baseline calculations to deliver the required CFM.

However, if the carburetor or throttle valve is too large, low-end performance and throttle response suffers and, if too small, engine output is limited. Following are sample calculations of baseline airflow requirements at 100 percent volumetric efficiency.

A 302 cubic inch engine with a 6,500 RPM limit:

6,500 x 302 ÷ 3456 = 568 CFM
Use a 600-CFM carburetor or throttle body

A 383 cubic inch engine with a 5,500 RPM limit:

5,500 x 383 ÷ 3456 = 610 CFM
Use a 650 CFM carburetor or throttle body

The required airflow at 100 percent VE can be calculated as shown based upon displacement and RPM. For quick reference, use the following table:

BASELINE AIRFLOW REQUIREMENTS IN CFM

| Displacement | | RPM | | | | | |
cm³	in³	5,000	5,500	6,000	6,500	7,000	7,500
1600	098	142	156	170	184	198	213
1800	110	159	175	191	207	223	239
2000	122	177	194	212	229	247	265
2200	134	194	213	233	252	271	291
2400	146	211	232	253	276	296	317
2600	159	230	253	276	299	322	345
2800	171	247	272	299	322	346	371
3000	183	265	291	318	344	371	397
3200	195	282	310	339	367	395	423
3400	207	299	329	359	389	419	449
3600	220	318	350	382	414	446	477
3800	232	336	369	403	436	470	503
4000	244	353	388	424	459	494	530
4200	256	370	407	444	481	519	556
4400	268	388	427	465	504	543	582
4600	281	407	447	488	529	569	610
4800	293	424	466	509	551	593	636
5000	305	441	485	530	574	618	662
5200	317	459	504	550	596	642	688
5400	329	476	524	571	619	666	714
5600	342	495	544	594	643	693	742
5800	354	512	563	615	666	717	768
6000	366	530	582	635	688	741	794
6200	378	547	602	656	711	766	820
6400	390	564	621	677	734	790	846
6600	403	583	641	700	758	816	875
6800	415	600	660	720	781	841	901
7000	427	618	680	741	803	865	927
7200	439	635	699	762	826	889	963
7400	451	652	718	783	848	913	979
7600	464	671	738	806	873	940	1007
7800	476	689	758	826	895	964	1033
8000	488	706	777	847	918	988	1059

Another place where calculations cannot be applied directly is with individual runner intake systems. There isn't very much volume in the runner and there is no sharing of a plenum. Vacuum is generally lower and the flow capacity of the throttle valve must be doubled or tripled to deliver the required airflow.

MATCHING EXHAUST SYSTEMS TO THE ENGINE

Fig.17-44 Four-into-one V8 headers (Hedman Headers)

In simple terms, a highly efficient performance exhaust system scavenges exhaust from the cylinders and helps initiate intake flow into the cylinder during overlap. Careful tuning takes advantage of reduced pressure in the exhaust port during overlap at the same time that atmospheric pressure is forcing air through the inlet valve. It is with this scavenging effect optimized that VE can approach or exceed 100 percent.

The dynamics of gas flow through exhaust systems are more complex than the explanation above suggests. While exhaust gas moves at approximately 250 feet per second through the system, there are simultaneous pressure waves traveling up and down the runners at approximately 1,600 feet per second.

The challenge in tuning exhaust is to synchronize the low point in the pressure wave with valve overlap at the beginning of the intake stroke and end of the exhaust stroke. As mentioned, tuning helps draw inlet air into the cylinder on the intake stroke and increases VE.

A basic requirement of an efficient system is that exhaust from one cylinder not be allowed to pressurize another. For this reason, exhaust ports and runners from individual cylinders remain separated for a distance of at least four times stroke length, beginning from the valve, prior to joining together. Further tuning requires even longer runners that remain equal in length before joining. Partly because of packaging constraints within the engine compartment, V-block engines are typically treated as two 3 cylinder or two 4 cylinder engines (see Fig.17-44).

Also keep in mind that when the exhaust valve opens, residual pressure in the cylinder of 90 PSI or more initiates the "blow-down cycle" through the port and runner. To facilitate this blow-down, the port and runner should be smooth and non-turbulent.

There are highly sophisticated approaches to exhaust tuning that take into consideration such factors as RPM, exhaust valve lift, valve timing and volumetric efficiency. However, for street applications with limited RPM, compression, valve duration and lift, there is a simpler approach that produces good results. This simplified approach involves looking at the basic relationship between cylinder displacement and runner volume.

For each configuration of headers, by this simplified approach, expect runner volume to remain constant at two times cylinder volume. This remains so even with changes in runner length or diameter. If runner lengths increase, diameters decrease, and if runner lengths decrease, diameters increase.

To determine volume, we begin with a runner cross-sectional area that matches the curtain area at a valve lift equal to one-quarter valve diameter; the point at which the area of the valve and curtain area are equal. This also makes runner diameter equal to valve diameter. If this seems small, keep in mind that the restriction below valve seat is only 85 to 90 percent of valve diameter. Since area times length equals volume, runner volume is calculated by multiplying runner cross-sectional area by the runner length including the port.

Although calculations require the runner inside diameter, tubing is sized by outside, not inside, diameters. It is necessary to either measure the inside diameter of the tubing or subtract two times the gage thickness from the tubing size and select the nearest size to optimum. Also, off-the-shelf tubing is made in a limited number of sizes, usually in 1/8in increments, making super precise selections difficult unless the builder is prepared to pay for custom fabrication (tubing "wrapped" and welded from flat sheets).

Most common is 20 gage tubing with a .035 thick wall. With coatings applied, wall thickness can go up to 040in. Thermal barrier coatings on the inside and thermal dispersant coatings on the outside are sometimes applied to improve efficiency and longevity. Gage thicknesses are as follows:

Tubing Gage	Thickness (Uncoated)
14	.083in
16	.065
18	.049
20	.035

Shown below is a sample calculation for curtain area, runner inside diameter and length based on valve diameter, valve lift and displacement per cylinder.

BASE ENGINE SPECS

Exhaust Valve Diameter	1.500in
Lift at 1/4 Valve Diameter	.375in
Cylinder Displacement	40in^3

Step 1 Calculate Curtain Area

$$\text{Curtain Area} = \text{Pi} \times \text{Valve Diameter} \times \text{Lift}$$
$$= 3.1416 \times 1.500 \times .375$$
$$= 1.767\text{in}^2$$

Step 2 Calculate Runner Diameter
(With runner area equal to curtain area)

$$\text{Area} = \text{Pi} \times r^2$$

$$r^2 = \text{Area} \div \text{Pi}$$

$$r = \sqrt{\text{Area} \div \text{Pi}}$$

$$\text{Diameter} = 2\sqrt{\text{Area} \div \text{Pi}}$$

$$= 2\sqrt{1.767 \div 3.1416}$$

$$= 2\sqrt{.5625}$$

$$= 2 \times .750$$

$$= 1.500\text{in Inside Diameter}$$

Step 3 Calculate Runner Length
(Runner area times one equals volume per inch)

$$\text{Length} = 2 \times \text{Displacement} \div \text{Curtain Area}$$
$$= 80\text{in}^3 \div 1.767\text{in}^2$$
$$= 45.3\text{in Runner Length Including Port}$$

Recall that exhaust tubing for runners is sized by outside diameters and in increments of 1/8 of an inch. This limited selection will likely make it necessary to select the nearest tube size matching the calculated curtain area with less than perfect sizing. If the tubing is larger, shorten the length proportionately and tuning will favor high RPM power. If smaller, lengthen the runner proportionately and tuning will favor low RPM power. Based on the earlier example, the possible selections with the nearest off-the-shelf 20 gage tube sizes, inside diameters, above and below calculations, are as follows:

Calculated Runner Inside Diameter
1.500in

ID of 1.625 Tubing
1.625 - .070in = 1.555in ID
2 x CID ÷ A = Length
Length = 42.1in Including the Port

ID of 1.500 Tubing
1.500 - .070in = 1.440in ID
2 x CID ÷ A = Length
Length = 49.1in Including the Port

Which compromise is best? Partly, it depends on where we want best average power, at the high end or the low end of the power curve. This is a good place to test each option using engine analysis software. Both combinations of diameters and lengths can be compared to each other and to the production system and perhaps further tuned.

Collectors and exhaust pipes make up the next section of the system. In a well-designed system, the collector tapers down from where the runners join to the exhaust pipe at an included angle of approximately 14 degrees (see Fig.17-45). Because of the need to fit collectors under the chassis, they may be formed into other shapes but transitions must remain smooth.

Pipe Diameter		HP	CID
Single	Dual		
2.00	2.00	100	150-200
2.25	2.00	150	150-200
2.50	2.00	200	150-200
2.25	2.00	150	200-250
2.50	2.00	200	200-250
2.50	2.25	250	200-250
2.50	2.00	200	250-300
2.25	2.25	250	250-300
3.00	2.50	300	250-300
3.00	2.25	250	300-350
3.00	2.50	300	300-350
3.00	2.50	300	350-400
3.50	2.50	350-400	350-450

Fig.17-45   Collectors taper down to the exhaust pipes (Hooker Headers)

Exhaust pipe diameter from the collector back will increase with displacement. A 2-inch diameter for a 100 CID four-cylinder engine may be adequate while a 450 CID V8 engine may benefit from a pair of 2 3/4-inch exhaust pipes. Diameters generally fall within this range in proportion to power and displacement. Following are some Flowmaster Mufflers Inc. recommendations for street systems:

In V-block engines, additional scavenging from one bank to the other is gained by adding a "crossover pipe" (see Fig.17-46). The crossover should be equal or greater in diameter and positioned ahead of the mufflers or converters as close as possible to the front of the system. The crossover enables scavenging with side-to-side pressure fluctuations in dual exhaust systems. Crossovers also greatly reduce resonant noise.

H-PIPE

Fig.17-46 A dual exhaust H-type crossover pipe (Hedman Headers)

As for mufflers and tailpipes, street vehicles must stay within reasonable noise limits. This typically requires backpressure not lower than 2-PSI under load. Good mufflers and crossover pipes for V-block engines are obviously important but tail pipes cannot be ignored. There can still be a lot of resonant noise in large diameter pipes especially when they exceed 6ft in length. It is sometimes necessary to add resonators for long wheel base vehicles or shorten the overall length possibly by exiting out the side of the vehicle.

In discussing exhaust systems, the word 'compromise" comes up frequently. However, choose components carefully and significant gains over production systems are likely

RUNNING COMPUTER SIMULATIONS

Considering the complexities of camshafts or intake and exhaust systems, it is beneficial to use computer engine analysis to plan projects prior to machining or selecting components. Computer simulation helps the builder achieve the desired results without needless and costly experimentation. Even when dyno tuning, computer analysis helps optimize the plan and reduces rework in testing. Without dyno tuning, such analysis is the only reasonably precise way of predicting output.

To run simulations, we need a program smart enough to ask the right questions and quality data to answer these questions. Once data is entered, the user can work through any number of "what if" scenarios. What if compression is increased, duration increased, or the cam advanced? What if another manifold or header is tried? After viewing the analysis on screen, it is not unusual for novice users to find that, with their first selections, power is too low or, on the other extreme, detonation is likely. However, when running simulations, it is best to focus on best average power and torque, not just peak numbers. Although not as impressive, you can readily give up a few HP at the peak if the average goes up.

While simulation is obviously important to race engine development, engines for the street

benefit possibly even more. Consider that octane for pump gas is limited and that computerized engine controls will not function properly with low manifold vacuum. Simulation helps predict manifold vacuum, the spark timing curve, port velocities and other factors important to drivability.

With predictions of horsepower and RPM, BMEP is also calculated and checked against the user specified fuel octane number. As can be seen in the formula for BMEP, pressure can be lowered by raising the RPM for peak torque. A shift of just 500 RPM can make a significant difference. With simulation, it is possible to experiment with later valve timing, shorter intake runner lengths, single versus dual plane manifolds and exhaust runners with larger diameters and shorter lengths. Adjustments in these areas can shift the RPM range upward without completely reconfiguring the engine.

Some engine analysis programs import camshaft and flow data directly from camshaft analysis and flow test equipment. The quality of this data greatly improves the accuracy of analysis. In analysis reports, aside from predictions of power output, there are a number of data points to examine for clues to where further improvements might be made. Some key data points include:

1. Effective compression ought to be in the 7 to 7.5:1 range for efficient engines to run safely on pump gasoline. If not in this range, intake duration or the intake valve closing point may require adjustment.

2. Manifold vacuum at peak power should not exceed 1 in/Hg. Higher vacuum indicates a restricted intake and the need for larger throttle bores.

3. Exhaust flow relative to intake flow should be in the 75-80 percent range. If low, exhaust duration could be low. If high, exhaust duration could be excessive or intake flow could be too low.

4. Peak volumetric efficiency ought to occur close to peak torque. If widely separated, intake and/or exhaust runner lengths might be better tuned.

5. Manifold vacuum at idle should be within acceptable limits for the engine application. It might be necessary to widen camshaft lobe centers to increase vacuum. Be careful to note the ignition timing curve in the analysis report. Unless the curve is set up as recommended, vacuum will not match the prediction.

6. Engines running full exhaust systems with exhaust back pressure below 2 PSI at peak power will likely be loud under load. A compromise between noise levels and power output is likely required.

7. Intake and exhaust velocities ought to average 250 feet per second. High velocity could indicate restrictive ports and runners. Low intake velocities could cause poor throttle response especially at lower engine speeds.

Making adjustments to the engine configuration based upon analysis helps the engine builder optimize the plan before machining or assembly. When it comes to tuning for power, recall that tuning for best average power is often more satisfactory in application than focusing on peak power alone.

TUNING PERFORMANCE ENGINES

Obtaining maximum efficiency and best drivability with modified engines also requires careful tuning and calibration of spark timing and fuel mixture. Changes in compression, duration and valve timing often call for recalibrating timing curves. Within limits, computer engine controls "learn" and recalibrate or, if the required recalibration is beyond this capability, they may be reprogrammed. For non-computer applications, distributor timing curves can be checked as follows:

1. Connect a tachometer to the ignition system and a vacuum gauge to a manifold source. Disconnect and plug the distributor vacuum advance hose for this procedure.

2. Start the engine and set the idle speed to the desired idle RPM.

3. Advance the ignition timing until the vacuum gauge reading reaches maximum and then retard the timing until the vacuum gauge drops between 1 and 2 in/Hg. Engine RPM changes so readjust idle (see pt.1, Fig.17-47). Since only a few degrees additional advance are permissible for emission-controlled vehicles, consider raising idle RPM slightly to improve idle quality.

4. Attach a dial reading timing light to number one cylinder and raise the engine speed to 3,000 RPM. With the vacuum advance disconnected, large bore engines run well with 28 to 32 degrees total spark advance, possibly less for small bore engines or those with spark plugs closer to center (see pt.2, Fig.17-47). If not within limits, calibrate the advance mechanism using aftermarket centrifugal weights and springs. Centrifugal distributors with no vacuum advance may run a few degrees more advance but should not reach the total until about 4,000 RPM (see pt.3, Fig.17-47).

5. Connect the vacuum advance and check total advance at 2,500 RPM. Under these conditions, a minimum of 40 degrees total advance is generally required for reasonable fuel economy (see pt.4, Fig.17-47).

6. Drive the vehicle and check for pinging. If minor adjustments up to 4 degrees in initial timing do not eliminate pinging, an adjustable vacuum advance unit might be necessary.

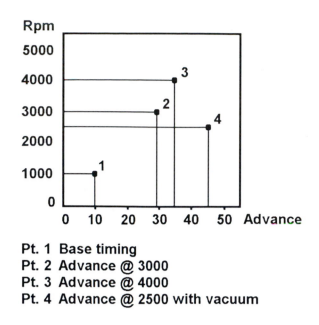

Pt. 1 Base timing
Pt. 2 Advance @ 3000
Pt. 3 Advance @ 4000
Pt. 4 Advance @ 2500 with vacuum

Fig.17-47 A base-line ignition timing curve

Especially in regard to spark timing, it is important to remember that more is not always better. Aside from the destructive effects of detonation, over advanced timing requires that pistons overcome expanding gases at the end of compression strokes thereby producing a power loss. Air-cooled engines are sometimes sensitive to heat and may tolerate less ignition advance. Small bore engines and those with centrally located spark plugs have a shorter length of flame travel to the outer edges of chambers and can develop peak power with as little as 24 degrees total advance. Engines with any form of forced induction have dense fuel mixtures that burn rapidly and much less advance is needed. In all engines, the time for burning shortens with RPM and the burning rate speeds up as mixtures richen under load. The constant is that all engines must vary ignition timing to achieve peak cylinder pressure by approximately 12 degrees ATDC.

Adjustments to fuel mixture are obviously critical. If mixtures are lean, the engine misfires at idle and overheats at high speeds. If mixtures are rich, the engine fouls spark plugs and has high exhaust emissions. A correct mixture yields good performance and also meets exhaust emission standards. Street driven engines ought to be tuned within mandated exhaust emission limits and all emission control devices kept functional.

Engines in a good state of tune run 14.7:1 air-fuel ratios at cruise and 12.5 to 13:1 at wide open throttle. Lean mixtures show up in emission testing as high readings of unburned hydrocarbons. This is because the engine misfires and unburned fuel goes out the exhaust. Rich mixtures show up as high readings of carbon monoxide because there is insufficient oxygen to complete combustion. In the author's experience, engines with reasonable compression and camshaft changes, and tuned in this way, maintain good idle quality and drivability.

PROJECT; PREPARE A PERFORMANCE ENGINE

As a case study, we are going to prepare a street driven performance engine under the guidance of a professional cam grinder and engine builder, Dimitri "Dema" Elgin of Elgin's Cams. Because most machining operations are similar to those covered earlier in the text, emphasis is placed upon points of particular importance to maximizing engine performance.

The first step was to decide upon the operating limits for the engine. The engine was the 350 Chevrolet small block from a restored 1973 Corvette with a Turbo 400 Hydramatic transmission and a 3.36:1 rear axle ratio that shifts by 5,500 RPM. Externally, the engine was to remain original including emission controls. Because of the automatic transmission, the objective was to produce maximum torque while maintaining acceptable idle quality.

It was decided that a 383 cubic inch engine based on the original small block 350-engine would produce the desired torque and acceleration without sacrificing drivability. Essentially, this combination uses all 350-engine parts except for a 400 small block crankshaft. Preparation of the 400 crankshaft was the first task beginning with cleaning and Magnaflux inspection. Following this, the main journals were ground .200in undersize to standard 350 dimensions including the rear flange of the thrust main (see Fig.17-48). At .200in undersize, particular attention was given to obtaining full, smooth fillet radii at each side of the bearing journals to prevent concentration in these critical areas. Also important was holding concentricity with timing sprocket and rear main seal locations to minimize the potential problems with timing sets or front and rear seal leaks.

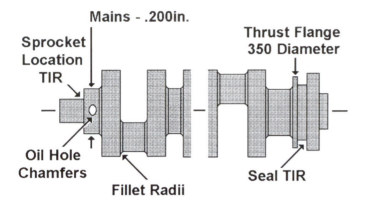

Fig.17-48 Points of detail in preparing a Chevrolet "383" crankshaft

It was decided after inspection that the crankpins required grinding .010in undersize. While crankpins in better condition might have been left standard, grinding the crankpins did provide the opportunity to equalize stroke lengths and improve fillet radii. After grinding, standard 350 main bearings and .010in undersize 350-rod bearings were required.

Next was the selection of connecting rods. The least expensive option was to use 5.565-inch long 400 rods with off-the-shelf replacement 350 pistons. Another option was to use longer 5.700-inch 350 rods with a "383" piston. It decided that the 350 rods had two advantages. First, because the piston remains for a longer period of time close to the top of its travel with a longer rod, cylinder pressure builds to higher levels before gas expansion forces the piston downward on the power stroke thus increasing output. The secondary benefit is a reduction in side thrust on the cylinder walls because of the smaller rod angle relative to the cylinder centerline.

Preparation of these rods began with cleaning and magnaflux inspection. The rod and cap parting surfaces were ground .002in and the rods reassembled with high strength rod bolts. The rod center-to-center lengths were then checked for variations (see Fig.17-49). Re-cutting the caps and repeating the resizing operation on the longer rods brought all rods within .002in of each other.

Of course, following any cap cutting, inspect bearing lock grooves for adequate length and depth.

Fig.17-49 Comparing rod center-to-center lengths

Balancing would normally follow rod preparation but it was decided to wait until rod clearance in the crankcase and at the camshaft was checked before removing any metal from the rods. With the long stroke, outside rod bolts can interfere with the crankcase and inside rod bolts with some cam lobes.

The pistons were forged "383" pistons designed with a 1.435in compression distance for use with 5.7in long Chevrolet 350 rods. The pistons were measured for size to confirm that the correct fit in .030in oversize cylinders. The weights of the pistons with pins were then checked and corrected within one gram. Machining of the engine block began with line honing the main bearing housing bores. This was done with the oil pump and the bolts for the four bolt caps tightened to specifications. To assure that the engine would not leak, the rear main seal flange was recut to restore concentricity to the new main centerline (see Fig.17-50).

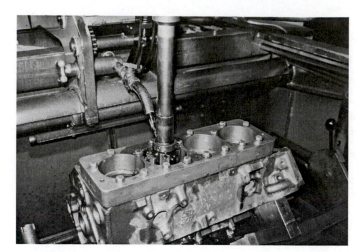

Fig.17-51 Cylinder honing with a torque plate simulates distortion from head bolts

Fig.17-50 Restoring main seal concentricity after line honing

The cylinders were then torque-plate honed to 4.030in and finished with stones producing a surface finish suitable for moly piston rings (see Fig.17-51). Since rod piston assemblies needed to be installed to check crankcase and cam clearance, surfacing the block could wait until deck was measured with the crankshaft, connecting rods, and 383 pistons in assembly.

The block was then turned upside down and one-half of a main bearing set installed in the crankcase. The bearing surfaces were oiled and the crankshaft lowered into place. One of the pistons, less rings, was assembled to a rod using a piston pin polished undersize to facilitate rapid assembly and disassembly during these checks. With this piston and rod assembly installed in one cylinder at a time, the crankshaft assembly was rotated to check for clearance in the crankcase. Because the outside rod bolts interfered with the crankcase, the block was set up in a vertical milling machine and the clearance increased using a ball end mill (see Fig.17-52).

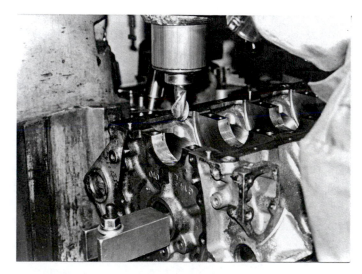

Fig.17-52 Cutting clearance for rods at the pan rails

The deck clearance was recorded while checking crankcase clearance and it was then determined that surfacing the block .010in would produce the .010in deck clearance specified by the piston manufacturer. The block was then set up for surfacing with a BHJ Bloc-True fixture (see Fig.17-53). Positioning from this fixture places the camshaft in the center of the "V" and places decks 45 degrees to each side of the crankshaft and camshaft centerline (see Fig.17-54).

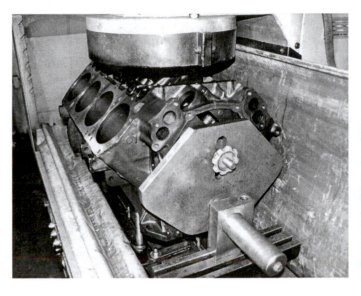

Fig.17-53 BHJ tooling attached to a V-Block

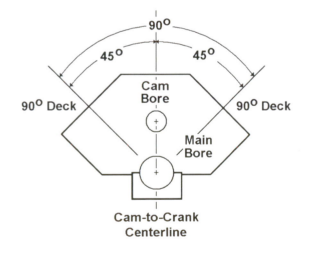

Fig.17-54 Alignment established by BHJ tooling

The check for connecting rod clearance at the camshaft required cleaning the block and installing cam bearings. The camshaft was then installed and cam lobe and connecting rod clearance checked as this was known to be a problem with this combination. With a piston and rod assembly in each cylinder, the crankshaft was rotated two revolutions to check cam clearance. There was interference and the heads of the inside rod bolts had be ground for clearance as shown in Figure 17-55. To maintain balance, all rods were modified in the same way. To be sure that clearance was adequate; it was checked again with the camshaft advanced then retarded 4 degrees. With some aftermarket cap-screw rods, interference is not an issue.

Fig.17-55 Grinding connecting rods to provide clearance for the camshaft

It was then time to complete the engine balancing. With the clearance modifications to the rods completed, they were then balanced. Because the 400 crankshaft was externally balanced, running the original 350 vibration damper would have required internal balancing with "heavy metal." An alternative was to run a 400 Chevrolet damper and flexplate with the original external weights (see Fig.17-56).

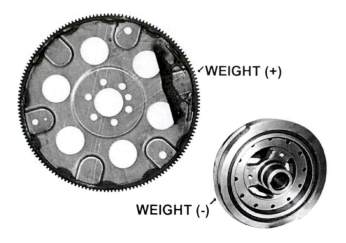

WEIGHT (+)

WEIGHT (-)

Fig.17-56 Externally weighted Chevrolet 400 damper and flexplate

Cylinder head preparation was next. Bronze guides and alloy exhaust valve inserts were installed. The port bowls were then matched to the high-flow 2.020 intake and 1.600in exhaust valves. The ports were cleaned up following the procedures discussed earlier in this chapter and the manifold runners and intake ports were reworked to correct their alignment to each other. The rocker arms were removed and stud bosses cut down to install screw-in studs and pushrod guide plates (see Fig.17-57).

Fig.17-57 Screw-in studs and pushrod guide plates require cutting guide bosses .360in

The combustion chambers volumes at the outside ends of each head were then measured or "CC'ed." From this it was found that each head required surfacing different amounts to maintain volumes within one cc of each other. Chamber volumes are reduced by one cc for each .005in removed from the heads. In this case, the volumes after surfacing came out at 76cc.

With combustion chamber volumes and deck clearance known, it was then possible to accurately calculate the static compression ratio:

Cylinder bore	4.030in (+.030in)
Stroke	3.75in (std. 400)
Chamber volume	76.0cc
Deck clearance volume	3.2cc (.010in deck)
Valve relief volume	3.2cc (specified)
Gasket volume	8.5cc (.038in thick)
Clearance volume	90.9cc (76+3.2+3.2+8.5)
Cylinder volume	784.2cc
Compression ratio	9.6:1 (784.2 + 90.9) ÷ 90.9

The finished cylinder heads were tested to determine the maximum valve lift required and the exhaust flow relative to intake flow. Based upon the test data below, it was decided that valve lift in the range of .425 to .450in would permit maximum flow. Testing also showed that exhaust flow maintained 75 percent of intake flow.

Flow Test Results
Chevrolet 350-400

Valve Lift	Intake CFM	Exhaust CFM	Percent Intake
.050	28.7	24.2	84.1
.100	62.1	50.5	81.2
.150	88.6	71.9	81.1
.200	121.5	98.5	81.1
.250	151.7	119.8	78.9
.300	175.5	138.2	78.7
.350	191.4	152.3	79.6
.400	206.2	161.2	78.2
.450	**214.4**	166.3	77.6
.500	214.4	**168.4**	78.5

Note: Flow at 28 in/water

The next task was to determine the intake closing point so that effective compression would come out between 7 and 7.5:1. Assembling on the high side of this range is recommended because of two factors that reduce cylinder pressure. First, normal timing chain stretch retards the camshaft at least two degrees, and second, the stock intake and exhaust manifolds reduce volumetric efficiency.

To find the correct intake closing point first required finding the cylinder displacement at 7.5:1 effective compression using one of the "Useful Formulas" listed earlier:

D = CV (CR-1) where;
D is displacement and CV is the clearance volume
= 90.9 (7.5-1)
= 90.9 x 6.5
= 590.9cc

Next, it was necessary to find the crankshaft position at this displacement. To do this requires finding the rod ratio for the 383 and the percentage of cylinder volume at IVC. With this information, the crankshaft position is found in the table of Cylinder Volumes at Intake Closing, page 17-22.

Rod Ratio:
= 5.700 Rod ÷ 3.750 Stroke = 1.52

Percent of cylinder volume at IVC @ 7.5:1
= 590.9 ÷ 784.2
= 75.3%

From the 1.5:1 rod ratio column in the table, the nearest to the actual 1.52, it was found that the intake valve must close by 69 degrees ABDC to maintain 75 percent of cylinder volume and 7.5:1 effective compression. The next task was to find a camshaft that would duplicate this intake timing and have adequate lift and acceptable overlap. Four off-the shelf camshafts were selected for comparison with specifications as follows:

SPECIFICATIONS	CAMS			
	#1	#2	#3	#4
Valve Lift, Intake	.443	.443	.435	.435
Valve Lift, Exhaust	.465	.443	.435	.435
Duration, Intake	280	280	288	288
Duration, Exhaust	290	280	288	288
Lobe Centers	112	112	114	112
Advance	3	3	4	2
Centerline, Intake	109	109	110	110
Centerline, Exhaust	115	115	118	114
Intake Opening BTDC	31	31	34	34
Intake Closing ABDC	69	69	74	74
Exhaust Opening BBDC	80	75	82	78
Exhaust Closing ATDC	30	25	26	30
Overlap	61	56	60	64

NOTE: Timing points retard at least 2 degrees due to chain stretch

The intake valve timing of all four possible camshafts was in an acceptable range so other considerations would make the difference in selection. With this in mind, the owner's requirements were recalled; good idle and drivability, low emissions, and strong low and mid range torque for acceleration. Additional exhaust duration was to be avoided as it would cause losses in vacuum and fuel economy. It was also decided that the intake closing point could be early since VE is reduced with the stock intake and exhaust systems. Given these considerations, camshaft number two was selected as the best compromise (see Fig.17-58).

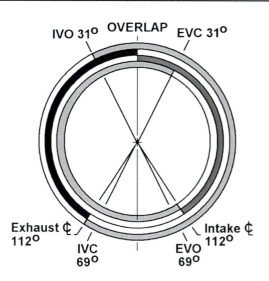

Fig.17-58 A valve timing diagram for the selected camshaft

The camshaft then had to be installed and timed so that the intake valve was wide open at close to 112 degrees ATDC. The first step in this procedure is to check the accuracy of the TDC mark on the vibration damper and the timing indicator by the following procedure:

1. Place a positive stop extending 1/2in into the cylinder over the center of number one piston (see Fig.17-59).

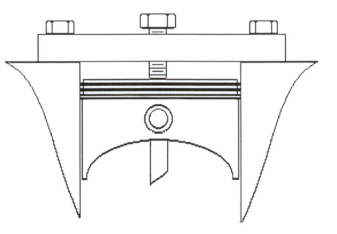

Fig.17-59 A positive stop on center over the piston

2. Rotate the crankshaft clockwise until the piston bumps the stop. Measure from the TDC mark on the damper to "0" on the timing indicator (see Fig.17-60).

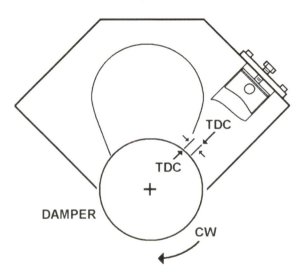

Fig.17-60 Measuring from TDC to the timing indicator, clockwise rotation

3. Repeat the procedure above rotating the crankshaft counter-clockwise. The measured distances should be equal. If not, the TDC mark is off by half the difference in measurements in the direction of the

longer measurement. Remark the timing indicator accordingly for the point of exact TDC (see Fig.17-61).

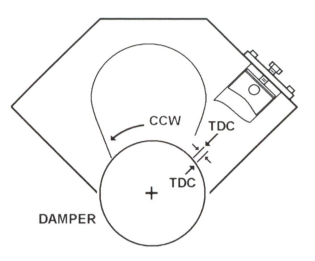

Fig.17-61 Measuring from TDC to the timing indicator, counter-clockwise rotation

To time the camshaft from the lobe center-line, the point in crankshaft rotation where the intake valve reaches wide open must be found. This procedure requires a degree wheel and a long stroke dial indicator as follows:

1. Attach a degree wheel to the damper and line up TDC on the wheel with TDC on the vibration damper (see Fig.17-62).

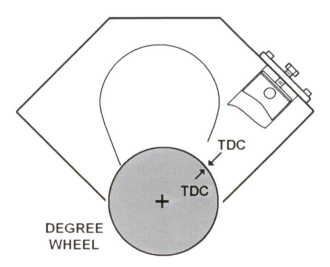

Fig.17-62 The degree wheel lined up with TDC

2. Place a dial indicator over the valve lifter of number one intake valve (see Fig.17-63). On overhead cam engines, the indicator can be placed on the spring retainer.

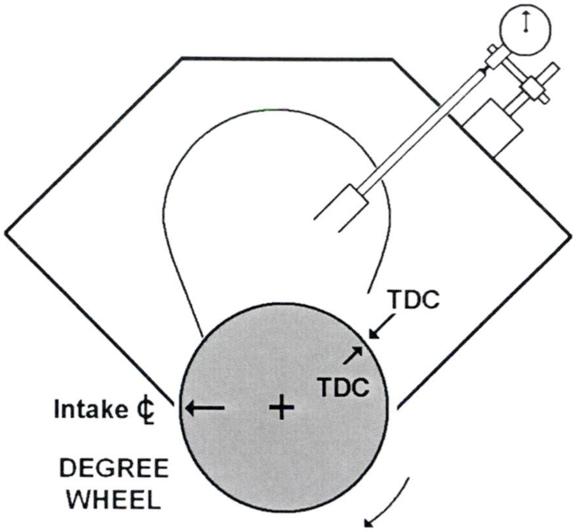

Fig.17-63 Positioning an indicator to read cam lift

3. Rotate the engine until the valve is wide open. Set the indicator to "0".

4. Backup the engine 1/4 turn and then rotate the crankshaft clockwise until the indicator is .050in short of "0" on the indicator. Write down the crankshaft position according to the degree wheel.

5. Continue rotating the engine until the indicator rises to "0" and drops .050in. Write down the crankshaft position according to the degree wheel.

6. Count the number of crankshaft degrees between positions one and two. The intake centerline is exactly half way between these two points.

In this case, the intake centerline checked out at 112 degrees ATDC but the cam was advanced two degrees just to compensate for timing chain stretch. Remember that with normal stretching of the chain during break-in, intake closing moves back at least two degrees. Advancing the camshaft required drilling the cam sprocket dowel pinhole

oversize so that a 2-degree (1-degree camshaft) offset bushing could be installed over the camshaft dowel pin (see Fig.17-64).

Fig.17-64 Advancing the camshaft with an offset bushing

To understand how these bushings work, keep in mind that the timing sprockets remain in normal alignment and that the camshaft rotates within the cam sprocket. After correcting the timing of the intake centerline, the point of intake closing can be checked using the degree wheel and dial indicator already set-up on the engine. To do this, rotation clockwise is continued from the centerline position until the rate of valve closing indicates the start of the closing ramp (see Fig.17-65). Depending on how "aggressive" the cam lobe profile, a .001in change in lift could require between one and three degrees of rotation. In this case, at 69 degrees ABDC, three degrees rotation should not produce more than .001' change in lift. Because this is not some precise point, and the difficulty in locating it especially with hydraulic valve trains, timing is more often checked at .050in timing points.

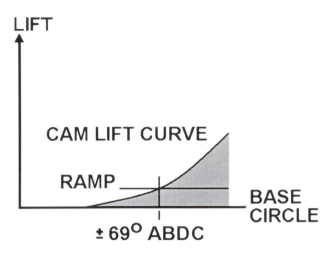

LIFT

CAM LIFT CURVE

RAMP

BASE CIRCLE

± 69° ABDC

Fig.17-65 The point at which lifters or followers encounter the closing ramp

Of course, the additional lift made it necessary to check spring travel for coil bind and clearance between the spring retainer and valve seal. These checks were made as follows:

1.   With .443in valve lift using 1.5:1 rockers, springs were checked to make certain that they compress .543in, or .100in past the wide-open point (see Fig.17-66).

Fig.17-66 Checking valve springs for coil bind at full lift plus .100in

2.   Next, retainers and keepers were assembled to the intake valve stem and the valve fully opened. The distance between the underside of the retainer and valve guide was then measured. There must be 3/16in extra travel to allow for the valve seal and spring compression during assembly (see Fig.17-67).

Fig.17-67 Checking for clearance over the valve guide and seal at full lift

This was also a convenient time for checking rocker arm geometry. To do this, the valve was opened half way and the point of contact on the valve stem checked. Half open and with correct geometry, the rocker arm contacts the valve stem on the centerline. To check geometry with hydraulic lifters, it is necessary to compress the lifter until the pushrod seat is approximately .040in (1mm) down in the lifter body. In this case, because there were minimal changes to camshaft base circles, cylinder head or block decks or valve stem lengths, the geometry checked out fine.

With aftermarket parts of different dimensions, it would also be necessary to check that the pushrod is perpendicular to its end of the rocker arm. This prevents operating through much of the arc of rotation "off-angle" causing lost motion through mid lift (see Fig. 17-68). In such assemblies, longer valves stems, pushrods and adjustments to the height of the rocker arm pivot might be needed.

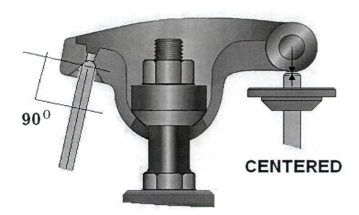

Fig.17-68 Pushrod and rocker arm alignment at mid lift

The last remaining check for clearance was made with the cylinder head installed over number one cylinder. Valves were installed for number one cylinder using very weak test springs (see Fig.17-69). The valve train assembly was completed for number one cylinder including lifters, pushrods, and rocker arms. After adjusting the rocker arms to zero lash, the piston to valve clearance was checked in 5 degrees increments between 15 degrees before and 15 degrees after TDC exhaust stroke. At each 5-degree increment, the piston to valve clearance was checked by making sure that .100in gauge passed between the rocker and valve stem.

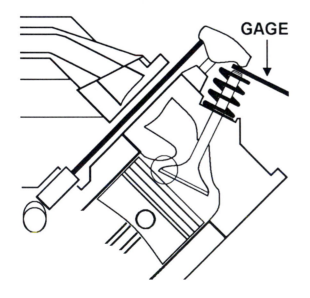

Fig.17-69 Checking for piston-to-valve clearance

With this series of checks and measurements completed, the engine was assembled, tested, installed and tuned. Computer analysis predicts the output of this engine at 90 percent VE with over 360 horsepower at 5,000 RPM and over 400 foot pounds of torque at 4000 RPM (see Fig.17-70). Of course, with the addition of a single plane intake manifold and headers, the RPM range would go up and power would climb another 10 percent. Even though VE is limited in this application by the stock intake and exhaust systems, output over the original still increases by over 100 foot pounds of torque and 100 horsepower. The engine also idles with 18-in/Hg manifold vacuum.

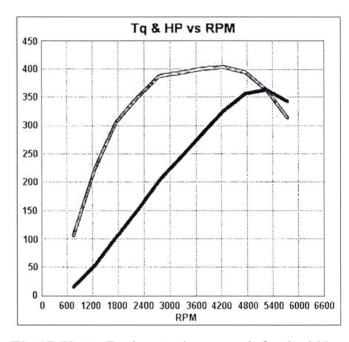

Fig.17-70 An Engine Analyzer graph for the 383 Chevrolet

The finished package produces everything expected; much improved torque with good idle quality and drivability. The project demonstrates the gains possible when appropriate attention is given to maximizing VE even within the parameters of a street driven engine.

SPORT COMPACT ENGINES

There is great interest in "sport compact" racing and even greater interest in the power output of street driven compacts. The majority of these engines are 4-valve per cylinder designs with pent roof or variations of pent roof combustion chambers (see Fig.17-71). Once an import-only market, these engines are now available from virtually all manufacturers, import and domestic.

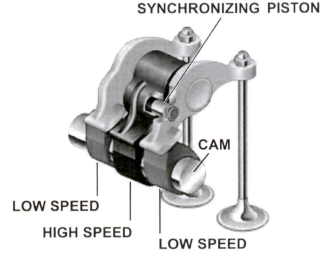

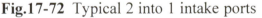

Fig.17-71 Honda V-TECH valve train

Tumble through the intake ports in these chambers creates mixture motion that helps improve mixture quality, reduce misfires, and complete combustion. With centrally located spark plugs and small bore diameters, the flame front need only travel a short distance to sweep across the full chamber. Through the normal RPM range for passenger cars, power is developed with as little as 24 degrees total spark advance. It is only above 6,000 RPM that more spark advance might be required. With such a short burn time, less power is lost compressing an expanding gas after ignition.

Ports in most of these cylinder heads flow very high numbers. Short of competition, beyond polishing and rounding sharp edges, major improvements are not likely (see Fig.17-72). For example, intake and exhaust port flow at 28 inches of water pressure in a production 1.8L VTEC Honda are 275 and 195 CFM at .500in lift. These are flow rates typical of efficient engines with double this cylinder displacement.

Fig.17-72 Typical 2 into 1 intake ports

An area of potential improvement is in the intake manifolds. Consider that these are small displacement engines capable of 8,000 RPM or more. With single length runners, it is difficult to tune outside of relatively narrow RPM bands. Performance can be improved for a particular band by fabricating a manifold with longer or shorter runners as needed but keep in mind that performance will likely suffer at opposite ends of the desired range. Computer simulations are helpful in determining the precise lengths and diameters for a given RPM range.

Some production engines use dual intake runners to widen the power band. This is done by running the engine on longer and smaller diameter runners up to mid-range RPM and then opening shorter and larger diameter runners. Dual runner intake systems work by keeping velocities elevated without limiting total airflow. In VTEC engines equipped with dual runner manifolds, runner changes are coordinated with a changeover to a higher lift and longer duration intake cam lobe (see Fig.17-73).

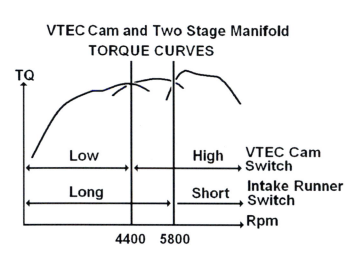

VTEC Cam and Two Stage Manifold
TORQUE CURVES

Fig.17-73 Changeovers of intake runners and VTEC cam

Exhaust tuning is similar to other engines except that there are two exhaust valves merging into one port. In a two valve cylinder, we typically begin with an exhaust runner diameter slightly larger than the valve. With 4-valve cylinders and only one exhaust runner for two valves, it is helpful to think in terms of total valve area. With 1.102 inch diameter valves in a VTEC Honda, twice the valve area is equal to 1.901 square inches. Another way of looking at this is to use twice the curtain area. At one-quarter valve lift, twice the curtain area is nearly the same. Based on this, the minimum exhaust runner inside diameter is 1.550 inches with a 1.901 square inch cross-section. With valve lifts exceeding 1/4 valve diameter, the runner diameter would be based on the larger curtain area.

A survey of available headers for this engine shows diameters of 1.500 and 1.625 inch diameters with some using runners stepped from 1.500 up to 1.625 inches a short distance from the cylinder head. Most of these headers also use a 4 into 2 into 1 design (see Fig.17-74 and 75). The runner lengths are long relative to diameter for these cylinder displacements however the extra length does improve low end torque.

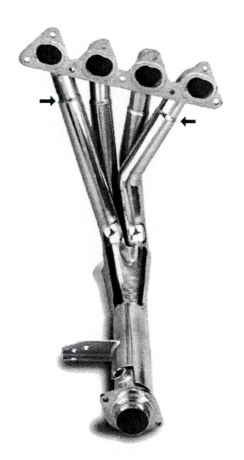

Fig.17-74 A 4 into 1 "stepped" header (Airmass)

Fig.17-75 A 4-2-1 header (Comptech)

As production engines go, specific output is already very high at 1.6 horsepower per cubic inch. Cylinder head port flow is high and compression is high at 10.6:1. Reciprocating forces at high RPM are low with very light rod and piston assemblies and relatively short stroke lengths. With careful attention to valve springs and camshaft profiles, the light weights of valve train components enable these engines to easily run 8,000 RPM. Improving output from the base numbers in these engines requires a very detailed look at intake and exhaust systems and camshafts and camshaft timing. Following are some variations in tuning possible with changes limited to intake and exhaust systems, valve timing, and camshafts.

COMPUTER PROJECTIONS
Honda 1.8L VTEC

Test #1   Base engine
Test #2   Base engine with optimized valve timing and header
Test #3   New cams only
Test #4   New cams with optimized valve timing
Test #5   New cams with optimized valve timing and header
Test #6   New cams with optimized valve timing, header, and intake

	Test #1	Test #2	Test #3	Test #4	Test #5	Test #6
Intake Runner Diameter	1.850	1.850	1.850	1.850	1.850	2.073
Intake Runner Length	8.5	8.5	8.5	8.5	8.5	16
Exhaust Runner Diameter	1.650	1.595	1.650	1.650	1.675	1.675
Exhaust Runner Length	34	28	34	34	31	31
Intake Centerline	105.5	107.5	101	106	105	105
Intake Duration @ .050	233	233	242	242	242	242
Intake Lobe Lift	.468	.468	.472	.472	.472	.472
Exhaust Centerline	99.5	99.5	107	114	111	111
Exhaust Duration	236	236	236	236	236	236
Exhaust Lobe Lift	.428	.428	.458	.458	.458	.458
Best Average HP *	166	174	169	173	173	174
Peak HP/RPM	174	189	181	188	189	194
	7500	7500	7500	7500	7500	7500
Average TQ	131	136	133	136	136	137
Peak TQ/RPM	149	144	147	143	144	146
	5500	5500	5500	5500	5500	5500

Note: Standard dyno conditions, 5500 to 8000RPM, 92 octane gasoline

For the tuning combinations above, best average power (*) was the objective. Note that the required engine changes are inter-related. Intake and exhaust runner diameters and length change and camshaft lobe centers are adjusted for overlap. More impressive numbers on the order of 200HP can be achieved by different tuning strategies but the peak RPM goes up and torque averages go down and overall performance in a street vehicle would likely be less satisfactory. Even at 194HP, output is 1.8 HP per cubic inch, not bad for a production engine with changes limited to "bolt-on" components. Tuning comes down to deciding what performance you are seeking; peak horsepower, peak torque, or a wide power band.

Important to engine builders is the value of computer modeling in arriving at these recommendations. With the inter-relationship of intake and exhaust tuning and camshafts, how many times would this engine have to go through dyno testing to arrive at this point? How many intake and intake systems and camshafts would require testing?

With the thought in mind that 1.6 to 2.0 liter "street" engines are developing 200 to 300 horsepower and turning up to 9,000 RPM, the major effort in competition is strengthening the engine block to withstand the stresses. The base Honda block, for example, is all aluminum, including cylinders, and has an "open" deck; that is, the deck does not support the top ends of the cylinders. Strengthening these blocks begins with replacing the cylinders with high strength sleeves and stabilizing the deck by enclosing the top end of the sleeves (see Figs. 17-76).

Fig.17-76 Deck closed with flanged sleeves (AEBS) including water ports

There is large number and variety of 16-valve four cylinder engines out there. Some have variable valve timing on one or both cams, some have multi-stage induction systems and some have forced induction. This discussion gives insight into the performance potential of these engines.

SUMMARY

In any engine, to improve efficiency requires increased airflow and cylinder pressure. Airflow increases are measured as a percentage of volumetric efficiency. Valve sizes, port flow, valve lift and duration are all areas that require study. Restrictions through intake manifolds and carburetors or throttle valves also require attention.

Compression ratios are increased to raise cylinder pressure and to enable increases in intake duration with the additional benefit of improved volumetric efficiency. Valve timing, in particular the intake valve closing point, is critical in controlling the effective compression ratio.

Tuned exhaust systems enhance intake airflow. Pressure and sound waves move through exhaust runners and, if tuned, they help draw air through the intake system.

Of course there are limits to address. Too much cylinder pressure causes detonation. Too much duration increases overlap and reduces manifold vacuum and low speed performance and idle quality suffer. Computer compatibility is another limit that may need to be considered. However, given the range of tolerance often used in production, efficiency in production engines can often be improved just by blueprinting to specifications.

Those involved in high performance engine preparation require all the basic skills of a technician or machinist. They must then learn the fine points of how to select valves and valve sizes, port cylinder heads, select camshafts, and rework or select intake and exhaust systems. For street engines, the most difficult challenge is to see that these components work together to provide both the performance and the drivability expected.

Chapter 17
PREPARING PERFORMANCE ENGINES

Review Questions

1. Technician A says that the probability of detonation increases with cylinder pressure. Technician B says that to increase output, average cylinder pressures must increase. Who is right?

   a. A only
   b. B only
   c. Both A and B
   d. Neither A or B

2. Technician A says calculate horsepower by multiplying torque times RPM and dividing by 5252. Technician B says that this requires dynamometer measurements of torque. Who is right?

   a. A only
   b. B only
   c. Both A and B
   d. Neither A or B

3. Technician A says that raising port rooflines and rounding short turn radii increases flow rates. Technician B says enlarge the bowl until it contacts the seat. Who is right?

   a. A only
   b. B only
   c. Both A and B
   d. Neither A or B

4. Efficiency requires that exhaust ports flow
   a. 75% of intake port flow
   b. 225 CFM
   c. 175 CFM
   d. equal to intake port flow

5. Valve "shrouding" refers to
   a. chamber or cylinder walls too close to valves
   b. the shape of valve heads
   c. the angle behind the valve face
   d. valve clearance in pistons

6. Modifications to reduce flow restrictions around valves include
   a. flattening the backside of the valve
   b. polishing the backside of the valve
   c. reducing the fillet radius
   d. combinations of these

7. Technician A says that maximum piston velocity occurs when the rod is 90 degrees to the crankshaft. Technician B says that, depending on the rod ratio, maximum piston velocity occurs at approximately 75 degrees ATDC. Who is right?

   a. A only
   b. B only
   c. Both A and B
   d. Neither A or B

8. Technician A says open exhaust valves as far as possible by 75 degrees ATDC. Technician B says open exhaust valves as far as possible by BDC. Who is right?

   a. A only
   b. B only
   c. Both A and B
   d. Neither A or B

9. Technician A says that increasing rocker arm ratios increases lift. Technician B says that higher rocker ratios open and close valves at a faster rate. Who is right?

   a. A only
   b. B only
   c. Both A and B
   d. Neither A or B

10. Technician A says calculate rod ratios by dividing stroke length by rod length. Technician B says calculate rod ratios by dividing rod length by stroke length. Who is right?

    a. A only
    b. B only
    c. Both A and B
    d. Neither A or B

11. Technician A says calculate static compression ratios by dividing displacement plus clearance volume by clearance volume. Technician B says that compression begins only when the intake valve closes. Who is right?

    a. A only
    b. B only
    c. Both A and B
    d. Neither A or B

12. Technician A says that with 100% VE and 91-octane fuel, limit effective compression to approximately 7:1. Technician B says ratios run slightly higher with less than 100% volumetric efficiency. Who is right?

    a. A only
    b. B only
    c. Both A and B
    d. Neither A or B

13. Technician A says that increasing compression ratios increases gas expansion during combustion. Technician B says that increasing compression ratios enables longer intake valve duration and thereby increases VE. Who is right?

    a. A only
    b. B only
    c. Both A and B
    d. Neither A or B

14. Technician A says that closing the intake valve early lowers cylinder pressure. Technician B says that closing the intake valve late causes reversions into the intake manifold. Who is right?

    a. A only
    b. B only
    c. Both A and B
    d. Neither A or B

15. Technician A says that opening the exhaust valve early wastes fuel. Technician B says that opening the exhaust valve late creates pressure at the end of the exhaust stroke. Who is right?

    a. A only
    b. B only
    c. Both A and B
    d. Neither A or B

16. Technician A says that opening intake valves too early dilutes the intake with exhaust gas. Technician B says that closing the exhaust valve too late reduces intake flow. Who is right?

    a. A only
    b. B only
    c. Both A and B
    d. Neither A or B

17. Technician A says that "lobe centers" refers to the distance between intake and exhaust cam lobes. Technician B says that this refers to the centerline of cam lobes. Who is right?

    a. A only
    b. B only
    c. Both A and B
    d. Neither A or B

18. Technician A says that lobe centers are measured in crankshaft degrees. Technician B says that duration is measured in camshaft degrees. Who is right?

    a. A only
    b. B only
    c. Both A and B
    d. Neither A or B

19. Technician A says that dual pattern camshafts have asymmetrical cam lobes. Technician B says that dual pattern cams have different intake and exhaust duration and lift. Who is right?

    a. A only
    b. B only
    c. Both A and B
    d. Neither A or B

20. Technician A says that asymmetrical cam lobes have different intake and exhaust grinds. Technician B says that asymmetrical cam lobes have different opening and closing contours. Who is right?

    a. A only
    b. B only
    c. Both A and B
    d. Neither A or B

21. Technician A says that engine displacement and RPM are used to calculate required inlet CFM. Technician B says that when valve overlap is extreme, increased air flow is necessary to achieve peak VE. Who is right?

   a. A only
   b. B only
   c. Both A and B
   d. Neither A or B

22. Technician A says that a 302 CID engine at 6,500 RPM and 100% VE requires less than 600 CFM. Technician B says that this engine is unlikely to achieve 100% VE with production inlet and exhaust systems. Who is right?

   a. A only
   b. B only
   c. Both A and B
   d. Neither A or B

23. Technician A says that single plane manifolds prevent cylinders from robbing airflow from one another. Technician B says that they are too restrictive for high RPM engines. Who is right?

   a. A only
   b. B only
   c. Both A and B
   d. Neither A or B

24. Technician A says that exhaust runner inside diameters are typically 1/8in smaller than the valves. Technician B says that runner volumes should equal cylinder displacement. Who is right?

   a. A only
   b. B only
   c. Both A and B
   d. Neither A or B

25. Technician A says that lean mixtures at cruise burn fast and less spark advance is needed. Technician B says that rich mixtures under load burn slow and more spark advance is needed. Who is right?

   a. A only
   b. B only
   c. Both A and B
   d. Neither A or B

FOR ADDITIONAL STUDY

1. A 5.0-liter engine develops 275 horsepower at 5,500 RPM. What is the Brake Mean Effective Pressure?

2. The engine above has a 3.8in cylinder bore. What intake valve diameter should this engine have? What approximate intake valve lift would be correct?

3. The engine above has a 6in long connecting rod and a 3.3in stroke length. What is the rod ratio? At what point will the piston reach maximum velocity?

4. What intake CFM is needed for 100% VE in the engine above?

5. What clearance volume is needed for 9.5:1 static compression in the engine above?

6. Describe three key steps in improving flow through ports.

7. What is the importance of the intake valve closing point? What happens if it is early? What if it is late?

8. What is the importance of the exhaust opening point? What happens if it is early? What if it is late?

9. What are the differences between the lobe centers and the lobe centerline of a camshaft?

10. What is the volume of a tuned exhaust runner? What happens to the length when the diameter is reduced? What happens to the diameter when the length is reduced?

11. How do changes in the length and diameter of an exhaust runner effect torque or horsepower?

12. At what point should an engine develop peak cylinder pressure?

13. The ignition timing for the engine above is set at 42 degrees BTDC at 3,000 RPM. What can be expected?

Chapter 18

ENGINE INSTALLATION AND BREAK-IN

Upon completion of this chapter, you will be able to:

- Describe some of the possible differences in engine removal and installation procedures for front wheel and rear wheel drive vehicles.
- List the steps in inspecting and servicing the cooling system and components.
- List the measures a heavy-duty mechanic can take on installation to prepare for engine tuning.
- Explain how the engine should be operated during break-in.
- List some of the key points to inspect on the first service following installation.

INTRODUCTION

While engines have evolved steadily, there have been even more dramatic changes in other technologies and production. The predominant change from rear to front wheel drive certainly affects how we service vehicles. More importantly, nearly all engines now have computerized engine management systems that require a new level of expertise for technicians. With these changes, technicians must now carefully prepare for any engine installation. Working on a computer controlled, front wheel drive car, requires more specialized tools, including computer scanners, and extra precautions to protect wiring, vacuum lines, and components of the fuel and electrical systems.

REMOVING AND INSTALLING THE ENGINE

Procedures for each vehicle vary. However, in regard to engine removal, there are disconnects that are typically the same or very similar. These include:

1. Negative battery terminal
2. Exhaust header pipe to manifold or manifold to engine
3. Throttle linkage or cable
4. Upper and lower radiator hoses
5. Automatic transmission cooler lines
6. Radiator
7. Heater hoses
8. Vacuum lines to body accessories
9. Oxygen sensor or wiring
10. Other sensors or wiring
11. Fuel lines and return hoses
12. Vapor recovery hoses
13. Speedometer cable or speed sensor at the transmission
14. Neutral safety switch or wiring
15. Transmission shift linkages
16. Manual transmission clutch linkage or hydraulics
17. Engine and transmission mounts

The key to working efficiently on any particular vehicle is planning. Do not rush into removal and installation without first checking service references. Some items to be especially watchful for include:

1. Is the engine removed from the top or the bottom?
2. Is the engine removed with or without the transmission? Does the transmission need to be supported while engine is removed?
3. If removing the transmission, how are front drive axles disconnected from the differential?
4. Does the engine need to be supported while the transmission is removed?
5. What special tools are require?

Other items of information may be helpful but unavailable in manuals. Check under the hood and consider possibilities for the following approaches:

1. Can the air-conditioning compressor stay connected and tied to one side of the engine compartment (see Fig.18-1)? If it is necessary to disconnect the system, evacuate and contain the Freon in a recycling system.

Fig.18-1   Pulling the engine with accessories tied to each side

2. On in-line engines, can the intake manifold be left under the hood without having to disconnect wiring, vacuum lines, or linkages?

3. On V-block engines, is it easier to leave the exhaust manifolds under the hood or pull them with the engine?

4. Are there oxygen sensors in exhaust manifolds or header pipes that need to be protected (see Fig.18-2)?

Fig.18-2   An oxygen sensor in an exhaust manifold

5. Are the vacuum hoses and electrical connectors color coded, numbered, or equipped with connectors that fit only one way (see Fig.18-3)? If not, number or color-code the hoses before removal.

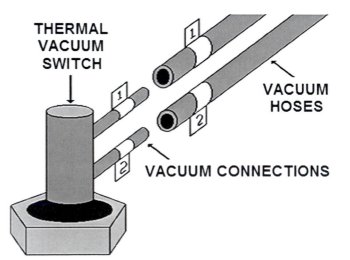

THERMAL VACUUM SWITCH

VACUUM HOSES

VACUUM CONNECTIONS

Fig.18-3   Marking vacuum and electric connections

6. Does the under hood decal show vacuum circuits and basic tune-up specifications (see Fig.18-4)? If not, check service references.

7. At what points are fuel lines and fuel return hoses disconnected? Be sure to plug them and to loosen the gas cap so that pressure does not build up in the system.

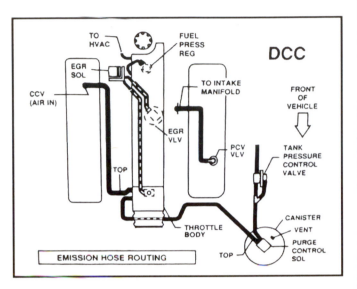

Fig.18-4 An under hood decal with vacuum circuits and basic tune-up specifications

Always look up both procedures and flat rate times for the engine installation. If this is your first installation in this model vehicle, expect that double the estimated time will be required. More important, high flat rate estimates are a clue that this installation is more complex than it first appears.

INSPECTING AND SERVICING THE COOLING SYSTEM

Considering the cost of rebuilding, many technicians prefer replacing all belts, hoses, and the water pump. At the very least, inspect belts for wear, cracking, or fraying, and inspect hoses inside and out for cracking or hardening. Check water pumps for stains suggesting that the pump has been leaking past the seal (see Fig.18-5). Also check for worn bearings by turning the pump by hand and feeling for roughness or wobble in the shaft. Water

pump impellers sometimes erode because of solids circulating within the coolant or because of cavitation (see Fig.18-6). If in doubt about the condition of any of these components, replace them.

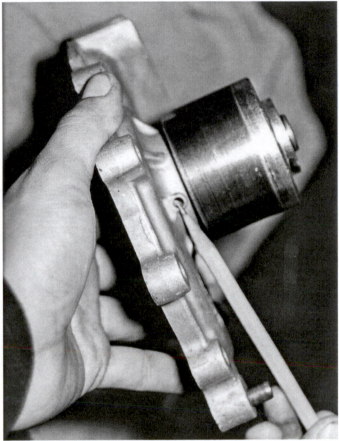

Fig.18-5 Looking for seepage past a water pump seal

Fig.18-6 Check for damage on water pump impeller

In replacing water pumps, there are occasional problems with belt and pulley alignment. Before installing a replacement pump, set the pump alongside the original and compare water fittings, pulley bolt patterns and the positioning of mounting hubs (see Fig.18-7). If not exactly the same, exchange the pump for one that is right.

Fig.18-7 Check hub position on new pumps

Most technicians do not like to take chances and sublet radiators to a radiator shop for cleaning, inspection and any required repairs. If a radiator is to be reused without a sublet inspection, look for deposit build up or clogged core tubes, flush the system and pressure test the radiator and cap (see Figs. 18-8 and 9).

Choices of coolant could include ethylene glycol, propylene glycol or Dexcool, each with different additive packages depending on the manufacturer and engine application. Use only the type of coolant specified by the manufacturer and do not mix coolants. Mixture concentration is typically in the 50-60 percent range. Thoroughly flush systems before refilling to prevent cross-contamination between coolants. Mixed or cross-contaminated coolants can cause clogging and a loss of protection.

Be sure to check automatic transmission cooler circuits built into radiator tanks. If restricted, the transmission fluid overheats and hydraulic pressures in torque converter increase as if the engine were under load. In some applications, pressures double or triple causing the torque converter to push forward against the crankshaft thrust bearing. Include coolers and cooler lines between the radiator and transmission in any cooling system inspection. If main bearing thrust wear was part of the original failure, consider installing an oil-to-air cooler in-line between the transmission and the radiator.

Fig.18-8 Pressure testing the cooling system

Fig.18-9 Pressure testing the radiator cap

ing causes bearing damage in pumps, alternators, compressors, idlers, and even crankshaft main bearings. Also be careful not to damage components by prying against improper points when tightening belts. Because new belts stretch immediately, check tension again after approximately 20 minutes operation.

Finally, add the specified coolant with a 50 to 60 percent concentration. It is necessary to purge air from the system to fill it completely. It helps to adjust the heater controls so that water circulates through the heater core when filling the system. Loosen the highest and most distant connection from radiator and then pressurize the system to allow air to escape before retightening the connection. Some systems have a bleed screw for this purpose. After this, add coolant as necessary. Check that hose clamps are tight and pressure test both the system and the radiator cap before start up. After warm up, compare inlet and outlet coolant temperatures. There should be a minimum 20 degree F difference.

PREPARING FOR EMISSIONS TESTING

To check fan clutches, first see if the fan turns freely when the engine is off and cold. Then bring the engine up to temperature, shut it off, and immediately check that clutch engages. Also check for wobble in the clutch assembly exceeding 1/4in (6mm) at the tips of the fan blades. Keep in mind that slight motion in the clutch is normal. Worn bearings cause excessive wobble and sometimes a growl when the fan coasts to a stop.

The thermostat is inexpensive and is normally replace. For troubleshooting, check the thermostat operation by suspending it in boiling water for a minute and visually watching for the opening. The thermostat should open fully in less than a minute.

Be sure to replace any V-belts in questionable condition. Also remember to replace double belts in sets and not individually. Tighten belts to the engine manufacturer's specifications or the lowest tension that does not slip at peak loads. Use care because, in addition to belt failure, over tighten-

Tuning the engine protects the shop from tune-up related engine failures and assures customer satisfaction from the start. Any technician can give examples of engine failures caused by tune-up or maintenance related problems. Examples include lean fuel mixtures causing valves to burn, rich fuel mixtures causing ring failure, and over advanced timing or non-functional EGR systems causing detonation. Consider that rebuilt engines typically go into high mileage vehicles with deteriorated emission control and cooling system components that require service.

While specific tuning requirements vary widely, making sure that each installed engine passes exhaust emission testing is one way of checking the state of tune. An engine that is out of tune simply will not pass such tests. It is also good customer relations to work with them in meeting legal and environmental responsibilities.

There are a number of basic services that heavy-duty technicians can perform during installations that improve chances of passing an emissions test. Some of these include:

1. On carbureted engines, clean exhaust passages that heat automatic chokes. Without sufficient heat, choke calibration is impossible and the engine will either stall when cold because of lean mixtures, or foul spark plugs because of rich mixtures.

2. Clean passages to the EGR valve. First, this is part of the emission control system and many states periodically check its operation. Second, without EGR, some engines detonate causing piston damage and head gasket failures.

3. Clean or replace the filter in the evaporative control carbon canister. This filter frequently goes without service and can prevent over rich fuel mixtures when "purging" the canister.

4. Replace fuel filters. Some vehicles use one filter between the tank and the engine and another filter at the engine. While the filter at the engine gets regular service, the filter nearest the tank sometimes goes without service until the engine stops.

5. Clean or replace fuel injectors. In addition to high mileage, fuel system components sit while the engine is rebuilt and dry fuel deposits can further contaminate injectors. Injector cleaning machines cycle cleaning solution through the injectors, spray patterns are checked and flow rates compared to match injectors (see Fig. 18-10 and 11). Unless injectors work properly, engine performance operation will be unsatisfactory. As a precaution while working with any fuel system, keep fuel lines plugged to prevent contaminating the system.

6. Check for clear idle circuits and accelerator pump operation on carburetor equipped engines. Idle circuits plug up from sitting with old fuel and accelerator pumps dry out. Remove limiter caps from idle mixture adjusting screws before installing the carburetor. With accumulated

Fig.18-10 A fuel injector cleaner

Fig.18-11 Checking injector spray patterns

mileage and engine rebuilding, the mixture is sure to require adjustment. With these problems present, idle quality is poor and the engine stumbles on acceleration

7. Carbureted and throttle body equipped engines use exhaust heat riser valves to direct exhaust across the floor of the intake manifold to heat the choke and to help vaporize fuel. Check this valve for free operation and clean or replace the assembly as necessary. Stuck open, these valves fail to properly heat choke housings and the floor of the intake manifold. Stuck closed, too much heat backs up into these areas.

8. Clean the positive crankcase ventilation system (PCV) including the valve or restricting orifice and hoses. Checks are too often limited to the PCV valve only but deposits built up elsewhere in the system that restrict flow volume leading to high crankcase pressure at highway speeds. With high crankcase pressure, engines leak and oil is forced into the inlet system through the PCV system into the inlet air.

9. Replace oxygen sensors or test on start-up and replace if needed. Oxygen sensors become contaminated by oil from worn engines.

10. Replace the oxygen sensor or clean it by heating the sensor end with a propane torch. These sensors are often contaminated by oil from the worn engine.

11. Check the routing and connection of vacuum hoses against the under-hood decal. This is easier in later model cars because most connectors fit only one way on each component.

12. If adjustable, set base timing, idle RPM, and the idle fuel mixture to specifications.

While the services above are readily done during installation, the vehicle often goes to an engine performance technician for fine-tuning. The technician normally checks for trouble codes with a computer scanner or for potential drivability complaints and makes additional adjustments or repairs as required. An emissions test assures that all engine parameters are within specifications and that both the shop and the customer are satisfied with the engine installation.

Basic clues to misadjustment or malfunctioning sensors or other fuel and ignition system components include the following:

1. Misfiring and high hydrocarbon emissions caused by excessively lean mixtures. Lean mixtures also cause engine overheating.

2. High carbon monoxide emissions caused by excessively rich mixtures. A rich air-fuel mixture fouls spark plugs, dilutes cylinder lubrication and contaminates engine oil.

3. Overheating and detonation caused by over advanced ignition timing.

4. Any number of possible malfunctions or misadjustments will set a check engine light or malfunction indicator on a computer controlled engine. If the check engine light comes on, it is necessary to extract "trouble codes" and trace the cause.

5. Marginal emission test results after tuning are often traced to contamination of the catalytic converter by oil or coolant. With the replacement engine in proper tune, driving the vehicle for a few hours at highway speeds often allows the converter to recover sufficiently to pass the test. With late model computerized engine management, a computer scanner can be used to test converter function. In earlier systems, check temperatures at the entry and exit to the converter with a pyrometer. The exit temperature should be 100 degrees F higher. If it does not recover sufficiently to pass emissions testing, replace it.

Be especially watchful for the indicators above, as these problems may be the very ones that caused the previous engine to fail. The challenge in diagnosis is the number of different systems used by manufacturers and the need for specialized product knowledge to correct problems in a reasonable time.

MAKING A FINAL INSPECTION

Make a final inspection of an engine installation to assure normal and safe operation of the reconditioned engine. Inspect each item on the following list before returning the vehicle to the owner.

1. Engine mounts for tightness or breaks
2. Exhaust leaks
3. Fuel leaks
4. Radiator hose condition
5. Heater hose condition
6. Hose clamp tightness
7. Condition and tightness of drive belts
8. Transmission linkage adjustment
9. Transmission oil cooler line leaks
10. Circulation of coolant through radiator
11. Pressure cap operation
12. Radiator and heater core leakage
13. Oil leaks
14. Oil pressure
15. For distributors:
    a. Condition of the distributor cap, rotor, secondary cables and spark plugs
    b. Adjustment of base timing and dwell if equipped with breaker points
16. For carburetors:
    a. Vacuum leaks
    b. Idle mixture and RPM
    c. Choke adjustment
17. Battery charge and cable connections
18. The proper routing of hoses and wiring
19. Computer trouble codes
20. Emissions testing
21. Cleanliness of the engine compartment and the inside and outside of the vehicle

STARTING AND BREAKING-IN THE ENGINE

While a properly assembled engine is lubricated on assembly, additional precautions against lubrication failures are recommended on start-up. This is especially important for many new engines with crankshaft driven oil pumps because it is not possible to pre-oil the engine by driving the pump only.

Use a pre-lubricator for these engines. If the engine has been sitting for some time, again use a pre-lubricator. At the very least, remove spark plugs, disable the ignition or fuel injection, and crank the engine for thirty seconds at a time until oil circulates. Remember that we want oil circulation to all moving parts before start up, not just pressure.

There is no "break-in" oil. With the exception of flat tappet engines, fill the crankcase with oil having the most current service rating for new car warranties. Because many emission controlled engines develop tremendous heat, oils with a lesser service rating break down and do not protect the engine. For flat tappet engines, select oils with improved scuff-inhibiting qualities to better protect camshafts and other scuff prone components.

Run the engine at high idle for the first few minutes of operation to ensure adequate oil pressure and circulation. Providing there is oil pressure, do not be alarmed by valve noise caused by hydraulic lifters or lash compensators. Air purges quickly and the noise goes away without harm to the valve train.

Drive the vehicle on the road for the initial seating of the piston rings. Seat rings by cycling the engine speed; first accelerate at half throttle, and then decelerate repeatedly through several cycles. Gas pressure forces compression rings against the cylinder walls on acceleration causing seating to take place more rapidly. On deceleration, vacuum draws oil up onto the cylinder walls and prevents cylinder scuffing or scoring. On carbureted engines, the idle speed sometimes increases after ring seating and has to be reset to specifications.

High speed driving overheats new bearings, which are relatively soft and deform easily when new. This softness allows bearings to conform to the shape of housing bores and rotating shafts. The bearings conform to shape and harden after approximately ten hours of engine operation. Avoid long periods at idle because of the high heat and reduced oil circulation.

Most shops recommend that the owner drive the car "normally" during the first 500 miles and caution them to avoid high speeds and long periods at idle. Advise owners that new engines run hotter because of the increased friction of new rings and renewed engine efficiency. It is especially important to caution drivers to watch temperature indicators and check engine lights.

Check the oil level in the crankcase frequently during break-in. Although new engine technologies allow for very little blow-by past rings, some engines may use oil initially on break in, especially if run at high speeds. After long periods at idle and low speeds or low temperatures, water and fuel vapors can collect in the crankcase and add to the oil level. At highway speeds, the oil heats up, contaminants evaporate, and the oil level drops.

FOLLOWING UP ON THE INSTALLATION

Change the oil and oil filter after break-in because metallic particles and possibly other contaminants come loose during initial operation. Where called for, retorque cylinder heads and intake manifolds to compensate for gasket compression caused by heating and cooling cycles. Carburetors may require readjustment after break-in. Choke operation changes because the exhaust passages used to heat the choke are clean and likely working normally for the first time in years. Idle speed increases during break-in as internal friction in the rebuilt engine diminishes. Fuel and timing calibration can be affected because ring seating increases intake manifold vacuum. Because valves sometimes recess slightly into valve seats during break-in, check valve adjustments in all adjustable valve trains. Readjusting valves with hydraulic adjusters is not necessary because plunger travel allows for minor changes.

Again, it is best to detect problems that might cause premature failure or customer dissatisfaction. With this in mind, pay careful attention to the customer regarding engine performance particularly the following:

1. Does the engine run smooth and have normal power?
2. Is there any abnormal engine noise?

3. Is there any exhaust noise?
4. Are oil pressure and water temperature indicators normal?
5. Are automatic transmission operation and shift points normal?
6. Are manual transmission and clutch operation satisfactory?
7. Are there any fuel, oil, or coolant leaks?
8. Are there any customer complaints?

SUMMARY

The installation and all that it includes is of crucial importance to engine performance and longevity. The procedures for removal, installation, and tuning are labor intensive. It is worthwhile to check shop references in advance of this work and search for ways of saving time.

The cooling system is especially important. The radiator must be clean and hold pressure and all belts and hoses must be in good condition and the fan and water pump require inspection.

Any or all of these components may require replacement. A number of preventive measures should be taken at this time such as the replacement of vacuum hoses, fuel filters, and secondary ignition cables. The engine requires tuning and testing to be sure that spark timing and fuel mixtures are within specifications. Consider that emissions testing will disclose any tuning problems likely to cause engine overheating or detonation.

The installation should also be professional in appearance. The engine should be clean with no water, oil, or exhaust leaks and with all hoses and wiring properly routed. As a starting point, checklists such as those in this chapter help in final inspection.

Chapter 18

ENGINE INSTALLATION AND BREAK-IN

Review Questions

1. Technician A says if possible, discharge the air conditioning systems for engine removal and installation. Technician B says if discharging the system, evacuate and recycle the Freon. Who is right?

   a. A only
   b. B only
   c. Both A and B
   d. Neither A or B

2. Technician A says that new engines run hotter because they are more efficient. Technician B says that lean fuel mixtures or advanced spark timing cause overheating. Who is right?

   a. A only
   b. B only
   c. Both A and B
   d. Neither A or B

3. Technician A says reuse radiators only after cleaning and testing for blockage or leaks. Technician B says reuse radiators so long as there was no prior overheating. Who is right?

   a. A only
   b. B only
   c. Both A and B
   d. Neither A or B

4. Technician A says that because automatic transmission fluid is cooled by engine coolant, transmission heat adds to engine heat. Technician B says that because the transmission fluid is cooled by engine coolant, engine heat adds to transmission heat. Who is right?

   a. A only
   b. B only
   c. Both A and B
   d. Neither A or B

5. Technician A says that hydraulic pressures increase in overheated automatic transmissions. Technician B says that this adds to crankshaft thrust wear. Who is right?

   a. A only
   b. B only
   c. Both A and B
   d. Neither A or B

6. Technician A says that new V-belts stretch and require adjustment after engine break-in. Technician B says readjust new V-belts after the first 20 minutes of operation. Who is right?

   a. A only
   b. B only
   c. Both A and B
   d. Neither A or B

7. Technician A says that fan clutches engage when cold. Technician B says that fan clutches disengage when hot. Who is right?

   a. A only
   b. B only
   c. Both A and B
   d. Neither A or B

8. Technician A says that ethylene glycol coolant is properly mixed 50 percent with water. Technician B says that raising the percentage of ethylene glycol increases cooling efficiency. Who is right?

   a. A only
   b. B only
   c. Both A and B
   d. Neither A or B

9. Technician A says that says that once engine oil is added to the crankcase, the new engine is ready to start. Technician B says first remove the spark plugs, disable the ignition, and crank the engine until there is oil pressure and circulation. Who is right?

a. A only
b. B only
c. Both A and B
d. Neither A or B

10. Technician A says match vacuum hose connections to under hood decals. Technician B says adjust idle RPM and spark timing to the specifications on under hood decals. Who is right?

a. A only
b. B only
c. Both A and B
d. Neither A or B

11. Technician A says that should choke or EGR systems fail to operate normally, the choke or EGR valve is defective. Technician B says check for clear exhaust passages leading to these devices. Who is right?

a. A only
b. B only
c. Both A and B
d. Neither A or B

12. Technician A says that when installing engines, clean the oxygen sensors in solvent. Technician B says replace them. Who is right?

a. A only
b. B only
c. Both A and B
d. Neither A or B

13. Technician A says that replacing the PCV valve assures normal system efficiency. Technician B says to assure normal operation, check or clean all hoses, traps, and fittings. Who is right?

a. A only
b. B only
c. Both A and B
d. Neither A or B

14. Technician A says that shops are legally required to run an emissions test on newly installed engines. Technician B says that making certain newly installed engines pass emissions testing helps protect against failures caused by incorrect spark timing and fuel mixtures. Who is right?

a. A only
b. B only
c. Both A and B
d. Neither A or B

15. Technician A says that if a new engine fails emission testing after making required adjustments, replace the catalytic converter. Technician B says that after driving a few hours with the new engine, catalytic converters often recover sufficiently to pass testing. Who is right?

a. A only
b. B only
c. Both A and B
d. Neither A or B

16. Technician A says that lash in adjustable valve trains loosen, not tighten, during break-in. Technician B says that new engines experience some valve recession during break-in. Who is right?

a. A only
b. B only
c. Both A and B
d. Neither A or B

17. Technician A says that after break-in, it is best practice to check valve lash and retorque heads and intake manifolds whenever possible. Technician B says that idle RPM, fuel mixtures and spark timing sometimes require readjustment after break-in. Who is right?

a. A only
b. B only
c. Both A and B
d. Neither A or B

18. Technician A says break in rebuilt engines on "break-in" oil. Technician B says break in rebuilt engines on oil with the most current service rating. Who is right?

   a. A only
   b. B only
   c. Both A and B
   d. Neither A or B

19. Technician A says that cycling engine speeds and loads soon after start up promotes ring seating. Technician B says that new bearings may require up to 10 hours under normal operating conditions to conform to shape and harden. Who is right?

   a. A only
   b. B only
   c. Both A and B
   d. Neither A or B

20. Technician A says that piston rings normally seat by 500 miles. Technician B says that during break-in, check for oil consumption especially in high speed driving. Who is right?

   a. A only
   b. B only
   c. Both A and B
   d. Neither A or B

FOR ADDITIONAL STUDY

1. How is discharging of an AC system avoided during engine removal and replacement? If it is discharged, what is done with the refrigerant?

2. How can you keep vacuum hoses and wires in order?

3. How is air purged from a cooling system?

4. An emissions test shows high hydrocarbons. What causes this?

5. An emissions test shows high carbon monoxide. What causes this?

6. The fuel mixture is lean and spark timing is advanced. What can be expected?

7. How can you prevent lubrication failures when starting a new engine?

8. What inspections should be performed on automatic transmissions during engine replacement?

Review Questions
ANSWER KEY

Chapter 1

1c	2c	3a	4c	5c	6c	7a	8c	9d	1d
11c	12a	13c	14c	15a	16c	17b	18c	19a	20c

Chapter 2

1b	2d	3c	4d	5b	6d	7d	8b	9c	10a
11d	12d	13b	14c	15c	16c	17d	18c	19b	20c
21d	22d	23a	24d	25c	26d	27c	28a	29c	30c

Chapter 3

1c	2b	3b	4a	5b	6a	7d	8c	9a	10c
11d	12a	13b	14a	15c	16d	17c	18a	19d	20a
21d	22b	23d	24c	25d					

Chapter 4

1c	2d	3c	4c	5a	6a	7a	8c	9d	10d
11c	12b	13a	14c	15c	16a	17c	18b	19c	20b
21b	22c	23b	24b	25a					

Chapter 5

1b	2b	3d	4a	5c	6c	7d	8d	9d	10c
11b	12b	13c	14a	15c	16c	17a	18b	19a	20d
21b	22a	23b	24c	25c	26c	27a	28d	29a	30d

Chapter 6

1c	2d	3c	4a	5a	6d	7b	8c	9a	10c
11c	12c	13c	14c	15b	16c	17a	18d	19b	20a
21d	22c	23c	24c	25b	26a	27c	28d	29c	30a

Chapter 7

1d	2a	3c	4b	5b	6d	7c	8b	9c	10b
11a	12b	13a	14d	15c					

Chapter 8

1b	2c	3d	4b	5a	6d	7c	8c	9b	10d
11c	12b	13a	14c	15c	16d	17a	18b	19d	20c
21c	22b	23d	24c	25b					

Chapter 9

1b	2b	3c	4b	5c	6b	7c	8c	9c	10a
11a	12a	13c	14d	15a	16c	17d	18b	19b	20c

Chapter 10

1c	2d	3d	4b	5b	6c	7a	8b	9a	10c
11a	12c	13a	14c	15b	16b	17d	18c	19a	20a

Chapter 11

1c	2d	3b	4c	5c	6c	7a	8a	9b	10b
11c	12c	13c	14a	15b					

Chapter 12

1c	2a	3a	4b	5b	6b	7b	8c	9d	10b
11b	12a	13c	14b	15c	16b	17d	18b	19b	20c
21a	22d	23d	24c	25d	26d	27d	28a	29c	30c
31a	32c	33a	34d	35a	36c	37a	38c	39d	40c
41d	42a	43d	44b	45c					

Chapter 13

1a	2b	3d	4c	5b	6c	7d	8d	9c	10d
11d	12a	13b	14a	15c	16a	17a	18c	19c	20d
21d	22c	23c	24a	25c	26d	27c	28c	29c	30c
31d	32d	33c	34c	35b					

Chapter 14

1c	2a	3d	4c	5b	6b	7b	8c	9c	10c
11a	12c	13c	14d	15c	16b	17c	18a	19c	20c

Chapter 15

1b	2a	3c	4d	5c	6c	7b	8c	9b	10b
11a	12a	13b	14a	15d					

Chapter 16

1c	2a	3c	4c	5b	6c	7b	8c	9c	10b
11c	12b	13d	14d	15c	16c	17c	18b	19a	20d
21b	22b	23a	24a	25a	26d	27c	28c	29c	30c
31d	32c	33b	34c	35c	36c	37c	38a	39c	40a
41c	42c	43a	44a	45c					

Chapter 17

1c	2c	3a	4a	5a	6d	7c	8b	9c	10b
11c	12c	13c	14d	15c	16c	17a	18d	19b	20b
21c	22c	23d	24d	25d					

Chapter 18

1b	2c	3a	4c	5a	6b	7d	8a	9b	10c
11b	12b	13b	14b	15b	16b	17c	18b	19c	20c

APPENDIX
TORQUE RECOMMENDATIONS

Recommendations are for clean threads lubricated with engine oil. Values are 33 percent below dry torque specifications.

Metric

Diameter	x	Pitch	Grade 8.8	Grade 10.9	Grade 12.9
6		1.00	56*	64*	88*
7		1.00	88*	116*	160*
8		1.00	160*	15	21
8		1.25	144*	166*	19
10		1.00	25	31	43
10		1.25	23	30	41
10		1.50	22	28	39
12		1.50	43	53	73
12		1.75	39	49	67

Inches

Diameter	x	Threads Per Inch	Grade 5	Grade 6	Grade 8
1/4		20	72*	88*	136*
		28	80*	104*	152*
5/16		18	135*	15	22
		24	160*	17	25
3/8		16	21	26	40
		24	23	30	45
7/16		14	33	43	64
		20	37	48	72
1/2		13	50	65	96
		20	57	73	110
9/16		12	73	93	140
		18	80	107	157
5/8		11	100	130	193
		18	113	147	220

Note: Any recommended torque value under 15 Foot/Pounds is given in Inch /Pounds *

PIPE PLUG TORQUE VALUES

Size	Actual O.D.	Torque Iron	Torque Aluminum
1/16	.320	10 ft/lbs	45 in/lbs
1/8	.410	15 ft/lbs	10 ft/lbs
1/4	.540	20 ft/lbs	15 ft/lbs
3/8	.680	25 ft/lbs	20 ft/lbs
1/2	.850	40 ft/lbs	25 ft/lbs
3/4	1.050	55 ft/lbs	45 ft/lbs

Note: torque values with sealer applied to threads.

TORQUE CONVERSION TABLE

Foot Pounds - Newton Meters - Meter Kilograms

FT/PDS	0	1	2	3	4	5	6	7	8	9
0	0	1.35	2.70	4.05	5.40	6.75	8.10	9.45	10.8	12.1
10	13.5	14.9	16.2	17.6	18.9	20.3	21.6	22.9	24.3	25.6
20	27.0	28.3	29.7	31.0	32.5	33.7	35.1	36.4	37.8	39.1
30	40.5	41.8	43.2	44.5	45.9	47.2	48.6	49.9	51.3	52.6
40	54.0	55.3	56.7	58.0	59.4	60.7	62.1	63.4	64.8	66.1
50	67.5	68.8	70.2	71.5	72.9	74.2	75.6	76.9	78.3	79.6
60	81.0	82.3	83.7	85.0	86.4	87.7	89.1	90.4	91.8	93.1
70	94.5	95.8	97.2	98.5	99.9	101	102	103	105	106
80	108	109	110	112	113	114	116	117	118	120
90	121	122	124	125	126	128	129	130	132	133
100	135	136	137	139	140	141	143	144	145	147
110	148	149	151	152	153	155	156	157	159	160
120	162	163	164	166	167	168	170	171	172	174
130	175	176	178	179	180	182	183	184	186	187

NOTE 1: The following formulas can be used to convert Foot/Pounds to Newton Meters:

Foot Pounds X 1.35 = NM and;
Foot Pounds X 7.23 = MKg

NOTE 2: Convert to Meter Kilograms within 5 percent by dividing Newton Meters by 10 (move the decimal point one place left).

DECIMAL EQUIVALENTS

Fractional, Wire, Gauge, And Letter-Size Drills

Drill Size	Decimal Size	Drill Size	Decimal Size	Drill Size	Decimal Size	Drill Size	Decimal Size
80	.0135	42	.0935	13/64	.2031	X	.3970
79	.0145	3/32	.0938	6	.2040	Y	.4040
1/64	.0156	41	.0960	5	.2055	13/32	.4062
78	.0160	40	.0980	4	.2090	Z	.4130
77	.0180	39	.0995	3	.2130	27/64	.4219
76	.0200	38	.1015	7/32	.2188	7/16	.4375
75	.0210	37	.1040	2	.2210	29/64	.4531
74	.0225	36	.1065	1	.2280	15/32	.4688
73	.0240	7/64	.1094	A	.2340	31/64	.4844
72	.0250	35	.1100	15/64	.2344	1/2	.5000
71	.0260	34	.1110	B	.2380	33/64	.5156
70	.0280	33	.1130	C	.2420	17/32	.5312
69	.0292	32	.1160	D	.2460	35/64	.5469
68	.0310	31	.1200	1/4	.2500	9/16	5625
1/32	.0312	1/8	.1250	E	.2500	37/64	.5781
67	.0320	30	.1285	F	.2570	19/32	.5938
66	.0330	29	.1360	G	.2610	39/64	.6094
65	.0350	28	.1405	17/64	.2656	5/8	.6250
64	.0360	9/64	.1406	H	.2660	41/64	.6406
63	.0370	27	.1440	I	.2720	21/32	.6562
62	.0380	26	.1470	J	.2770	43/64	.6719
61	.0390	25	.1495	K	.2810	11/16	.6875
60	.0400	24	.1520	9/32	.2812	45/64	.7031
59	.0410	23	.1540	L	.2900	23/32	.7188
58	.0420	5/32	.1562	M	.2950	47/64	.7344
57	.0430	22	.1570	19/64	.2969	3/4	.7500
56	.0465	21	.1590	N	.3020	49/64	.7656
3/64	.0469	20	.1610	5/16	.3125	25/32	.7812
55	.0520	19	.1660	0	.3160	51/64	.7969
54	.0550	18	.1695	P	.3230	13/16	.8125
53	.0595	11/64	.1719	21/64	.3281	53/64	.8281
1/16	.0625	17	.1730	Q	.3320	27/32	.8438
52	.0635	16	.1770	R	.3390	55/64	.8594
51	.0670	15	.1800	11/32	.3438	7/8	.8750
50	.0700	14	.1820	S	.3480	57/64	.8906
49	.0730	13	.1850	T	.3580	29/32	.9062
48	.0760	3/16	.1875	23/64	.3594	59/64	.9219
5/64	.0781	12	.1890	U	.3680	15/16	.9375
47	.0785	11	.1910	3/8	.3750	61/64	.9531
46	.0810	10	.1935	V	.3770	31/32	.9688
45	.0820	9	.1960	W	.3860	63/64	.9344
44	.0860	8	.1990	25/64	.3906	1	1.0000
43	.0890	7	.2010				

TAP DRILL SIZES
BASED ON APPROXIMATELY 75% FULL THREAD
Coarse and Fine Threads

Thread	Drill		Thread	Drill		Tapered Pipe Thread	Drill
0-8	3/64		7/16-14	U		1/8-27	R
1-64	53		7/16-20	25/64		1/4-18	7/16
1-72	53		1/2 -12	7/64		3/8-18	37/64
2-56	50		1/2 -13	27/64		1/2-14	23/32
2-64	50		1/2 -20	29/64		3/4-14	59/64
3-48	47		9/16-12	31/64		1-11	1 5/32
3-56	45		9/16-18	33/64			
4-40	43		5/8-11	17/32			
4-48	42		5/8-18	37/64			
5-40	38		3/4-10	21/32			
5-44	37		3/4-16	11/16			
6-32	36		7/8- 9	49/64		Straight Pipe	
6-40	33		7/8-14	13/16		Thread	Drill
8-32	29		1-8	7/8			
8-36	29		1-12	59/64		1/8-27	S
10-24	25					1/4-18	29/64
10-32	21					3/8-18	19/32
12-24	16					1/2-14	47/64
12-28	14					3/4-14	15/16
1/4-20	7					1-11	33/64
1/4-28	3						
5/l 6-l 8	F						
5/16-24	1						
3/8-16	5/16						
3/8-24	Q						

METRIC TAP DRILL SIZES

Diameter and Pitch	Metric Drill	Inch Drill
5 x .80	4.20	11/64
6 x 1.00	5.00	13/64
7 x 1.00	6.00	15/64
8 x 1.25	6.75	17/64
10 x 1.50	8.50	11/32
12 x 1.75	10.25	13/32

HELICOIL TAP DRILL SIZES

Inches	Tap Drill
1/4-20	17/64
5/16-18	21/64
3/8-16	25/64
7/16-14	29/64
1/2-13	17/32

Metric	Tap Drill
5 x .80	13/64
6 x 1.00	1/4
7 x 1.00	9/32
8 x 1.25	21/64
10 x 1.50	13/32
12 x 1.75	31/64

Note: All tap-drill sizes calculated for 75% contact area by subtracting the pitch
From the outside diameter.

ENGLISH-METRIC CONVERSIONS

Decimal	mm	Decimal	mm	Fraction	Decimal	mm	Fraction	Decimal	mm
.010	.254	.510	12.954	1/64	.0156	.3969	33/64	.5156	13.097
.020	.508	.520	13.208	1/32	.0313	.7938	17/32	.5313	13.494
.030	.762	.530	13.462	3/64	.0469	1.191	35/64	.5469	13.891
.040	1.016	.540	13.716	1/16	.0625	1.588	9/16	.5625	14.288
.050	1.270	.550	13.970	5/64	.0781	1.984	37/64	.5781	14.684
.060	1.624	.560	14.224	3/32	.0938	2.381	19/32	.5936	15.081
.070	1.778	.570	14.478	7/64	.1094	2.778	39/64	.6094	15.478
.080	2.032	.580	14.732	1/8	.1250	3.175	5/8	.6250	15.875
.090	2.286	.590	14.986	9/64	.1406	3.572	41/64	.6406	16.272
.100	2.540	.600	15.240	5/32	.1563	3.969	21/32	.6563	16.669
.110	2.794	.610	15.494	11/64	.1719	4.366	43/64	.6719	17.066
.120	3.048	.620	15.748	3/16	.1875	4.763	11/16	.6875	17.463
.130	3.302	.630	16.002	13/64	.2031	5.159	45/64	.7031	17.859
.140	3.556	.640	16.256	7/32	.2188	5.556	23/32	.7188	18.256
.150	3.810	.650	16.610	15/64	.2344	5.953	47/64	.7344	18.653
.160	4.064	.660	16.764	1/4	.2500	6.350	3/4	.7500	19.050
.170	4.318	.670	17.018	17/64	.2656	6.747	49/64	.7656	19.447
.180	4.572	.680	17.272	9/32	.2813	7.144	25/32	.7813	19.844
.190	4.826	.690	17.526	19/64	.2969	7.541	51/64	.7969	20.241
.200	5.080	.700	17.780	5/16	.3125	7.938	13/16	.8125	20.636
.210	5.334	.710	18.034	21/64	.3281	8.334	53/64	.8281	21.034
.220	5.588	.720	18.288	11/32	.3438	8.731	27/32	.8438	21.431
.230	5.842	.730	18.542	23/64	.3594	9.128	55/64	.8594	21.828
.240	6.096	.740	18.796	3/8	.3750	9.525	'7/8	.8750	22.098
.250	6.350	.750	19.050	25/64	.3906	9.922	57/64	.0891	22.622
.260	6.604	.760	19.304	13/32	.4063	10.319	29/32	.9063	23.019
.270	6.858	.770	19.558	27/64	.4219	10.716	59/64	.9219	23.416
.280	7.112	.780	19.812	7/16	.4375	11.113	15/16	.9375	23.813
.290	7.366	.790	20.066	29/64	.4531	11.509	61/64	.9531	24.209
.300	7.620	.800	20.320	15/32	.4688	11.906	31/32	.9688	24.606
.310	7.874	.810	20.574	31/64	.4844	12.303	63/64	.9844	25.003
.320	8.128	.820	20.828	1/2	.5000	12.700	1	1.0000	25.400
.330	8.382	.830	21.082						
.340	8.636	.840	21.336						
.350	8.890	.850	21.590						
.360	9.144	.860	21.844						
.380	9.652	.880	22.352						
.390	9.906	.890	22.606						
.400	10.160	.900	22.860						
.410	10.414	.910	23.114						
.420	10.668	.920	23.368						
.430	10.922	.930	23.622						
.440	11.176	.940	23.876						
.450	11.430	.950	24.130						
.460	11.684	.960	24.384						
.470	11.938	.970	24.638						
.480	12.192	.980	24.892						
.490	12.446	.990	25.146						
.500	12.700	1.000	25.400						

METRIC-ENGLISH CONVERSION TABLE

mm	Inches	mm	Inches	mm	Inches	mm	Inches	mm	Inches
0.01	.00039	0.41	.01614	0.81	.03189	21	.82677	61	2.40157
0.02	.00079	0.42	.01654	0.82	.03228	22	.86614	62	2.44094
0.03	.00118	0.43	.01693	0.83	.03268	23	.90551	63	2.48031
0.04	.00157	0.44	.01732	0.84	.03307	24	.94438	64	2.51968
0.05	.00197	0.45	.01772	0.85	.03346	25	.98425	65	2.55905
0.06	.00236	0.46	.01811	0.86	.03386	26	1.02362	66	2.59842
0.07	.00276	0.47	.01850	0.87	.03425	27	1.06299	67	2.63779
0.08	.00315	0.48	.01890	0.88	.03465	28	1.10236	68	2.67716
0.09	.00354	0.49	.01929	0.89	.03504	29	1.14173	69	2.71653
0.10	.00394	0.50	.01969	0.90	.03543	30	1.18110	70	2.75590
0.11	.00433	0.51	.02008	0.91	.03583	31	1.22047	71	2.79527
0.12	.00472	0.52	.02047	0.92	.03622	32	1.25984	72	2.63464
0.13	.00512	0.53	.02087	0.93	.03661	33	1.29921	73	2.87401
0.14	.00551	0.54	.02126	0.94	.03701	34	1.33858	74	2.91338
0.15	.00591	0.55	.02165	0.95	.03740	35	1.37795	75	2.95275
0.16	.00630	0.56	.02205	0.96	.03780	36	1.41732	76	2.99212
0.17	.00669	0.57	.02244	0.97	.03819	37	1.45669	77	3.03149
0.18	.00709	0.58	.02283	0.98	.03858	38	1.49606	78	3.07086
0.19	.00748	0.59	.02323	0.99	.03898	39	1.53543	79	3.11023
0.20	.00787	0.60	.02362	1.00	.03937	40	1.57480	80	3.14960
0.21	.00827	0.61	.02402	1	.03937	41	1.61417	81	3.18897
0.22	.00866	0.62	.02441	2	.07874	42	1.65354	82	3.22834
0.23	.00906	0.63	.02480	3	.11811	43	1.69291	83	3.26771
0.24	.00945	0.64	.02520	4	.15748	44	1.73228	84	3.30708
0.25	.00984	0.65	.02559	5	.19685	45	1.77165	85	3.34645
0.26	.01024	0.66	.02598	6	.23622	46	1.81102	86	3.38582
0.27	.01063	0.67	.02638	7	.27559	47	1.85039	87	3.42519
0.28	.01102	0.68	.02677	8	.31496	48	1.88976	88	3.46456
0.29	.01142	0.69	.02717	9	.35433	49	1.92913	89	3.50393
0.30	.01181	0.70	.02756	10	.39370	50	1.96850	90	3.54330
0.31	.01220	0.71	.02795	11	.43307	51	2.00787	91	3.58267
0.32	.01260	0.72	.02835	12	.47244	52	2.04724	92	3.62204
0.33	.01299	0.73	.02874	13	.51181	53	2.08661	93	3.66141
0.34	.01339	0.74	.02913	14	.55118	54	2.12598	94	3.70078
0.35	.01378	0.75	.02953	15	.59055	55	2.16535	95	3.74015
0.36	.01417	0.76	.02992	16	.62992	56	2.20472	96	3.77952
0.37	.01457	0.77	.03032	17	.66929	57	2.24409	97	3.81889
0.38	.01496	0.78	.03071	18	.70866	58	2.28346	98	3.85826
0.39	.01535	0.79	.03110	19	.74803	59	2.32283	99	3.89763
0.40	.01575	0.80	.03150	20	.78740	60	2.36220	100	3.93700

CONVERSION FACTORS

Multiply	By	To obtain
LENGTH		
Millimeters (mm)	.03937	Inches
	.1	Centimeters (cm)
Kilometers (km)	.6214	Miles
	3281	Feet
Inches	25.4	Millimeters (mm)
Miles	1.6093	Kilometers (km)
AREA		
Inches^2	645.16	Millimeters^2 (mm^2)
	6.452	Centimeters^2 (cm^2)
Feet	.0929	Meter^2 (M^2)
	144	Inches^2 (in^2)
VOLUME		
Cubic Centimeters (cc)	0.06102	Cubic inches (in^3)
	1000	Liters (L)
Liters (L)	61.024	Cubic inches (in^3)
	.2624	Gallons
	1.0567	Quarts
Cubic inches (in^3)	16.387	Cubic centimeters (cc)
	231	Gallons
Cubic feet	1728	Cubic inches (in^3)
	28.32	Liters (L)
Fluid ounces (oz).	29.57	Milliliters (mL)
MASS		
Gram (g)	.03527	Ounce
Kilograms (kg)	2.2046	Pounds
	35.274	Ounces
FORCE		
Ounce	.278	Newton (N)
Pound	4.448	Newton (N)
Kilogram	9.807	Newton (N)
TORQUE		
Foot-pounds	1.3568	Newton-meters (N-m)
	.1383	Kilogram/meter (Kg/m)
Inch-pounds	.11298	Newton-meters (N-m)
	.0833	Foot-pounds
Kilogram-meters (Kg/m)	7.23	Foot-pounds
	9.30665	Newton-meters (N-m)

PRESSURE

Atmospheres	14.7	Pounds/sq inch (PSI)
	29.92	Inches of mercury (in/Hg)
Inches of mercury (in/Hg)	.49116	Pounds/sq inch (PSI)
	13.579	Inches of water
	3.377	Kilopascals (kPa)
Inches of water (in/H$_2$0)	.07364	Inches of mercury (in/Hg)
Bars	100	Kilopascals (kPa)
	14.5	Pounds/sq inch (PSI)
Kilograms/cm^2 (Kg/cm^2)	14.22	Pounds/sq inch (PSI)
	98.07	Kilopascals (kPa)
Kilopascals (kPa)	.145	Pounds/sq inch (PSI)

FUEL PERFORMANCE

Miles/gallon	.4251	Kilometers/liter (km/L)

VELOCITY

Miles/hour	1.467	Feet/second
	88	Feet/minute
	1.6093	Kilometers/hr (km/h)
Kilometers/hr (km/h)	.27778	Meters/second (m/s)